VACCINE NATION

VACCINE NATION

|||||||||||

AMERICA'S CHANGING
RELATIONSHIP
WITH IMMUNIZATION

ELENA CONIS

The University of Chicago Press

Chicago and London

ELENA CONIS is assistant professor of history at Emory University.

The University of Chicago Press, Chicago 60637
The University of Chicago Press, Ltd., London
© 2015 by The University of Chicago
All rights reserved. Published 2015.
Printed in the United States of America

24 23 22 21 20 19 18 17 16 15   1 2 3 4 5

ISBN-13: 978-0-226-92376-5 (cloth)
ISBN-13: 978-0-226-92377-2 (e-book)
DOI: 10.7208/chicago/9780226923772.001.0001

Library of Congress Cataloging-in-Publication Data
Conis, Elena, author.
Vaccine nation: America's changing relationship
with immunization / Elena Conis.
pages cm
Includes bibliographical references and index.
ISBN 978-0-226-92376-5 (cloth : alkaline paper) —
ISBN 978-0-226-92377-2 (e-book)
1. Vaccination—United States—History—20th century. I. Title.
RA638.C66 2015
614.4'70973—dc23   2014009846

Portions of chapter 8 appeared in the *Journal of Medical Humanities*:
"'Do We Really Need Hepatitis B on the Second Day of Life?':
Vaccination Mandates and Shifting Representations of Hepatitis B,"
Journal of Medical Humanities 32, no. 2 (2011): 155–66.

A version of chapter 5 previously appeared in the *Bulletin of the
History of Medicine*. Copyright © 2013 The Johns Hopkins University
Press: "A Mother's Responsibility: Women, Medicine, and the Rise of
Contemporary Vaccine Skepticism in the U.S.," *Bulletin of the History
of Medicine* 87, no. 3 (2013): 407–35.

♾ This paper meets the requirements of ANSI/NISO z39.48-1992
(Permanence of Paper).

For JVR

CONTENTS

| | | | | | | | | | |

INTRODUCTION
| | | | | | | | | |

At the turn of the new millennium, childhood vaccines were constantly in the news. A new vaccine against rotavirus was recalled because it caused a severe bowel disorder. A long-used vaccine preservative was banned because it exposed children to the toxic metal mercury. Congressional hearings investigated the safety of the hepatitis B vaccine, following news reports that it caused multiple sclerosis and autoimmune disorders. A growing number of parents worried that the vaccine against measles, mumps, and rubella was behind rising rates of autism. And a new vaccine, against human papillomavirus (HPV), sparked a firestorm of debate when lawmakers attempted to require it for sixth-grade girls. As these episodes unfolded, blogs, message boards, and best-selling books buzzed with advice about how to circumvent vaccine requirements and undo feared damage caused by vaccines. Celebrities spoke out against vaccines on talk shows and the nightly news. Angry parents marched on Washington. Even presidential hopefuls weighed in from the 2012 campaign trail. Vaccines for children had become one of the biggest controversies of the day.

In response, medical and health experts did all they could to assure Americans that vaccines were safe. The nation's prestigious Institute of Medicine convened expert panels that reviewed and confirmed the safety of vaccines against hepatitis B and measles, mumps, and rubella. Scientists launched massive studies that examined the causes of autism and exonerated vaccines. Physicians and health officials took to the airwaves to remind parents that the diseases prevented by vaccines were both real and far worse than the exceedingly rare — and frequently unproven — risks increasingly attributed to vaccines.

But such reassurances may well have been beside the point. For the larger debate over vaccination that took place in the early 2000s wasn't just about vaccine risks. At a deeper level, it was a debate about the role of children in our society, our health care politics, gender relations, chronic disease risks, and more.[1] Moreover, the ongoing debate obscured the fact that in the new millennium, vaccination was widely accepted by the American public. By 2009 America's youngest children were more vaccinated than ever: 90 to 95 percent were protected against over a dozen infections, including diphtheria, pertussis, tetanus, polio, measles, rubella, mumps, *Haemophilus influenzae* b (Hib), hepatitis B, and chicken pox. Only a small fraction — one-half of 1 percent — had received no vaccines at all.[2] Vaccine controversies may have grabbed Americans' attention at the start of the twenty-first century, but the broad acceptance of vaccination deserved just as much scrutiny. For the controversies were just one small piece of the much larger, more complicated, and less-examined story of the rise of childhood vaccination in modern times.

This book follows that story from the 1960s to today, in order to answer two questions: How did we come to place such importance on the mass vaccination of children against such a varied array of infections over the last half century? And why did vaccination become so contentious as the new millennium began? The story begins in the 1960s because that's when a new era of vaccination dawned.[3] Government scientists, triumphant over polio and smallpox, considered how the country might deploy new vaccines against what they called the "milder" diseases, including measles, rubella, and mumps. While they grappled with these new technologies, federal support for childhood vaccination expanded (especially when it served the political interests of White House occupants). The vaccines themselves — and others that came after — drew attention to their target infections, changing how Americans saw them and thought about the urgency of their prevention. Meanwhile, major New Left social movements, including second-wave feminism and environmentalism, helped cultivate criticism of vaccines and vaccination. And a wildly diverse set of influences — Cold War anxiety, the growing value of children, the emergence of HIV/AIDS, changing fashion trends, and immigration worries, to name just a few—helped prop up vaccine acceptance, as the chapters in this book reveal. If the stories herein reveal just one thing, it is that we have never vaccinated for strictly medical reasons. Vaccination was, and is, thoroughly infused with our politics, our social values, and our cultural norms. By acknowledging and understanding the divergent reasons why we've vaccinated in the past, however, we just may ensure the continued success of vaccination in the future.

VACCINATION BEFORE THE SIXTIES

Before the 1960s—from the advent of the first vaccine against smallpox at the turn of the nineteenth century up to the middle of the twentieth century—vaccination in the United States was typically an individual, local, and reactive affair. Parents vaccinated children to protect them from feared diseases. Towns, cities, and states vaccinated their citizens to prevent epidemics. Often they did so because an epidemic loomed in a neighboring region or port. Local government involvement in vaccination was patchy and usually sporadic. And the federal government, on the whole, had little to nothing to do with the local and individual civilian matters of who was vaccinated, against which diseases, and when or how they received their vaccines.

Because vaccination protected both individuals and communities from disease, however, it was both a private activity and a public one. Over the first half of the twentieth century, as historian James Colgrove has shown, cities, states, and the federal government increasingly invested in vaccination in pursuit of its community benefits, and state power over vaccination expanded.[4] In 1902, with passage of the Biologics Control Act, for example, Congress gave the federal government the authority to license and inspect vaccine manufacturers to ensure safe vaccines for all. In 1905, and again in 1922, the Supreme Court upheld the right of states to enforce health through compulsory vaccination.[5]

By then, the first vaccine, against smallpox, had been joined by vaccines against diphtheria and tetanus, and scientists began to amass a wealth of evidence in favor of enforcing vaccination outside the context

3

of raging epidemics. Medical and public health journals reported approvingly on the successes of compulsory vaccination measures in Europe, and they bemoaned the proximity of the U.S. smallpox rate to the rate of disease in China, India, and Russia.[6] State health departments, meanwhile, compiled data demonstrating that vaccines had an impressive capacity to drive down rates of long-established infectious killers when widely used.[7] To many, the data illustrated the merits of mandatory vaccination in particular: New York—which required schoolchildren, teachers, and janitors to be vaccinated against smallpox, for instance—saw roughly 3,000 cases of the disease among 10.3 million residents in the 1920s, noted one early twentieth-century health official. Utah, which prohibited compulsory vaccination, saw more than 13,000 cases among less than half a million residents that same decade.[8]

Over time, the sum total of these events and observations gradually elevated the position of vaccination in American science, policy, and society. But the rise of vaccination in this period was neither entirely smooth nor inexorable, and compulsory vaccination in particular faced challenges. Anti-vaccination groups, first mobilized in the late nineteenth century, decried vaccination as inherently dangerous and *compulsory* vaccination as un-American for its violation of individual freedoms. During the Progressive Era, anti-vaccinationists tackled local and state compulsory vaccination statutes and managed to win a handful of state bans on such laws.[9] Threats to compulsory vaccination came from within the scientific professions as well. Vaccination became a flashpoint in an ongoing territorial dispute between the professions of medicine and public health over where the line between treatment (medicine's traditional domain) and prevention (public health's domain) was drawn.[10] The nation's economy, too, posed its own challenges to the expansion of vaccination: the Depression diminished the capacity of local governments to launch vaccination campaigns and resulted in declining use of medical services overall.

That contraction would turn out to be short-lived, however. Scientific and popular support for vaccination swelled to an unprecedented extent during the Second World War. Thanks to vaccines, wartime news reports declared, American troops were safe from tetanus, pneumonia, Rocky Mountain spotted fever, yellow fever, and more.[11] In Europe, headlines announced, vaccines had thwarted the spread of epidemics among Allied troops in France and saved 2 million civilians from typhoid in war-torn Warsaw.[12] When a preliminary measles vaccine was discovered in 1940, the American press described it as a potent tool to protect soldiers from otherwise devastating epidemics of pneumonia.[13] Vaccines, in short, were subtly but deftly woven into the narrative Americans told—and were told—about how scientific and technological prowess had made the United States the leader of the free world.

Vaccination was also deemed necessary to maintain a robust and war-ready population back home. Health officials urged Americans to get immunized against smallpox and diphtheria to protect themselves from infections spread by migrating war workers and, later, soldiers returning home from combat.[14] Americans were happy to oblige. In 1943 New York health official Leona Baumgartner reported results of a poll on American immunization attitudes: more than 90 percent trusted in vaccines to protect against death and serious disease, prompting Baumgartner to declare Americans "ready" for mass immunization.[15] That readiness was vividly apparent in 1947, when a smallpox outbreak threatened New York City and more than 6 million New Yorkers lined up to get voluntarily vaccinated, displaying an alacrity that historian Judith Leavitt attributes to the spirit of cooperation and sacrifice engendered by the war.[16]

The war made vaccination a symbol of American exceptionalism and a patriotic duty; it also helped set the stage for the American public's clamor for the polio vaccine in the 1950s. The 1955 announcement that Jonas Salk had developed the first effective vaccine against polio was met with fanfare and demand so intense that it led to a national outcry over how and to whom the vaccine would be distributed.[17] Nervous and overwhelmed state health officials and members of the public called on President Dwight Eisenhower to settle the matter, which he did by signing the Poliomyelitis Vaccination Assistance Act into law later that year.[18] The act represented a watershed moment in U.S. vaccination history, as it carved out, for the first time, an active role for the federal government in the funding and dissemination of a vaccine to everyday Americans. The act also established a foothold for federal health officials, who would later use it to argue for further federal involvement in vaccination promotion.[19]

The nation's experience with polio also strengthened federal authority in the realm of vaccination by boosting the power and prestige of the Atlanta-based federal Communicable Disease Center (CDC; now known as the Centers for Disease Control and Prevention). When a faulty lot of polio vaccine began causing cases of paralytic polio and deaths in late April 1955, the CDC played a critical role in stemming the outbreak and maintaining public confidence in the vaccine; these acts solidified the agency's reputation as an authority on the prevention of communicable diseases generally, as historian Elizabeth Etheridge has shown.[20] Over the following few years, CDC officials increasingly pressed for federal support for broad-based immunization campaigns, not just those targeting polio. When, in 1957, Congress asked the surgeon general and the head of epidemiology at the CDC to predict when polio vaccine programming would conclude, the health experts responded by pointing out the need for ongoing *smallpox* vaccination, despite 150 years of immunization against the disease.[21] The remark provided insight into a still-nascent political agenda being crafted in Washington, D.C., and Atlanta: short of disease eradication, vaccination was a perpetual enterprise. Policies, therefore, had to support one approach or the other.

A NEW ERA OF VACCINATION

Polio's dramatic decline — from a high of nearly 60,000 cases in 1952 to just 3,000 in 1960 — helped push eradication to the top of the nation's public health agenda in the 1960s.[22] In that decade, an unprecedented number of effective new vaccines were developed, including a new polio vaccine (developed by Albert Sabin), followed by several vaccines against measles, and vaccines against rubella and mumps. The rapid-fire development of these new weapons against infection — combined with technological hubris, the expansion of the welfare state, and other factors — prompted health experts to forecast a future free of infectious disease. Health officials soon began a push for universal vaccination against what they called the "milder" infections, namely, measles, mumps, and rubella, which they saw as less severe than — and therefore categorically different from — polio, smallpox, and diphtheria, the previous targets of mass vaccination. This push was largely accomplished through the vaccination of children, and increasingly in the post-sixties era, it was supported by the force of law.

A new era of vaccination thus dawned, marked by federal leadership, an emphasis on disease eradication, and a focus on the enforced vaccination of children against a list of diseases ranging in severity. The story of "Sarah W.," a baby girl born in Minneapolis in 1964, illustrates the prospects and challenges of this new era.[23] In the first two years of her life, Sarah was vaccinated against smallpox, diphtheria, whooping cough, pertussis, tetanus, polio, and measles. She received not Salk's but Sabin's polio vaccine, which was licensed in 1961 and placed at the forefront of anti-polio efforts in the 1960s. Her protection against diphtheria, pertussis, and tetanus came from a combined vaccine that scientists had developed in the 1940s. The measles vaccine she received was brand-new, first approved for use the year before she was born. And the smallpox vaccine she received would be rendered obsolete before she left grade school, thanks to an eradication campaign that began in the 1960s and effectively wiped the virus from the face of the earth before the 1970s were through.

But many of the very questions that doctors, health professionals, cities, states, and Washington politicians had grappled with in the first half of the century remained unanswered as Sarah's life began. When should a vaccine be made compulsory, and when should a city or state limit itself to vaccination promotion? When should a vaccine's use be left to the discretion of physicians, or of parents? Which diseases should be priorities for prevention via vaccination? How should the medical and public health professions divide responsibility for ensuring widespread vaccination? Who should cover the costs of vaccination? And how best should worries about vaccine safety (since all vaccines carried some small risk of harm) be addressed?

In the first few decades of Sarah's life, some answers would be forthcoming. When Sarah was born, Minnesota state law didn't require any of the vaccines she received, either for children generally or those ready to enroll in school. But in the late 1960s, the state began requiring measles vaccine for schoolchildren, and in the 1970s, Minnesota along with nearly every other state would adopt laws making all childhood vaccines a requirement for school.[24] The ease with which such laws were adopted across the country in the 1970s created a new norm for vaccination policies in the 1980s, 1990s, and 2000s, when laws requiring the vaccination of children for school (or day-care) enrollment became the default policy approach for many new vaccines.[25]

With the exception of the brand-new measles vaccine, the shots Sarah received as a child were recommended by the American Academy of Pediatrics.[26] The year that she was born, however, a newly formed federal committee—the CDC's Advisory Committee on Immunization Practices (ACIP)—began to assume the role of issuing consolidated vaccination recommendations for all Americans. The ACIP's recommendations have never been intended as enforceable rules, but in the decades that followed, they began to serve as the impetus for legislative acts and regulations making vaccines mandatory for children at the local level. State laws and regulations requiring children's immunization against a litany of targeted diseases—hepatitis B, rotavirus, diphtheria, tetanus, pertussis, Hib, pneumococcus, meningococcus, polio, flu, measles, mumps, rubella, and chicken pox—are now common. As a result, American children today receive significantly more vaccines and vaccine doses than Sarah did: at least thirty-two recommended vaccinations, which protect children against at least thirteen different infections before the age of six.[27]

Sarah's parents recorded the dates of all of her vaccines in her "baby book"; unsurprisingly, they didn't note how they paid for them. They likely paid out of pocket, but thanks to legislation passed under President John F. Kennedy two years before Sarah's birth, new federal funds were beginning to support the vaccination of children against all vaccine-preventable infections. Kennedy's vaccination law, described in chapter 1, laid the foundation for broadscale federal involvement in childhood immunization. Two of his successors—Jimmy Carter, whose Childhood Immunization Initiative is the subject of chapter 4, and Bill Clinton, discussed in chapter 7—later expanded federal support for childhood vaccination even further. For all three administrations, the mass vaccination of children was an expedient means of achieving budgetary, social welfare, and health reform goals. Each president's support for the cause had significant, lasting consequences for childhood immunization in the United States, even as it reflected his own political philosophies and the national priorities of his time.

During the decade Sarah was born, popular, political, and scientific faith in vaccines ran so high that the United States announced the objectives of completely eradicating not just one but two vaccine-preventable diseases: smallpox and measles. But despite a decade of strong federal support for vaccination, at the start of the seventies, vaccine-preventable diseases such as polio, measles, and diphtheria were proving intractable in certain pockets across the country. "If we cannot, by putting our collective minds to it, arrive at workable policies for a technology as fundamentally simple and effective as immunization," one health administrator lamented, "heaven help us in any future efforts to deal with more unsettled and ambiguous issues all around us in biomedical science and technology."[28] Vaccination may have been simple, but its effectiveness relied on uniform application to a heterogeneous and ever-changing population. And over time, the promise of the sixties—that vaccines would eliminate infectious diseases as public health problems, if not eradicate them entirely—proved elusive.

Each period offered its own explanation for the persistence of vaccine-preventable diseases. In the 1960s it was the nation's poverty-stricken who bore much of the blame. In the early 1970s and 1980s, it was a stingy federal government. From the 1980s into the 2000s, the broken health care system, greedy pharmaceutical companies, selfish upper-middle-class liberals, and an irresponsible media all took turns shouldering the blame for the country's inability to wipe out vaccine-preventable diseases for good. The reasons offered to parents to encourage them to vaccinate also shifted over time. In the 1960s and 1970s, health officials emphasized the gruesome complications of the "milder" diseases and encouraged parents to vaccinate for the good of their children. Decades later, in the 2000s, health officials increasingly encouraged parents to vaccinate their children in order to protect the community at large from disease. The benefits of vaccination hadn't changed, but each era produced its own call to arms.

Each era also produced its own interpretation of the diseases targeted by new vaccines. New vaccines brought fresh scientific and popular attention to their disease targets; the diseases were then scrutinized, analyzed, and understood anew within the context of the time.[29] Mumps (the subject of chapter 3), for example, caused the same symptoms and complications in 1980 as it had in 1960, but cultural and scientific descriptions of the infection at these two points in time suggested two very different diseases: one was comical and largely harmless, the other devastating if not deadly. The mumps vaccine released in 1967 and a series of policy and marketing approaches to encourage its use spurred this transformation. Mumps was not unique in this sense. Measles (discussed in chapter 2), chicken pox (chapter 6), hepatitis B (chapter 8), and HPV (chapter 10) were all framed very differently after their vaccines were introduced. Sometimes the new light shed on these infections supported the cause of vaccination; at other times it worked against it.

Vaccines themselves were also cast in the image of their own time.[30] The measles vaccine was cast as a great equalizer at a time — the 1960s — marked by the war on poverty. Mumps vaccine was framed as a modern, middle-class convenience in the 1970s, hepatitis B vaccine as a triumph over AIDS-like infections in the 1980s. In the 1990s and 2000s, the chicken pox vaccine served as a fulcrum for cultural debates over the beneficence or malevolence of nature. In the later 2000s, the HPV vaccine served as a political flashpoint for a nation marked by deepening ideological divides. Each vaccine was colored by the most prominent anxieties and disputes of its historical moment. In turn, the way in which each vaccine was framed by contemporary preoccupations influenced both the policies that took shape to govern its use and how the public received them.

The decades from the 1960s to today marked a new era of vaccination resistance, as well. As new vaccination policies took shape in the 1960s and 1970s, feminism, environmentalism, and other social movements — the subjects of chapters 5 and 6 — challenged scientific and governmental authority, with profound implications for how some Americans came to think about vaccines. The recent history of vaccination is thus characterized by a deep paradox. Beginning in the 1960s, a liberal view of government expanded federal vaccination activity in order to ensure equitable access to vaccines for Americans of every class and race. At the same time, however, liberal social movements gave rise to modern critiques of vaccination as sexist, racist, corporatist, and toxic. These movements also opened the door for a full-fledged revival of older forms of anti-vaccinationism. Modern vaccine skeptics thus still included traditional anti-vaccinationists, who rejected all vaccines for religious or safety reasons, or out of a principled devotion to unfettered individual liberty. But in the late twentieth century, vaccine resisters counted among their ranks critics who adopted tenets of the New Left social movements to critique the political and social hegemonies they saw embodied in vaccination. And as the number of childhood vaccines climbed even higher at the turn of the twenty-first century, the ranks of vaccine resisters also included those who discerned *among* vaccines, accepting some and forgoing others they deemed too risky or just unnecessary.[31]

Modern vaccination resistance has also, importantly, often taken the form of a child-protection movement, precisely because children became the primary target of enforced vaccination in the latter half of the twentieth century. On a population level, effective vaccination relies on a social contract among citizens; if most everyone gets vaccinated, everyone is protected. From the sixties to today, the burden of upholding this social contract has fallen increasingly on the shoulders of children. It was not always a given that the vaccines against rubella, mumps, and hepatitis B, for example, would be administered to young children instead of adults. But policies requiring vaccines for children built on conceptualizations of children as reservoirs of infection in their communities. They also built on the practical and political expedience of vaccinating children, who with the rise of pediatric care in the first half of the twentieth century had more regimented and malleable contact with health care professionals than civilian adults have ever had. Children were vaccinated to protect them from disease, but also to protect their communities from disease, and the state and nation from the medical and economic burdens of disease. The prevention of epidemics through vaccination thus became a key health-citizenship responsibility of children at the same time that the vaccination of children (and ever-younger children) against a growing number of communicable infections expanded.[32]

Children's very participation in public life is now contingent to a notable extent on their immunization status. Vaccination is largely a right of children in this country; federal social insurance programs, dating back to Kennedy's 1962 act, ensure that they don't go without required vaccines for an inability to pay. But it is also their duty, thanks to laws requiring vaccines for school. This health responsibility was born of a particular moment in our collective history, when the nation had high confidence in its experts and optimism about its future wealth, power, and health. Over time, however, both vaccination policies and vaccine acceptance have reflected changing cultural valuations of— and beliefs about—children as family members and citizens. Over the last fifty years, the state's notion of children as health citizens grew increasingly out of sync with parents' ideas about the value of children as family members. Parents in the seventies and later articulated both a responsibility and a right to care for their children in a manner that highlighted allegiance to family, not community. In part, this attitude had been shaped by the public and private vaccination promotion efforts of the preceding decades, which deliberately emphasized the dangers that vaccine-preventable diseases posed to the individual child, and which encouraged parents to vaccinate their children for their own good. Today's public health exhortation to vaccinate for community good is thus not surprisingly sometimes lost on parents long socially, culturally, medically, and politically conditioned to make a decision deemed best for their own family.

Public health pleas to vaccinate also call on parents to abide by technocratic risk calculations. Today's high child immunization rates suggest that most Americans concur with the majority of health experts that the benefits of vaccines substantially outweigh their minimal risks. To some vaccine critics, however, the benefits of certain vaccines are simply not worth their risks, no matter how minimal or disputed those risks may be. To other vaccine critics, vaccine risks epitomize the risks attendant with modern technological hubris, because they are diffuse and universal, and because we won't recognize the harm they've done until it's too late to undo.[33] As many scholars of vaccination have noted, vaccine risks become more salient as the diseases prevented by vaccines disappear; vaccines are, in this sense, "victims of their own success."[34] But perceptions of the risks posed by both vaccines *and* their target infections are constructed in far more complicated ways as well. Social movements, for instance, have helped amplify the perceived risks of vaccines. The rise of epidemiology, on the other hand, has amplified the perceived risks of diseases prevented by vaccines. So has economic growth, inasmuch as it has increased the value of individual lives in affluent countries like the United States, with implications for the types of risks modern parents are willing to subject their children to. The perception and calculation of acceptable risks is thus both context dependent and value driven.

For the last half-century, our political values have strongly supported the expansion of childhood vaccination; generally, our cultural and social values have, too. On occasion, these values have prompted objections to the rise and expansion of childhood vaccination, but on the whole these objections have done relatively little to impede it.

HOW THIS BOOK IS ORGANIZED

This book is divided into three parts. Part I opens with the introduction of President Kennedy's Vaccination Assistance Act, which ushered in a period marked by growing federal involvement in childhood vaccination over the subsequent decade. The act provided the resources and political will for a joint attack on poverty and disease within the nation's changing urban landscape, which is the subject of chapter 2. It also laid the groundwork for federal health officials to become more actively involved in vaccination promotion against diseases both "mild" and severe. One of these mild diseases, mumps—and the process by which it was transformed by its vaccine and vaccination policies in the seventies—is the subject of chapter 3.

Part II focuses on the emergence of popular vaccination resistance as a significant social force in the seventies and eighties. It begins with an analysis of the Carter administration's Childhood Immunization Initiative, which made school vaccination regulations more uniform across the country and took the nation's childhood vaccination rates to new heights. The initiative took place amid some of the most transformative social movements of the seventies, including the women's movement and the environmental movement. The impact that changing ideas about gender and the environment had on popular vaccine reception is explored in chapters 5 and 6.

The final section of the book, part III, opens with the vaccine initiative spearheaded by President Clinton (chapter 7). Clinton's Vaccines for Children Program (VFC) went into effect as the nation's health experts were deliberating on the use of the hepatitis B vaccine, the subject of chapter 8. In the 2000s, the hepatitis B vaccine came to signify for many vaccine skeptics all that was wrong with the nation's vaccination approach, and long-simmering popular misgivings about vaccines exploded into a full-fledged, multidimensional controversy. Chapter 9 looks at how one of the biggest vaccine controversies of the 2000s — over a contested link between vaccines and autism — unfolded in the changing media environment of the new millennium. The new media environment also played a central role in national disputes over HPV vaccination later that same decade, which is the subject of chapter 10.

Much, inevitably, is left out of this account. The book focuses on the vaccines developed for and deployed most widely in U.S. children in the second half of the twentieth century and the beginning of the twenty-first. The book's focus on domestic vaccine use and policies means that discussion of foreign or global vaccination efforts, even those that had an impact on U.S. vaccination, is limited. Many good histories of vaccination in international settings have been written, and I refer readers to these works to put this American story in broader context. The book also focuses most closely on those vaccines subject to school mandates, but it does not address all of these. I discuss the Hib, meningococcal, and pneumococcal vaccines only in passing, for example. I do not discuss the flu vaccine, despite the fact that it is recommended for children and has recently become mandatory for some children in a few states and for some health care workers. I have also excluded vaccines long under development that might someday be the object of school mandates, such as a vaccine against HIV. Lastly, I use the term "mandate" to refer to laws that require vaccination for school enrollment, even though exemption clauses permit some students to be exempt from these requirements in all fifty states.[35]

A word on chronology: the chapters proceed in roughly chronological order, from the sixties to today. On occasion, the chronology of chapters overlaps or wraps back on itself. In some cases, I have done this deliberately in order to trace a single idea over an extended period of time in a single chapter, such as the influence of the women's health movement or the nation's changing environmental ethics on vaccination beliefs. In other cases, the chronology of separate chapters overlaps because the events discussed—such as the debate over vaccines and autism and debate over the HPV vaccine, discussed in chapters 9 and 10—overlapped themselves.

The story outlined on the pages that follow highlights several unresolved questions about the very nature and purpose of vaccines. Vaccines are developed in response to disease threats, but they also fundamentally change how we think about disease. They are an important medical technology, but they have also been important political tools in an era plagued by tensions over the cost and provision of health care. Resistance to vaccination and vaccine policy in recent decades has been complicated and deeply layered, and informed by significant social trends, the depths of whose influence hasn't been fully examined. But just as interesting—and important—a line of historical inquiry concerns the question of how and why Americans came to accept the universal vaccination of their children against such a long and varied list of infections in the last decades of the twentieth century. The answer, as this story illustrates, has varied by vaccine and by historical moment, but it has consistently reflected prevailing cultural attitudes toward technology, state power, social hegemonies, women, children, and disease itself.

PART I

1

KENNEDY'S VACCINATION
ASSISTANCE ACT

A crush of reporters gathered for a presidential press conference late one afternoon in April 1961. They were poised with pressing questions about the national economy, "Red" China, Communists in Cuba, and the Soviets' recent success launching the first human into outer space. But President John F. Kennedy put their questions on hold to make a brief announcement: "Today is the sixteenth anniversary of the death of President Franklin D. Roosevelt," said the president. "It is also the anniversary of the announcement of the vaccine which has been discovered to prevent paralytic polio." He told the reporters that there was a renewed national effort to vaccinate all Americans against polio, and that he hoped "every parent in America" would take advantage of "this miraculous drug."[1]

It was a succinct but carefully timed and scripted announcement. By mentioning Roosevelt, Kennedy invoked the kind of bold, progressive Democrat he intended to be; Roosevelt also happened to be the nation's most famous polio victim and survivor. By mentioning the "miraculous" polio vaccine, Kennedy invoked the nation's impressive biomedical enterprise and its recent victory against a disease that had gripped Americans with fear in the 1940s and 1950s. As Kennedy went on to field questions from the reporters, he found no further opportunity to mention polio or its vaccine; the nation was no longer worried about polio, and the press was fully preoccupied with the economy and the Communist threat. But even as the president spoke, federal regulators were reviewing the evidence on a newer and more effective polio vaccine. And his administration was crafting a bill that would give the federal government an unprecedented role in supporting and shaping immunization all across the United States, against an unprecedented number of diseases.

That bill was initially conceived as a continuation of an Eisenhower-era program of the mid-1950s that had helped provide the first polio vaccine to tens of millions of Americans. But the Kennedy administration proposed taking things further. The resulting Vaccination Assistance Act of 1962 was a quietly significant piece of legislation.[2] It gave states federal dollars and guidance to protect all Americans from a list of diseases—diphtheria, pertussis, tetanus, and polio—chosen not for the number of lives they took, or for the public attention they commanded, but for the existence of their vaccines. Policy, here, took its shape from existing technologies. Indeed, the act was remarkable because it gave Americans something they hadn't asked for: immunization for all, against all vaccine-preventable diseases.

But why did Kennedy promote legislation that few Americans had asked for, and how did he succeed in getting it enacted? As this chapter shows, the vaccination act squared neatly with the president's domestic policy focus on social welfare and health reforms, and it reflected his predilection for cost-conscious and "compassionate" legislation. It bore the influence of his family's personal interest in children's health, and it helped counterbalance the administration's focus on health care for the elderly—a fact that Kennedy's sister Eunice Shriver was blunt in pointing out. It was propelled through Congress by Cold War fears of disease as a national security threat, a zenith of cultural faith in biomedicine and medical technologies, the nation's recent experience with polio, and widening acceptance of routine medical care and vaccination for children.

Kennedy's act wasn't just remarkable for its time; as subsequent chapters of this book illustrate, it would also turn out to have enduring significance. In fact, the story of vaccination and society in modern times begins with Kennedy's proposal precisely because it laid a new foundation for federal immunization policy and set a precedent for federal involvement in vaccination promotion that subsequent Democratic administrations—such as those described in chapters 4 and 7—would mimic and expand upon. Decades later, health experts continued to refer to it as "one of the most successful prevention programs in public health."[3] To this day, in fact, it remains a "mainstay" of public support for child vaccination—even as it remains a product of the Kennedy administration's priorities and the national preoccupations of the early sixties.[4]

POLIO POLITICS

In many ways, today's vaccination enterprise is a legacy of the Vaccination Assistance Act of 1962. And that law, in its own time, was made possible by the nation's experience with polio and the success of a federal polio vaccination act passed in the 1950s. Polio, a viral infection that can cause permanent crippling and fatal paralysis, appeared somewhat suddenly in the United States around the turn of the twentieth century. Roosevelt was struck in 1921 and left paralyzed at the age of thirty-nine.[5] He found some relief at a crumbling mineral springs resort in Georgia, which he bought and turned into a nonprofit institute for polio patients. By the 1940s, the small institute had been transformed into a massively successful philanthropy and fund-raising machine, the National Foundation for Infantile Paralysis, later renamed the March of Dimes. From the 1930s through the 1950s, the foundation inundated Americans with information about polio through ubiquitous and celebrity-studded publicity blitzes. Its campaigns popularized the "poster child," a young patient whose image tugged at heartstrings and helped prompt millions of Americans to give time and money to the cause.[6] They also put polio firmly on the national agenda and made the disease's defeat a national priority (even as heart disease, cancer, accidents, and pneumonia and flu claimed more lives).

The foundation also heavily financed the vaccine research of several scientists, including Jonas Salk and Albert Sabin. In 1954 the foundation financed a field trial of epic proportions to test Salk's killed-virus vaccine in children. With some 1.4 million "Polio Pioneers" participating in forty-four states, it was the largest medical experiment ever conducted. A year later, on the tenth anniversary of Roosevelt's death, the trial results were announced before a bank of reporters and television cameras.[7] The Salk vaccine worked, and the country was jubilant—and then desperate for access to the long-awaited miracle drug. The foundation had arranged to buy the first $9 million worth of vaccine for the country's first- and second-graders, who suffered the highest rates of the disease, but it was not at all clear how and when others would get the vaccine. Rumors spread of black markets and doctors charging exorbitant fees, and panicked health officials, parents, and others turned to Washington for help.[8]

Federal assistance came, but it was hard-won. Representatives of the American Medical Association (AMA) and the American Drug Manufacturers' Association argued that the government had no place determining how to distribute or pay for a vaccine. President Dwight D. Eisenhower himself was a moderate conservative with little tolerance for federal involvement in health care.[9] The Democratic Senate wanted to make the polio vaccine available to all children, but the Republican-controlled House favored a bill to provide the vaccine to needy children only. The lawmakers' partisan dispute made headlines, drew charges of socialism, and at one point forced the House Speaker to demand that legislators take a "cooling-off" recess.[10] The compromise bill, which Eisenhower signed in August 1955, allotted $30 million for states to vaccinate children under twenty and pregnant women—as they saw fit.[11] As the Polio Vaccination Assistance Act was being hammered out, and upon its approval, the press made much of the fact that although a federally devised formula had determined how much vaccine each state received, ultimately the authority remained with each state, and not the federal government, to decide who would get the shots and where.[12] The message was clear: the act was not to be taken for socialized medicine.

By the late 1950s, as historian James Colgrove has shown, the National Foundation for Infantile Paralysis had helped move vaccination from local health departments and medical societies to the national stage. By setting in motion the series of events that prompted passage of the 1955 act, the foundation had also (inadvertently) helped nudge the federal government into the "reluctant" role of director.[13] This new arrangement appeared to be effective. In the year Kennedy entered office, 1961, the United States saw just 1,000 polio cases, down from more than 58,000

SPACEKNIGHT™
Armor, Spaceknight, Warrior

Galador Spaceknight can use **Charge**, **Force Blast**, and
Shot.

SPACE TRANSPORT (Phasing/Teleport)

NCED STRENGTH (Super **Strength**)

RED STAFF (Blades/Claws/**Fangs**)

RALIZER (Incapacitate)

DANIUM ARMOR (Invuler **ability**)

INT VALUE: 100

MARVEL
© MARVEL   marvel.com

han half of all Americans had received the requi-
vaccine; many more had received one or two.[15]
os, health experts had begun to notice a new pat-
men were still vulnerable to the disease, a finding
ational campaign urging that "Babies and Bread-
d.[16] In the course of the nation's battle against
emed, had come to think that vaccines were for
oolchildren only.

A CAPTIVE AUDIENCE

f the National Foundation for Infantile Paraly-
t the notion that vaccines were for children
firmly onto the American psyche. School-age children were most at risk
of polio, but they were not exclusively at risk, as starkly evidenced by
former president Roosevelt's paralytic polio. Nonetheless, the ubiquitous images of poster children, the recruitment of over a million young
Polio Pioneers, and the foundation's plan to ration limited vaccine to
five- to nine-year-olds placed children at the center of the polio vaccination crusade, and as a result vaccination as a service for adults was
increasingly overlooked. But the nation's experience with polio wasn't
the only factor driving home the impression that vaccination was for
children; it had much to do with the rise of pediatric care, as well.

Pediatrics as a medical specialty had emerged in part from a social
movement, originating in the latter nineteenth century, that focused on
child health and welfare.[17] This child welfare movement spawned laws
governing compulsory schooling for children, child labor laws, delin-
quency reform, and more. In the 1900s and 1910s, Progressive reformers
made schools the site of compulsory medical examinations to test hearing, vision, and dental health, taking the position that schools provided
the most expedient means of improving not just children's minds, but
their bodies as well. To social reformers of the time, the comprehensive
welfare of all children promised a strong future for the state. This same
belief inspired increased federal involvement in children's lives in the
early twentieth century, dubbed the "Century of the Child" by reform-

ers.[18] The U.S. Children's Bureau was established in 1912 to "investigate and report upon all matters pertaining to the welfare of children and child life," including infant mortality, child labor, orphanages, adoption, accidents affecting children, "degeneracy," and childhood diseases.[19] The Sheppard-Towner Maternity and Infancy Protection Act, passed in 1921, gave states federal dollars to provide health and welfare services for children and their mothers.[20]

Pediatrics took shape in the context of this broader social movement. In the first few decades of the twentieth century, the practice of pediatrics—the medical care of children—was increasingly characterized by regular, repeated medical examinations of healthy children. Children were X-rayed, weighed, and measured, and their blood and urine were tested, all in an effort to ensure their "normal" development and keep them free of illness and disease.[21] This type of preventive pediatrics was initially designed for the working class, as sociologist Sydney Halpern has noted.[22] And originally, the care of the well child (as opposed to the sick child) often took place in health centers supported by the federal U.S. Children's Bureau and public clinics supported by the Sheppard-Towner Act. But by the 1940s, this had changed.

In part, that's because pediatrics had become firmly established as an independent medical specialty in the 1930s. The American Academy of Pediatrics was formed in 1930, and the organization issued its first set of immunization recommendations for children in 1934. Meanwhile, as publicly supported sites for well-child care disappeared (Congress stopped funding Sheppard-Towner in 1929), pediatricians' offices became the most logical setting for all well-child services, including vaccination; a series of shots followed by boosters could be easily administered at the same time as a child's regular barrage of tests and measurements. But as a result of this shift from public clinics to private offices, well-child care—and vaccination—increasingly became the domain of the middle class. Over the next few decades, pediatric care offered a convenient setting for the administration of vaccines, a set of professionals eager to own the practice, and a captive audience—but it was an audience defined by its ability to afford preventive care. As members of Congress considered federal legislation to make vaccines available to all Americans in the 1960s, they remarked on precisely this fact: middle-class children received vaccines as a matter of course, but the same was not true of poor children.

THE VACCINATION ASSISTANCE ACT OF 1962

On June 26, 1962, members of the House of Representatives debated a mass immunization bill that had come directly from President Kennedy's office. The administration's bill proposed to give states grants to carry out intensive immunization programs against diphtheria, tetanus, pertussis, polio, and—as new vaccines became available—any other disease deemed a health threat by the surgeon general. It was a problem, the assembled representatives agreed, that so many Americans were not taking advantage of vaccines to prevent disease: 80 percent of all adults and two-thirds of children under the age of five, the majority of them from "low-income" families, were not fully immunized against polio, diphtheria, pertussis, and tetanus. But as members of the House debated the bill's merits, many made it clear that they saw the president's proposal as an imperfect solution. To Peter Dominick of Colorado and Barry Goldwater of Arizona, the root of the problem was not a lack of facilities, vaccine doses, or dollars, but inadequate communication and education. To Harold Collier of Illinois, the bill was redundant, as states already had access to general public health funds from Washington that they could use to vaccinate their populations. To Representative Dave Martin of Nebraska, the bill was simply a worrisome and unnecessary expansion of federal activity.[23]

The administration's proposal provided for the immunization of all Americans, but it placed special emphasis on preschoolers, as did the House debate and other hearings on the bill. The problem of inadequate immunization was "not confined to children," noted Secretary of Health, Education, and Welfare Abraham Ribicoff, but children were at "greatest risk" from the specified diseases and thus constituted a "community epidemic hazard."[24] The bill did not place special emphasis on poor Americans—of any age—but discussions in Congress returned again and again to the findings of a recently commissioned Public Health Service report that "the lower the socioeconomic class, the lower the level of immunization."[25] Though some of the bill's proponents, like Representative Leonor Sullivan of Missouri, endorsed the idea of providing all children, of all means, with vaccines against any disease, this was not the view that held sway with most members of Congress in the spring of 1962. A House committee narrowed the bill's scope to the four diseases for which vaccines already existed and altered its emphasis to cover children in need above the age of five. In that revised form, the bill passed the House. In October the bill passed the Senate, and on the twenty-third of that month, in the midst of the Cuban Missile Crisis, President Kennedy signed it into law. Its title: "An Act to assist States and communities to carry out intensive vaccination programs designed to protect their populations, particularly all preschool children, against poliomyelitis, diphtheria, whooping cough, and tetanus."

The Vaccination Assistance Act added a new section, section 317, to the federal Public Health Service Act, which had been enacted in 1944 to consolidate the nation's public health legislation. Like the Polio Vaccination Assistance Act before it, the 1962 act gave states funds to conduct short-term mass immunization campaigns, over just a few years; it did not support ongoing vaccination efforts. Under the act, the federal government could cover both the cost of vaccines and the personnel needed to carry out immunization programs. The government could also buy vaccines directly from manufacturers and give them to states in lieu of cash. States that applied for funds had to assure the surgeon general that they would make vaccines available to private physicians and not just public health departments. The act also specified that states were not required to force vaccines on people who didn't want them for themselves or their children. The White House held fast to the line that, like the polio program before it, the new program was not a form of socialized medicine: "While the program would provide Federal stimulation and assistance, the actual work would be carried out by State and local people," Ribicoff stressed.[26]

The idea of federal support for such a broad mass immunization plan took shape quietly and was initially announced almost as an aside, just one small component of a much larger vision to improve American health and welfare. Kennedy first unveiled the plan, along with a list of proposed welfare measures, in his State of the Union address early in 1962.[27] In that speech, Kennedy's unanticipated description of a program "aimed at the virtual elimination of such ancient enemies of our children as polio, diphtheria, whooping cough, and tetanus" caught even some of his own administration officials off guard, prompting the *Wall Street Journal* to equate the "mystifying" announcement to the dropping of a "bomb."[28] A month later, when Kennedy gave a special address on national health needs to unveil his proposal for what would become Medicare, he simultaneously announced several measures to improve the "health and vitality" of the overall population. Here, his immunization plan joined a list of proposed measures including a major budget increase for the National Institutes of Health and added funds for medical schools and medical education, mental health and mental retardation, air pollution prevention, and children's physical fitness.[29]

Medicare was crucial for protecting the nation's elderly; the rest of the proposals Kennedy outlined that day were described as crucial for safeguarding the health of the nation's youths. By 1962, the tail end of the baby boom years, close to 40 percent of the nation's population was under twenty.[30] This large stratum of the population, key to the country's future security and prosperity, was in less-than-ideal shape, noted the president: they were physically unfit generally and handicapped by preventable communicable diseases and venereal infections. Because of these "correctable health defects," a full fifth of applicants for U.S. military service were being rejected.[31] To hear the president expound on the state of the nation's children before Congress, the poor health of young Americans was no small threat to the United States' position as a world power, especially as the Cold War escalated.[32] The idea of unhealthy youths constituting a national security threat was likely familiar to many members of Congress, as it had been expressed before, when the military took stock of the poor health of its recruits in World War I and World War II.[33]

But why focus on a set of diseases that were already in steady and sharp decline? Thanks to the widespread administration of the Salk polio vaccine, cases of polio, for instance, had fallen from over 30,000 in 1955 to less than 900 in 1961.[34] Nevertheless, polio was still a familiar and feared disease in the early 1960s. The National Foundation for Infantile Paralysis had spent the late 1950s reminding Americans that polio wasn't "licked yet," and when Albert Sabin's new orally administered polio vaccine, protective against all three types of polio virus, became available in 1961, Americans turned out in droves once again to get immunized. On "Sabin Sundays" held across the United States in 1962 and 1963, 85 to 95 percent of the residents in rural counties, small towns, and even cities took the anxiously awaited vaccine, whose development (like that of the Salk vaccine before it) had been dramatically chronicled in the press.[35]

Moreover, polio wasn't the only "ancient enemy" on American's minds in 1962. That year smallpox returned to the United States for the first time since the 1940s when an infected boy, traveling with his family from Brazil to Canada, landed in New York's Idlewild Airport, rode in a taxi, and hopped on board a train at Grand Central Station. Media outlets reported on CDC officials' efforts to track down individuals who had come into contact with the boy and place them in quarantine—a seemingly ancient remedy for modern times. When smallpox broke out aboard an ocean liner from Italy that dropped off passengers in Washington, D.C., later that year, the press again chronicled CDC scientists' efforts to track down the exposed.[36] In 1962 and 1963, health departments across the country reminded citizens of the importance of vaccination in an era when "rapid international travel" could reintroduce smallpox to the country at any time from foreign ports.[37]

Smallpox wasn't included in the Vaccination Assistance Act, though several observers asked whether it should have been.[38] (Child-care guru Dr. Spock, for one, still called the vaccine a "must for all babies."[39]) The other three infections included in Kennedy's immunization proposal, meanwhile, were well off the public's radar in 1962. About 8,000 children a year still suffered from pertussis, the medical term for whooping cough.[40] Cases of diphtheria and tetanus—preventable with a combined vaccine available since 1947—had meanwhile dropped to a few hundred each. But none of the three infections had been the subject of high-profile campaigns. Nor did they attract much press coverage. In fact, throughout 1962, the three infections had received only the rarest of mentions in the media.[41]

Outside the United States, however, the infections had become the focus of renewed scientific attention following World War II. Cases of diphtheria, long in decline, had spiked across Europe during the war, inspiring new research on the disease. When the World Health Organization (WHO) was established shortly after the war, it quickly set to studying diphtheria and its prevention via immunization. By then, a combined vaccine against diphtheria, pertussis, and tetanus was available. As the WHO continued its diphtheria research in the 1950s, it increasingly studied the disease as one of the several vaccine-preventable diseases of childhood. When the Salk and Sabin polio vaccines were developed, European nations used anti-polio drives as opportunities to immunize against diphtheria, pertussis, and tetanus, too. Combined vaccines that protected against all four infections soon streamlined the process even further. Scientific journals in Europe, the United States, and Canada began to predict that combined vaccines then under development would soon facilitate prevention of smallpox and measles as well. Across North America and Europe, the coincidence of war and the development of new technologies had cultivated scientific interest in childhood immunization generally, and a few "ancient" diseases specifically.

But as one observer remarked in the mid-1950s, combined vaccines did nothing to address an enduring vaccination question: just how childhood immunization was to be carried out, "whether by legal compulsion or by persuasion . . . is still a matter of debate in most countries in the temperate zones."[42] Wartime France, for one, had chosen to make diphtheria and tetanus immunization compulsory in 1940; laws in Sweden and the German Democratic Republic, by contrast, supported immunization on a voluntary basis. The United States was one of a handful of Western countries (including Finland, the Netherlands, Norway, and others) in which authorities had the power to enforce vaccination as needed.[43] But as one textbook of the day noted, U.S. health officials opted to use compulsion "as sparingly as possible."[44] Persuasion was preferred, and congressional proponents of the Vaccination Assistance Act appeared to have taken note of this.

SELLING THE ACT

Though Americans showed little concern for diphtheria, tetanus, and pertussis in the early 1960s, congressional supporters of the president's proposal cast the infections as pressing public health matters. Against the backdrop of the Cold War, it was easy to argue that even seemingly defeated infections posed a threat to all Americans. Although the Vaccination Assistance Act was "not designed as a national defense measure . . . [it] might have a substantial impact in this regard," the bill's House sponsor, Oren Harris, pointed out.[45] In the event of a nuclear attack, roughly 70 percent of those affected would suffer injuries including "penetrating wounds, contaminated with dirt." In such a scenario, tetanus would be rampant, and less than 50 percent of those infected could be expected to survive.[46] Diphtheria, too, could be expected to spread wildly in a time of disaster. The infection would thrive inside crowded bomb shelters; Germany had faced major outbreaks during the Second World War, noted the bill's Senate sponsor, Lister Hill. Vaccinating all Americans against these diseases—and Hill focused on all Americans, not just children— would not only save lives in times of disaster; it would also serve the economy in times of peace, by preserving "the productive capacity of individuals who are now handicapped as a result of these diseases."[47]

If these justifications were abstract or hypothetical, Secretary Ribicoff's arguments in favor of the bill were practical and immediate. As some of its critics, like Nebraska Representative Martin, noted, the Vaccination Assistance Act expanded federal involvement in immunization efforts beyond the level set by the polio vaccination act. It established a leadership role for the Public Health Service, and the CDC in particular, in coordinating and overseeing nationwide immunization efforts.[48] Indeed, this was the administration's objective; the bill's purpose, according Ribicoff, was to "provide Federal leadership in assuring that [vaccines] will be so utilized as to achieve the maximum benefits and protection to the public."[49] The timing was ideal for such an effort, argued Ribicoff: across the United States, communities were already carrying out campaigns to stamp out polio with Sabin's new oral vaccine. With "comparatively little extra cost and effort," the same campaigns could administer the combined vaccine against diphtheria, pertussis, and tetanus (DPT) and wipe out four diseases at once, much as was being done in parts of Europe. Federal coordination of such immunization efforts would yield enormous benefits: a national program could make use of national organizations and resources not accessible to local programs.

Equipment, supplies, and educational materials could be standardized and used more efficiently. Federal leadership would not just be a boon; it was a necessity. While a handful of states were immunizing all citizens against all four diseases, they were at constant risk of new infections coming in from other states. The U.S. population, noted Ribicoff, was increasingly mobile, and they took their infectious diseases with them. It was an age-old refrain with new urgency in the age of "jet travel": diseases did not recognize state borders.[50] So, the maxim was put forth as justification for increased federal power.

The idea that the nation's health and medical needs would be best served by stronger federal leadership was a matter of political philosophy, of course, and it was one that Kennedy and his selected advisers and cabinet members espoused. Shortly after his election in November 1960, Kennedy had assembled a task force to recommend health and welfare legislation. He appointed Wilbur Cohen, a founder of the Social Security system, as its chair. It was the task force that initially recommended that the new administration invest in medical education and manpower, medical research, and medical care facilities, and establish a program to provide medical care for the aged. The task force did not recommend a vaccination program, but many of the characteristics of the program it outlined for the aged were reflected in the administration's mass immunization proposal. The group's final report advised the federal government to take a "more vigorous role in the financing, organization, and stimulation of health and medical care." The program should not "socialize" medicine, nor should it "supervise" or "control" the practice of medicine, they advised, but a national advisory council should set policy and recommendations to guide it.[51]

The immunization program, likewise, did not tell states how to carry out their campaigns or which vaccines to use. (Members of the administration stressed that local programs could decide for themselves whether to use Salk's or Sabin's polio vaccine, for example.) But even though state activities wouldn't be supervised or controlled by the federal government, the CDC's role in streamlining state activity quickly expanded. In 1963 the CDC was given responsibility for administering the new federally funded immunization grants to states.[52] The agency's immunization operations grew as it provided not only money, but also promotional and educational materials, courses, seminars, and even government-trained personnel to states and metropolitan areas, giving immunization programs across the country what one agency official called "a unity of purpose."[53] By nationally coordinating immunization efforts, the CDC in effect began to do for all vaccines what the March of Dimes had once done for the polio vaccine.

In 1964 the agency took on an even more authoritative role. That year Surgeon General Luther Terry appointed the Advisory Committee on Immunization Practices, a group of experts chaired by the head of the CDC and charged with producing recommendations on the "most effective" use of vaccines in "communicable disease control." The committee was the first of its kind; prior to its establishment, separate immunization recommendations were issued by the armed forces, the American Academy of Pediatrics (AAP), and the American Public Health Association (APHA). But the armed forces concerned itself with only the needs of its troops; AAP's recommendations were limited to the private pediatric care setting; and APHA's recommendations were issued every five years, and thus failed to keep pace with the "rapidly changing field" of vaccine development.[54] With the formation of the ACIP, the United States had, for the first time, a singular, centralized governmental authority advising the nation's health and medical professionals on who should be vaccinated against what and when. The federal government no longer inhabited the role of director reluctantly, but with a newfound sense of permanence.

CHILDREN AND YOUTHS

The immunization program outlined in the Vaccination Assistance Act bore superficial resemblance to the administration's plan to provide care for the aged. But its emphasis on children (a cause célèbre of some members of the Kennedy family) was also likely proposed as a counterbalance to Medicare. President Kennedy came from a politically liberal, politically active, and wealthy family, and when he entered the White House, his sister Eunice Kennedy Shriver was running the family's philanthropy, the Joseph P. Kennedy, Jr. Foundation. Their father had founded the philanthropy in the 1940s, with somewhat dubious intentions, but under Shriver's direction in the 1960s, the foundation's focus narrowed in on the cause of mental retardation. John and Eunice's sister Rosemary had been born "mentally retarded" (an accepted diagnosis at the time); she suffered ineffective treatments as a child and had been institutionalized since her twenties.[55] When Shriver heard that Social Security alum Cohen would chair the task force on health, she worried that the health needs of the young, including mental retardation, "would be lost sight of with Wilbur Cohen's singular concern for the elderly."[56]

Shriver pressed her brother to include Nobel Prize–winning geneticist Joshua Lederberg and Johns Hopkins University pediatrician Robert Cooke on the task force.[57] With Lederberg and Cooke's contributions, Shriver hoped the panel would come up with cutting-edge and "creative" proposals to serve all children, but particularly the mentally retarded. In the end, the task force recommended that the administration move the Children's Bureau and maternal and child health grant programs to elevated positions in new agencies; create an institute of family and child welfare research; and establish a new National Institute of Health devoted to children's health. The last proposal—which faced resistance for its focus on a stage of life rather than a specific organ or disease, like the other National Institutes—was the administration's hardest won.[58] Kennedy also championed legislation specific to mental retardation, and as he signed the first such bill into law in 1963, he pointed to the near-elimination of diphtheria and pertussis as proof that mental retardation would soon meet a similar fate.[59]

Shriver insisted that her brother's support for the cause of children's health was driven not by the Kennedy Foundation or by Rosemary's experiences, but by his own interest in the health and welfare of children. In an interview later in life, Shriver claimed that the death of Kennedy's youngest son, Patrick, from neonatal respiratory distress just a few days after birth, convinced the president to push for meaningful investments in children's health.[60] But Patrick died in August 1963, well after the administration proposed both the National Institute of Child Health and Development and the Vaccination Assistance Act. Kennedy did have his own personal health issues to draw on. He suffered ill health from a very young age; he had debilitating gastrointestinal problems and back pain that worsened when he was a teenager, and into adulthood medicine failed repeatedly to provide him with definitive cause or cure. On top of these chronic maladies, Kennedy suffered a long string of childhood illnesses while growing up: chicken pox, rubella, measles, mumps, scarlet fever, whooping cough, and bronchitis. Of all these ailments, only whooping cough was preventable with a vaccine by the time he proposed his mass immunization bill. And at any rate, according to Kennedy biographer Robert Dallek, Kennedy had "pleasant" memories of reading and being read to by his mother while confined to bed with the common childhood diseases.[61]

More important than personal health issues, perhaps, were the midterm elections the president would face in 1962, his second year in office. According to Dallek, Kennedy's key political objectives that year were to strengthen his support in Congress by prioritizing domestic issues, bolstering the economy, and proving that America remained a strong but "compassionate nation."[62] A far-reaching vaccination program would seem to fit the bill; it would improve American health, which would, as Hill pointed out, improve productivity, and it would be cost-efficient to boot. The "mass immunization" bill, as Kennedy staffers called it, was placed on the president's list of major legislative proposals for 1962 alongside bills on trade expansion, tax reform, the Peace Corps, and a farm bill.[63] One of the president's congressional liaisons called it a "major point of President Kennedy's legislative program."[64]

The act aligned with the progressive, "compassionate" agenda for which the Kennedy administration became known; it also aligned with the administration's focus on the nation's youths, who were often held up as a key measure of the state's future strength. "If our young men and women are to attain the social, scientific and economic goals of which they are capable," said Kennedy, "they must all possess the strength, the energy and the good health to pursue them vigorously."[65] The act thus joined other social policies that originated in the White House to support youth physical fitness, education, and employment. As in an earlier era characterized by progressive politics, the interests of state and children were equated as one and the same.

On announcing the Vaccination Assistance Act, Kennedy remarked that there "was no longer any reason" why American children should have to suffer from polio, diphtheria, whooping cough, or tetanus.[66] On one level, the act was meant to equalize access to health services across social strata, but there was also more to it than that. Vaccines both symbolized and secured American scientific and economic exceptionalism. The United States, "with the most vigorous, productive, and imaginative total research effort that has ever been witnessed by the world," was leading the way in a "golden age of medical research," noted Kennedy's Assistant Secretary of Health, Education, and Welfare Boisfeuillet Jones.

Jones, among others, predicted an imminent end to the "major disease of mankind."[67] In this golden age, the vaccine-preventable infections, even those that didn't appear to be major health threats, ought to be regarded at the very least as unnecessary; modern medicine had made it so. Vaccines, in this view, didn't just prevent unnecessary deaths; they supported productive, prosperous, modern, and convenient American lifestyles. They safeguarded communities, protected workers in all industries, shielded overseas travelers, and kept pupils safe in burgeoning schools, enabling them to become productive citizens in the future. Moreover, because vaccines were a product of American research efforts, to the Kennedy administration, they were the province of every American and *needed* to be the province of every American to ensure a great future for a great country.

In the context of national pride in the biomedical sciences, mass immunization was a safe political bet. And in fact, objections to the Vaccination Assistance Act were few and far between. Christian Scientists, who eschewed medical interventions on principle, saw in the act an unprecedented expansion of mandatory "mass medication."[68] A group that called itself the Commission on Constitutional Government bemoaned the law's "encroachment upon responsibilities that rightfully should be shouldered by the states, the localities, and by private volunteer efforts."[69] A few individuals sent letters directly to the White House to express the view that in a "free country," health decisions such as vaccination should be a "personal matter."[70] "People should be free to refuse vaccination if they do not believe in it (no matter what their reason is)," Mr. and Mrs. James De Haan, of Chicago, wrote in a letter to the president.[71] Such objections were near-perfect echoes of the libertarian protests of Progressive Era vaccination opponents (but were too faint to hamper the advent of pro-immunization policy).

Doctors, however, also wrote to Washington to oppose the new vaccination law. Pediatricians in particular saw the proposal as a threat. "A Government immunization program that would wrest from the physician the responsibility for the prevention of disease would deny, or seriously compromise, the concurrent supervisory health care that attends such immunization in the doctor's hands," wrote the head of the Cincinnati Pediatric Society, which viewed the bill "with grave concern."[72] The AMA, meanwhile, supported the act—with several caveats. In particular, they wanted to make sure the bill was limited to just four infections. They also wanted it to require states to match federal funds and be limited to supporting state-run activities, not more local campaigns.[73]

The need for diplomacy with the medical profession brought CDC chief James L. Goddard before a meeting of the AMA in 1963. Immunizations, Goddard told the assembled physicians, were part of a multipronged approach to infectious disease control. Vaccines alone had never eradicated a disease, he acknowledged, and in the case of smallpox, improvements in environmental conditions and standard of living were crucial to the disease's disappearance. But changing times meant that vaccines were becoming increasingly important as a means of protecting Americans against these sometimes-forgotten diseases. For nothing but immunization would have guaranteed New Yorkers protection from the smallpox imported by the boy from Brazil, he said. And nothing but vaccines would protect the wives and mothers who now—"with the shifting patterns of society"—accompanied their families on camping and fishing trips, for example, where the risk of tetanus was ubiquitous.

Goddard went on to diplomatically outline a sort of division of labor between public health and medicine, one that secured a clear role for federal health authorities without stepping on medicine's toes. The CDC would advise, conduct surveillance, respond to outbreaks, and maintain the expertise to confirm diagnosis of infections, like smallpox, that were seen less and less often in doctors' offices. Organized medicine, meanwhile, needed to counteract Americans' prevailing notions regarding vaccines: that they were for children; that they were needed only upon school entry; and, thanks to the nation's experience fighting polio, that a one-time vaccination campaign was all it took to wipe out an infectious disease. Goddard, a physician himself, pressed the assembled doctors to urge their patients to get vaccinated against all four of the diseases targeted by the Vaccination Assistance Act. Although these diseases were "of relatively minor importance in this country," said Goddard, they were — thanks to vaccines — "diseases about which we could say that *one* is too many."[74]

Under Kennedy, the federal government outlined a clear set of reasons for consolidating centralized authority in the realm of immunization. These reasons merged with national interests in defense and scientific advance, as well as the Kennedy administration's specific interests in social equity, the needs of youths, and liberal governance. The Vaccination Assistance Act likely had another point of origin as well. In early 1961, prior to the approval of Sabin's new oral polio vaccine, Kennedy had asked Congress to appropriate funds to stockpile 3 million doses of polio vaccine in the event of an outbreak. A congressional committee held hearings on the matter, but instead of approving the request, the committee asked the Department of Health, Education, and Welfare to prepare a report, mentioned earlier, on "the vaccination picture in the Nation."[75] It was this report, conducted by researchers with the national Public Health Service, the parent agency of the CDC, that pointed out that immunization rates were particularly low among preschoolers, in areas of poverty, for all vaccine-preventable diseases. The report also noted that a vaccine against measles — which "causes more deaths than polio" — was likely to be available soon, as well as vaccines against other infectious diseases. But such new technologies would do little to improve health if they weren't universally used, which unfortunately seemed to be the pattern. "Vaccines are effective," the authors wrote, "but not reaching all in need."[76] The report made a convincing case for expanded federal support for immunization, an agenda federal health officials had been hoping to advance. With the support of the Kennedy administration, this agenda moved swiftly forward.

The administration's original draft of the Vaccination Assistance Act was in fact written with measles in mind. The vision, articulated by Ribicoff before Congress, was that the act would position the nation to immediately launch a nationwide immunization campaign against the disease as soon as the vaccine was ready.[77] With Congress's adjustments to the bill, this vision was not realized. But the vision of federal leadership was. Unlike its legislative predecessor, the polio vaccine act of 1955, the 1962 law represented a forward-thinking effort to lay the groundwork for ongoing federal leadership for local, state, and municipal vaccination activities. And, indeed, that's how it came to be used.[78] In 1965 the Vaccination Assistance Act would be renewed. It would be expanded to cover all immunization activities, not just "intensive" ones, and it would be extended to cover measles. The following year, the CDC would lead the nation in an unprecedented effort to completely eradicate measles from the country by vaccinating all children against the disease. States would still retain the sovereignty to launch immunization campaigns and purchase and distribute vaccine as they saw fit, but from 1963 on they would increasingly do so with strong federal guidance and resources.

2

POLIO, MEASLES, AND THE
"DIRTY DISEASE GANG"

In September 1962, about a month before Kennedy signed the Vaccination Assistance Act into law, a white eleven-year-old girl was admitted to a hospital in Sioux City, Iowa. Her face was swollen, her cervix was inflamed, and her enlarged tonsils were coated with a "grayish membrane" that covered the upper part of her throat. The swelling obstructed her breathing, so doctors cut an incision in her trachea to bring oxygen to her lungs. But the girl only worsened: the wound turned purple and red and then bled, her kidneys failed, and she seized repeatedly. She fell into a coma, and on her fifth day in the hospital she died. Cultures of bacteria from her throat confirmed the cause of death: diphtheria.

Alarmed by the severity of the case and the prospect of an outbreak, scientists from the state health department and the CDC descended on Sioux City. The outbreak turned out to be mild: it caused seventeen cases, claimed no other lives, and was over in a couple of weeks. But investigators were stumped as to how it had begun in the first place. Ninety-one percent of children in the district were vaccinated, so it was surprising that an outbreak had taken hold. There was also no obvious source of infection. The bacterium wasn't present in nearby schools where the affected children had siblings or friends. The outbreak wasn't caused by raw milk, once a common source of diphtheria, because all of the city's milk was pasteurized. Moreover, noted investigators, Sioux City had no "true skid-row section," nor was there any "district catering to transients" nearby.[1]

The start of the 1960s was characterized by an optimism about the conquest of infectious disease, as described in the previous chapter. New vaccines and new federal resources had led a growing number of experts to predict that vaccine-preventable infections would soon be wiped out for good. But in Sioux City and elsewhere, outbreaks of preventable disease persisted. Health experts who attempted to explain the trend often revived age-old assumptions about the ignorance and disease-breeding proclivities of the poor; these were the very ideas insinuated by the Sioux City investigators' comments about "skid row" and "transients." This tendency to hold the poor accountable for outbreaks also reflected a new pattern of disease that emerged over the course of the sixties. With record numbers of middle- and upper-class parents vaccinating their children, preventable infections began to concentrate in new populations. This was particularly true for polio and measles, both targets of federally sponsored vaccination campaigns that overshadowed diphtheria prevention. In the wake of national immunization efforts, polio, once a middle-class disease, became a disease of the "slums" and, in some areas, of minorities. Measles, which once struck all children, became a disease of the disadvantaged.

The complicated relationship between vaccination and social class and the ways in which vaccination transformed disease in the sixties is the subject of this chapter. The decade's campaigns against polio and measles took place in the context of a national war on poverty, widespread anxiety over the decline of American cities, and the civil rights movement; worries about poverty, urban transformation, and race were thus subtly inscribed upon the nation's efforts to immunize against these infections. The decade was also marked by growing scientific enthusiasm for disease eradication, which inspired a push to not just vaccinate against diseases, but eliminate them entirely. The decade's most high-profile vaccination campaigns both shifted their target diseases' epidemiology—the pattern of who got sick where and when—and provoked changes in the diseases' popular reputations. Measles eradication proponents, for instance, urged Americans to see measles not as a familiar part of childhood, but as a fate worse than polio, drawing upon middle-class anxieties about poverty and urban decay as they did so. As one health educator put it, measles immunization programs needed to highlight the disease's "dramatic aspects" in order to make Americans fear the disease, for only then would the country stand a chance at wiping out a disease still harbored in its "ghettos" and "slums."[2] This approach reinscribed vaccination as a middle-class concern, even as the decade's social welfare programs aimed to ensure vaccination's equitable distribution across class lines.

POLIO AND POVERTY

In 1959 Surgeon General Leroy Burney penned a letter to health departments across the country, encouraging them to redouble their efforts against polio. "We in the Public Health Service share with you a deep concern that there was more paralytic poliomyelitis in 1958 than in the previous year," he wrote. When Salk's polio vaccine was first introduced in 1955, demand for it was so overwhelming, and vaccination rates climbed so quickly, that cases of the disease quickly plummeted. But demand then quickly slackened, noted Burney. And while rates of polio were still far lower than they had been a decade before, the sudden decline in cases showed troubling signs of reversal. There were roughly 5,500 cases in 1957 and close to 6,000 cases in 1958. If health departments didn't act quickly to halt the trend, things would only get worse, because more than half the population under forty was either unvaccinated or incompletely vaccinated, and close to a third of children under five weren't protected at all against the disease.[3]

Burney's letter also pointed out that polio wasn't affecting the same segments of the population as it once had. Before 1955, children between the ages of five and nine were at greatest risk of the disease, but by the end of the decade, paralytic cases were concentrated in children *under* five, and attack rates were highest among one-year-olds. Cases of polio were also appearing clustered in urban areas, with attack rates concentrated in the poorest of urban districts. When polio struck Rhode Island in the summer of 1960—the state's first epidemic in five years— CDC epidemiologists noted that the "pattern of polio" was "quite different from that generally seen in the past." Prior to the introduction of the Salk vaccine, polio cases were scattered throughout the capital

city of Providence, "without preference for any socioeconomic group."
But in 1960, the cases were almost entirely confined to the city's lower
socioeconomic census tracts, especially in areas with housing projects,
through which the disease seemed to spread without resistance. By stark
contrast, noted investigators, "the degree to which the upper economic
areas were spared is quite remarkable."[4]

These observations were of no minor consequence; the fact that polio
was now concentrated in urban areas and among the poor meant that its
"epidemic pattern" had changed in the wake of the vaccine, according to
epidemiologists.[5] When outbreaks hit rural areas, the cases were "scattered" and the disease struck the usual school-age children. But in cities,
where most outbreaks now occurred, epidemics were concentrated in
"lower socioeconomic" areas and younger children were affected more
frequently than older schoolkids.[6] "A definite trend seems to be developing whereby poliomyelitis is appearing more in lower socio-economic
groups and among pre-school children," noted a Public Health Service
fact sheet that described polio's new predilection for urban areas.[7] There
was no mistaking that the trend had been sparked by the advent of Salk's
polio vaccine. But the trend was particularly troubling against the backdrop of the country's changing urban landscape.

In the 1950s, most of the country's largest cities had begun losing
population and wealth to new developments of detached homes on their
edges and outskirts. "The city is not growing; she is disintegrating: into
metropolitan complexes, conurbations, statistical areas, or whatever
one chooses to call them," said Burney's successor, Luther Terry, at a national meeting on health in American cities in 1961. He went on: "Those
who love her despair as they perceive new gaps in the beloved façade;
a raffishness in her grooming; her growing relief rolls; another respect-
able old neighborhood turned 'honkey tonk'; a sudden show of violence.
Those who despise her feel vindicated as they move their homes farther
from these unpleasant surroundings in which, nevertheless, they work,
or study, or trade."[8] In the decade and a half since World War II, millions
of Americans with the means to do so had moved to new suburbs that
had been built to address a national postwar housing shortage. Federal
housing loans and investment in the interstate highway system accelerated the migration. By the start of the sixties, central cities were mark-

edly poorer than their suburbs and continued to grow even more so.[9] And in public health literature and correspondence on the new patterns of polio distribution, "urban" and "poor" often came to be used synonymously. The conflation of the two categories was a testament to changing realities and presumptions about American cities and the people who called those cities home.

The nation's cities in this period were poor; they were also, increasingly, black. From the end of the Second World War through the 1950s, vast numbers of African Americans moved out of the rural South and into cities in the North and West. In total, some 3.6 million whites moved to the suburbs while 4.5 million non-whites moved to the nation's largest cities over the course of the decade. By 1960 most Americans lived in the suburbs, but only 5 percent of African Americans did; the vast majority of them lived in these "disintegrating" areas marked by "raffishness," violence, and poverty.[10] Thus, when public health officials began examining new urban outbreaks of polio, they also documented rates among "Negroes" and other non-whites that far exceeded those among whites living in suburban settings. In the Providence, Rhode Island, outbreak, there were few "Negro" cases, but that was consistent, investigators noted, with the black population in the city and state as a whole.[11] In Baltimore, by contrast, a polio outbreak affected "Negroes" at twice the rate of whites and left the suburbs untouched. In an epidemic in northern New Jersey, non-whites were six times as likely to contract polio compared to whites. In Detroit, non-whites were eighteen times as likely to contract paralytic polio compared to whites, and cases clustered in the center of the city, "paralleling the known distribution of negroes in the city." Whites who got sick lived either on the edge of "Negro areas" or in "mixed areas," investigators observed.[12]

The history of American medicine is riddled with instances of the poor and non-white being held culpable for their own infirmities and the diseases of their communities. In nineteenth-century New York, cholera epidemics were blamed on the intemperance, filth, and godlessness of the city's poor. In turn-of-the-century California, responsibility for outbreaks of plague was pinned on the habits and customs of Chinese and other Asian immigrants. Not long after, public health workers blamed poor eastern and southern European immigrants for the spread of polio through American cities (its means of transmission still unknown). Through the middle of the twentieth century, scientists held that "syphilis soaked" blacks were uniquely susceptible to the disease and its effects because of racially determined inferiorities.[13] In the late 1950s and early 1960s, however, a new, seemingly objective explanation was offered for the appearance and persistence of polio among urban non-white populations: the cities were populated by the poor, and the poor simply weren't vaccinated, whether they were black or white. In Harrisburg, Pennsylvania, 65 percent of the city's uppermost socioeconomic group was fully vaccinated, but just 35 percent of the city's poorest residents were protected. In Atlanta, Georgia, 78 percent of the city's wealthiest residents were immunized, and only 30 percent of the poor were.[14] Immunization surveys in the city concluded that this differential had more to do with class than with race, as the "upper echelons in both the white and Negro population" had similarly robust rates of vaccination.[15]

Not surprisingly, the dramatic income gap between the vaccinated and the unvaccinated prompted many to speculate that "economics" was the root cause of failure to vaccinate; those who could afford to vaccinate did, and that was that.[16] But thanks to the Polio Vaccination Assistance Act and the March of Dimes, polio vaccines had been administered free of cost in a great many communities — so there must have been some other root cause of the failure to vaccinate. Some observers fell upon explanations that blamed the poor and minority communities for their purported ignorance, apathy, and procrastination habits. Others pointed to their lack of motivation and imperviousness to the high-profile publicity of the polio immunization campaigns of the previous decade. Often this attribution of blame was subtle. With such widespread publicity and access to free vaccines, the poor's failure to vaccinate was perceived to have been to some extent deliberate. "We need a doorbell ringing campaign in poorer sections where vaccination is ignored," said Donald Henderson, head of surveillance at the CDC, in 1964.[17]

A growing body of research on the sociology of poverty influenced investigations as to why the poor "ignored" pleas to immunize. The Welfare Administration's 1966 report *Low-Income Life Styles* epitomized the tone of such research; the report summarized the poor's "typical attitudes" about life, courtship, marital relations, education, money management, and health. "Literally hundreds" of recent research studies, said one review, were showing how "poverty-linked attitudes and patterns of behavior . . . set the poor apart from other groups."[18] At the Public Health Service, a team of behavioral scientists argued that educated mothers with "white-collar" spouses were more likely than mothers with less education and "blue-collar" spouses to vaccinate their children because it was more convenient for them and because peers and community groups encouraged them. Moreover, when lower-income people read the papers, they read only about "crime, disaster, and sports" and therefore missed the news on public and social affairs, science, and politics—and, it followed, free polio vaccines.[19] The vaccination status of the poor was thus chalked up to a combination of misplaced social pressures, inconvenience, and a lack of education that made them different from the middle-class norm. In short, those in poverty—and their "Life Styles"—were responsible for their newfound vulnerability to polio.[20]

In the early sixties, it was clear that polio vaccination campaigns had thus far simply "'skimmed off the cream' of people who are well motivated toward immunization programs," as one health official put it.[21] The families who lined up in droves to vaccinate their children early and completely were the middle-class families who, from the start, felt most threatened by the disease and most invested in the search for a means of treatment or prevention.[22] Future immunization campaigns would have to try harder to reach those who had been missed the first time around, advised health officials. But even this advice had its limitations. An adherence to middle-class values and a sense of trepidation about the dangers of urban poverty areas were evident in federal guidelines that instructed health departments on how to identify the unvaccinated in their communities. The guidelines advocated going from house to house to identify pockets of the unvaccinated, even though it might be difficult to find people at home in areas where "both parents are working during the day." In such cases, it was best to leave a note on the door and call back in the evening—though it was expressly not recommended to return to "slum areas" after dark.[23]

A TRINITY OF SOCIAL PROBLEMS

Such fear of the slums was driven by urban decay that would reach a crisis point in the mid-1960s, fueled by urban deprivation and sparked by movements demanding civil rights and economic justice. In the postwar period, the United States had become not only the most powerful but the wealthiest nation on earth. The proportion of people living in poverty had declined since the end of World War II, but in the 1950s the decline had slowed, the gulf between the rich and the poor had widened, and the poor had become increasingly segregated from the middle class and the wealthy; in the words of one prominent writer, they had become "invisible." At the end of the decade, the issue of persistent mass poverty became the subject of intellectual and political debate. Social critics and economists declared American poverty "chronic," a more intractable and widespread problem than most Americans realized or cared to admit.[24] When Kennedy was elected in 1960, it was difficult to disagree. By most estimates, some 40 million Americans — 22 percent of the population — were living in poverty, and for some segments of the population, the statistics were even worse: 40 percent of the elderly and more than half of all African Americans were poor.[25]

In light of such figures, the Kennedy administration promised prosperity, equality for all, an end to poverty — and better health. "Let it not be said that the world's richest nation in all history failed to meet the people's health needs," said Luther Terry, Kennedy's surgeon general.[26] For health officials under Kennedy, and subsequently under President Lyndon B. Johnson, poverty, education, and disease were tightly intertwined; it was folly to address one while ignoring the others, and it was unjust that any of the three should exist in a nation that had come to be defined by its wealth. "Too many of our families are experiencing privation in the midst of plenty. We cannot accept poverty as inevitable, any more than we accept illiteracy or disease as inevitable," said Secretary of Health, Education, and Welfare Anthony Celebrezze.[27] The sentiment was embraced by Johnson, who believed it fell to government to "battle the ancient enemies of mankind, illiteracy and poverty and disease."[28]

Kennedy's promises of civil rights and an anti-poverty program were realized under Johnson, who rolled them into a vision of a "Great Society" built on "abundance and liberty for all" and an "end to poverty and racial injustice." As Johnson made his case for massive societal investments, he frequently referenced the nation's changing urban landscape, the "despoiled" suburbs and city centers marked by "decay" and "blight." For Johnson, the achievement of a great future rested in large part on remedying the lack of jobs, opportunities, good schools, and health and social services in urban areas. "Our society will never be great," he declared, "until our cities are great." His Great Society would launch a war on poverty, improve education, control crime, clean up the environment, preserve voting rights, protect civil rights, support the arts, launch a "massive attack on crippling and killing diseases"—and "make the American city a better and more stimulating place to live."[29]

As the Kennedy and Johnson administrations expounded on the trinity of poverty, education, and disease, health officials increasingly spoke of the connection between urbanization and health. Cities were losing population to suburbs, but the metropolitan areas they formed when combined were growing by leaps and bounds. By 1970 three-fourths of the U.S. population was projected to be living in "great complexes of city and suburb."[30] Early in the sixties, Surgeon General Terry had identified metropolitan growth as a major factor challenging health programs in the future. Later he cited the plight of American cities as justification for more federal involvement in addressing local problems. "The continued exodus to suburbia is draining the central city of resources and complicating the delivery of health services. . . . Great concentrations of people increase the hazards and complicate the control of communicable diseases," he remarked.[31] Under Johnson's Great Society, both the Public Health Service and the Department of Health, Education, and Welfare launched "an attack on urban ills." As the nation's "urban crisis" devolved into inner-city riots, they would cite a new triumvirate in need of addressing the "problems of poverty, the problems of the Negro, and the problems of the city," in the words of Johnson's Health, Education, and Welfare Secretary John W. Gardner.[32] If "poor" and "urban" were synonymous before, now "black, urban, and problem" came to be used interchangeably, too.[33] The shape of this "problem," for public health specifically, had been sketched by the shift in polio epidemiology in the late fifties, when polio epidemics appeared concentrated among urban populations of color. In the sixties, urban decay and civil unrest etched the "problem" of race, urban poverty, and disease into stone. This view of the problem of vaccine-preventable disease persistence would be subtly reflected in the national anti-measles campaign launched later that decade.

EXPANDING THE VACCINATION ASSISTANCE ACT

Great Society legislation expanded federal involvement in American lives generally, and it paralleled an expansion of federal involvement in health that had been under way for more than a decade. Between 1953 and 1965, more than fifty new programs were added to the Public Health Service, and its budget increased almost tenfold, from $250 million to $2.4 billion. In 1966 Johnson reorganized the burgeoning Public Health Service, increasing its number of bureaus from three to five.[34] This expansion was evident in the area of immunization, specifically. In 1965 the expiring Vaccination Assistance Act came up for renewal. Congress chose not only to extend it for another five years, but to dramatically broaden its scope. Now it covered not just "intensive" immunization programs but all "community" immunization programs. Measles was added to the act's list of target diseases—along with any other infectious disease for which a vaccine became available in the future, provided the surgeon general deemed it a "major" problem and one "susceptible of practical elimination . . . through immunization."[35]

Even before the Vaccination Assistance Act was expanded, it had an arguably impressive track record. Between 1962 and 1964, 50 million children and adults were vaccinated against polio, and 7 million children were immunized against diphtheria, pertussis, and tetanus. As a result, annual cases of polio had dropped from 910 to 121, and cases of diphtheria had fallen from 404 to 306. Communities had also established programs to ensure that rates continued to decline in the future.[36]

In its first two years, the Vaccination Assistance Act had supported a massive federally sponsored educational effort spearheaded by the CDC. At the creative heart of this effort was "Wellbee," a smiling, round-faced cartoon bumblebee designed by a Hollywood artist. Wellbee was designed to be "the personification of good health," a standardized symbol that state and local public health agencies across the country would use to communicate the importance of immunization widely. Wellbee urged children in Atlanta and Tampa to "drink the free polio vaccine" and appeared on billboards and pin-on buttons in Chicago. He posed with the Red Sox and with Mayor John F. Collins (a former polio patient) in Boston, went from school to school in Honolulu, rode a dogsled in Anchorage, and warned against "Illbee" in Dallas. His main cause was promoting oral polio vaccine: "Tastes Good • Works Fast • Prevents Polio," read posters with the smiling bee.[37] Local health departments improvised with Wellbee, but his message of preventive health and, above all, the importance of immunization was consistent across the nation.

The expansion of federal involvement in health generally was justified by the nation's rapid growth and the changing demographics of the population. The numbers of the very old and very young were increasing particularly fast, and it was these groups that consumed the most health resources and medical attention.[38] The expansion of federal involvement in immunization specifically—well captured by Wellbee's far-flung appearances—was justified quite differently, in terms that made reference to a world in which germs knew no geographic boundaries, whether between two nations, two states, or between city and suburb. "Modern air travel" meant that communicable disease was "no longer a *local* problem," noted the Chicago Board of Health (adding, "For this reason, we have welcomed the aid of the United States Public Health Service in our overall control program").[39] A Children's Bureau film made more explicit reference to the diseases latent in poverty-stricken cities and the threat they posed to the suburbs: "When large numbers of unimmunized persons live together in close proximity, as they now tend to do in low-income neighborhoods—the danger is a concern to everyone. Public health officials warn that under such circumstances, an individual case could quickly grow into an epidemic—and once started, epidemics do not respect neighborhood boundaries."[40] The only way to protect everyone from an infectious disease was to make sure no one was at risk—and that was best done by eliminating the disease entirely.

This philosophy lay at the heart of the Vaccination Assistance Act's expansion to cover measles and the first federal attempt to eradicate the long-familiar childhood infection.

TARGETING MEASLES

In its pre-vaccine days, measles was inescapable: nearly all children caught the infection, which causes fever, fatigue, and often a cough or runny nose followed by a blotchy, full-body rash. Most children were protected for the first year of life by antibodies they received from their mothers, but when that wore off, they were susceptible to infection. In the 1930s and 1940s, children who were exposed to someone with measles sometimes received a shot of "convalescent serum," blood serum taken from people who had previously had measles. The antibodies in the serum reduced children's risk of coming down with the disease and lessened their symptoms if they did get sick.[41] A measles vaccine containing live, weakened (attenuated) virus was licensed for use in 1963. A vaccine containing a killed version of the same strain of measles virus, the so-called Edmonston strain, was licensed that same year. Before the decade was through, two more live, further attenuated measles vaccines were also licensed for use.[42]

Within just a few years of the first measles vaccine hitting the market, 20 million doses had been administered to American children.[43] The number of measles cases declined, but the disease's national morbidity rate—the amount of total illness and disability it caused—didn't budge nearly as much.[44] In large part, that was because of who was getting the vaccine, and who wasn't. From 1963 through 1965, most of the measles vaccine in the country had been administered by physicians in private practice, noted the CDC. Little was given to children living in poverty, and the consequence was a new pattern of disease. In Los Angeles, for example, measles declined in predominantly white, "middle class and upper socioeconomic areas" and "increased in areas with large populations of the lower socioeconomic group," a UCLA epidemiology professor noted. As a result, "measles became . . . a disease of the Negro population and the population with Spanish surnames."[45] Across the country, measles-susceptible children were now concentrated "in a central core, lower socioeconomic area of the city," said CDC epidemiologist Robert Warren. As with polio, the new vaccine had shifted the disease's epidemiology. If the country was going to eradicate measles, noted Warren, it would have to ensure the vaccine was used more "homogenously" in the future.[46]

But even before the measles vaccine had become available, an Oklahoma physician was among those who had anticipated problems encouraging parents of *any* race or class to vaccinate their children against the disease. "Before the ultimate benefits from vaccination can be realized," he said, "it is necessary that the public be educated to a realization of the great hazards of measles."[47] Measles had been a part of childhood for so long, explained a California physician, that the disease was widely considered as inevitable as "wornout shoes" and scraped knees.[48] It was no wonder, then, that the vaccine hadn't been met with a fraction of the enthusiasm shown for the polio vaccine. Measles affected all children, almost without exception, and most of them recovered without incident. That and the long-standing use of immune globulin to prevent outbreaks and serious cases meant that for most parents, measles wasn't a prominent health worry. "It was only measles, doctor," said a reportedly baffled parent whose infant daughter died after catching the disease from a sibling. "No one worries about measles anymore."[49]

If that was true, it was also because measles was deeply overshadowed by polio. The threat of communicable diseases of childhood, measles included, was already fading when polio began to appear in truly epidemic proportions in the 1940s. By midcentury, parental anxieties, research dollars, and scientific attention all focused tightly on polio, at the expense of other (albeit still present) infections. The development of new measles vaccines, however, prompted both medical and public health professionals to take a new and closer look at the familiar infection. "Now that means are at hand to prevent measles and poliomyelitis, it is interesting to compare the mortality associated with the two," wrote the editors of the *Journal of the American Medical Association*. The exercise may have been partly intellectual, but it ended up yielding a host of reasons for immunizing the population widely and "homogenously" against measles, as had been done against polio. In 1964 polio killed seven people; measles killed four hundred. Polio had the potential to cause paralysis, measles the potential to cause ear infections, bronchopneumonia, and encephalitis, which could result in deafness, brain damage, or death. "Physicians would do well to ponder these facts," the editors concluded.[50]

Measles' long-ignored risks and litany of serious complications formed a refrain that surfaced as the first new measles vaccines were still being tested. In 1961 Surgeon General Terry told physicians it was time to think differently about measles, which each year killed more children than any other childhood disease.[51] In 1963 Merck, which had just licensed a live-virus measles vaccine, pointed out that the disease threatened 20 million Americans, a figure that, with the birth rate as high as it was, would grow by another 4 million each year.[52] CDC scientists began to tally measles' threats even more precisely: one in every thousand cases suffered encephalitis, one in three encephalitis cases died, and another third suffered permanent damage to the central nervous system.[53] Such findings encouraged Merck to chime in on measles complications: "All too often, measles is a killer or the cause of mental crippling so severe that the victim survives only as a mental defective," the vaccine maker propounded.[54]

Polio had caused national panic and roused national demand for a widely available vaccine, as noted in chapter 1. But measles was another story. Many doctors and health officials agreed that the disease had the potential to pose serious complications, but it was never a given that mass immunization was the appropriate response. Pediatricians debated their role in promoting the vaccine, the price of vaccine, how measles shots complicated the process of scheduling the growing number of vaccines for children, and whether measles was even "worth worrying about," as one doctor put it.[55] A New York physician griped about having to "sell" mothers on the prospect of more painful shots in their "baby's tender bottom"; an Alabama physician, by contrast, lamented that vaccine maker Merck was the only one promoting the vaccine.[56] At the CDC, immunization officials' position on the matter evolved. In 1964 the Advisory Committee on Immunization Practices concluded that "rarely would there appear to be a need in the United States for mass community immunization programs" targeting measles.[57] A year later, the committee revised its position: experience with the measles vaccine to date indicated a need for high levels of mass immunization, as well as a way to achieve such levels.[58]

OPTIMISM AND ERADICATION

The same year that Johnson renewed the Vaccination Assistance Act, he also signed a bill to greatly expand federal support for medical and health sciences research. On signing the bill, Johnson gave an "evangelistic" speech on the nation's role in leading a "worldwide war on disease." Polio had nearly been eliminated in the United States, and the new measles vaccine promised another victory over disease in the future. With the advances that scientists had already made against these diseases as well as cancer, kidney disease, and heart disease, a "staggering era for medicine" lay ahead.[59] The president's remarks captured the scientific spirit of the day. Scientists announced that the medical sciences were experiencing a revolution akin to the one in the physical sciences two decades before. It was an era that prompted scientists in government, the academy, and industry alike to expound on a future in which people would live longer, happier, healthier, and even more intelligent lives.[60] American science had the proven capacity to conquer disease, and the nation was witnessing just the beginning of what it would ultimately achieve.

In this era of medical optimism, disease eradication superseded disease control. Control had already proven within reach, remarked American Society of Tropical Medicine and Hygiene president E. Harold Hinman. With the help of drugs, pesticides, and vaccines, developed countries including the United States had all but eliminated from within their borders a long list of diseases: typhoid fever, tuberculosis, diphtheria, smallpox, whooping cough, typhus, cholera, plague, malaria, yellow fever, and more.[61] The next step was to move beyond national borders, and beyond mere control. In 1963 the World Health Assembly had pledged its support for a push to vaccinate 80 percent of the earth's population against smallpox.[62] In 1966 the assembly formally adopted a resolution to eliminate the disease from the face of the earth, and the CDC donated experts, equipment, and vaccine to the effort.[63] A diplomatic plea from the Soviet minister of health played an important role in pushing the world's nations to sign on to the project; in the USSR as in the United States, said the minister, smallpox had been eliminated, but the disease's presence in other countries posed a perpetual threat of imported outbreaks.[64] A few years into the program, containment of the disease in poor, populous countries with high rates of smallpox demonstrated that with cooperation and the right technology, the ancient and horrifying disease could be eradicated even "under the most difficult conditions expected."[65] The accomplishment suggested a model that the United States could follow in its own intractable urban poverty areas.

The United States' commitment to smallpox eradication abroad was in fact mirrored by the campaign to eradicate measles at home, announced officially by President Johnson early the following year.[66] Measles was markedly different from smallpox. It was a disease of "only mild severity," which caused "infrequent complications" and only rarely caused deaths, noted CDC head David J. Sencer and chief epidemiologist Alexander Langmuir.[67] Over the last half century, they continued, "man" had developed a "deep respect for the biological balance of the human race with measles virus." But this accepted doctrine, they argued, was ready to be overturned. They outlined three bases for the disease's eradication. The measles virus had been isolated and studied, and, like smallpox, was now known to infect only humans, creating neither chronic carriers (as diphtheria could) nor "inapparent" infections (as polio sometimes did). The isolation of the measles virus, furthermore, had led to several "potent" vaccines. Lastly, epidemiological studies of the disease had demonstrated both how it spread and how many "susceptibles" were needed to sustain an outbreak. As with smallpox, it was clear that the disease could be eliminated with less than 100 percent vaccination coverage. Based on this knowledge, the CDC officials confidently declared measles could—and would—be eradicated from the United States before the end of 1967.[68]

Scientific reasons for measles eradication partially belied cultural ones. In discussions about measles eradication (and disease eradication generally), attitudes toward "nature" (of which measles virus was a part) reflected a modern intolerance for ancient woes and challenges. Nature was something to be overcome with technological breakthroughs and scientific expertise; it was not something to be sentimentally respected.

"To those who ask me, 'Why do you wish to eradicate measles?' I reply with the same answer that [Sir Edmund] Hillary used when asked why he wished to climb Mount Everest," said Langmuir. "He said, 'Because it is there.' To this may be added, '. . . and it can be done.'"[69] Tackling nature was the modern way. But it required an all-or-nothing approach. Experience had already shown that partial vaccination changed a disease's epidemiology, concentrating cases in certain segments of the population. There were other concerns, too: members of a CDC immunization committee noted that because the research community was uncertain about how long the "artificial immunity" created by vaccines would last, it was possible that vaccination of children was creating a situation in which, "in the not too distant future, many adults . . . may become once again susceptible to such diseases as diphtheria, pertussis, and measles." The prospect prompted the committee members to remark on the "sensitive balance of nature"—and to conclude that the only answer was to immunize early and repeatedly, until nature had effectively been conquered.[70] In the 1960s, vaccination against disease created an imperative to eradicate disease. Modern confidence in humankind's capacity to triumph over nature helped this idea take shape and take hold.[71]

The public, meanwhile, was given a very different reason why measles needed to be eradicated in 1967. Among themselves, health experts may have reflected on the balance of nature and referred to measles as "mild," but publicly they described measles as a serious disease with horrific and sometimes fatal complications. The CDC-led anti-measles campaign— designed to reach 8 to 10 million unvaccinated children—produced public service announcements, billboard ads, films, comic strips, coloring books, and more with the message that measles was a treacherous threat. Radio and television stations across the country notified the public that "measles is a serious disease that sometimes causes pneumonia, deafness, encephalitis and even death."[72] Personalities from the surgeon general to Ann Landers spread the word that measles could leave children blind, deaf, and mentally impaired.[73] The federal government didn't go it alone in this effort; new vaccine-maker Merck, hoping to make good on the resources it had invested in developing its measles vaccine, launched a marketing campaign with the slogan "Measles Only Gave Her Spots—Will Your Child Be As Lucky?"[74] The national measles campaign, borrowing an idea from the 1950s polio campaigns, selected a white poster child, ten-year-old Kim Fisher, who at age four suffered a case of measles so severe it left her mentally impaired, partially blind, and partly deaf.[75] Kim vividly illustrated the idea that measles *could* be serious, even if parents didn't see it as such. And if the possibility of complications seemed remote, that was beside the point. Measles immunization was necessary because "one death, one brain-damaged child, or even one child who needs hospitalization is one too many," as one supporter of the campaign put it.[76]

Notably, the national campaign also exploited urbanization anxieties of the suburban middle class. In "Spot Prevention," a cartoon film and coloring book developed by the Department of Health, Education, and Welfare, "measles" was a red-faced, yellow-eyed fiend who popped out of metal garbage cans left on the sidewalk and climbed through nursery- and elementary-school windows, where innocent white children sat unaware and unprotected. At the end of the story, young "Billy" received his vaccine at Happyville Clinic, set in a wide expanse of lawn surrounded by trees, the city where measles once lurked darkly visible in the background.[77] In another cartoon developed for the campaign, young "Emmy Immunity" fought off the "Dirty Disease Gang," a group of social misfits and offenders including Mean-ole Measles, Rolly Polio, and Dippy Diphtheria. "Can Emmy Immunity protect the 'happy little family' against the plotting, conniving, evil-doing 'Disease Gang?'" ran the teaser.[78] The cartoons explicitly linked disease and degeneracy; in the world they constructed, measles and other infections arose from filth and social decay. Health officials may have argued that the measles vaccine needed to be used more homogenously, but promotional materials clearly played to white middle- and upper-class anxieties about degenerate and disease-breeding urban cores.

MEAN-OLE MEASLES IN THE SLUMS

The public wasn't asked to eradicate a mild disease in 1967; they were asked to eradicate a serious one. But they weren't quick to see it that way, and neither were all health professionals. The disconnect between measles' public image and its private one was addressed in a memo that CDC head James L. Goddard sent to rally immunization personnel around the country to the cause. "Measles has not suddenly become a more serious disease. It has always been a scourge of childhood," Goddard wrote. "It commands special attention now because modern medical research has provided us with vaccines which can prevent the disease."[79] As in the case of diphtheria, tetanus, and pertussis, technology warranted renewed focus on a long-familiar infection.

With health experts unevenly enthusiastic, it only followed that the public responded similarly. In Delaware's anti-measles campaign, the message "that measles can be a dangerous disease with widespread complications was met by general public apathy," noted a state health officer. "Measles, unlike poliomyelitis, is not a 'glamorous' disease. . . . It was found that for the most part the public still considered it a minor, childhood disease."[80] Officials in Washington, D.C., and Atlanta may have believed that all the pieces were in place to conquer yet another childhood scourge, but the buy-in they needed from both local health professionals and the public was only halfhearted, at best.

Nonetheless, an internal report published by the CDC late in 1967 stated that measles cases and deaths had markedly declined. In addition, *reporting* of measles cases and outbreaks had greatly increased. As a result, a new picture of the disease began to form. Measles complications, once considered rare, now seemed common. Deaths, once thought to occur in 1 of every 100,000 cases, now seemed to occur in 1 of every 10,000 cases.[81] Studies had also begun to suggest that even mild cases affected the brain, and that measles might increase susceptibility to polio, cause pulmonary emphysema, and harm the fetus in pregnant women.[82] Indeed, the calculated representation of measles as a dangerous infection began to seem like a self-fulfilling prophecy. The measles vaccine became possible only with increased knowledge of the disease; with deployment of the vaccine (combined with disease surveillance) came even more knowledge about the disease and its dangers, which ultimately validated the impression that public health officials had cultivated in order to encourage even further vaccination. Within a decade, measles would no longer be referred to as a minor or mild disease, either by lay observers or health and medical professionals. And its vaccine would come to be widely accepted as a routine part of childhood health care.

In the meantime, measles failed to be eradicated, and at the end of the decade, any progress achieved toward eradication started to come undone.[83] The number of measles cases nationwide fell from more than 260,000 in 1965 to just 22,000 in 1968, but soon thereafter the number began to tick upward again.[84] At the end of the sixties, new vaccines against mumps and rubella drew efforts and attention away from measles, and immunization funding shortfalls compounded the problem. The Vaccination Assistance Act expired in 1969, and a 1970 extension was left unfunded by President Richard Nixon's new administration.[85] In the wake of the abandoned measles campaign, old patterns became visible again. "Things in Atlanta continue to go as usual," wrote a CDC immunization officer to a colleague in 1970. "Measles is up 50 percent over last year with epidemics in a number of urban ghettos."[86] In Indiana, too, measles was up, much of it "coming from central city indigent areas," reported a health officer in that state.[87] The same was true in Los Angeles, Chicago, and elsewhere. The CDC was both "chagrined and concerned" about measles resurgence, and the fact that much of the increase was due to "outbreaks in poverty areas of large cities."[88] Despite their best efforts, measles persisted as a disease of the slums.

Explanations for the upsurge in measles cases piled on top of one another. Some blamed the poor and the conditions in which they lived. Some blamed the web of social problems bound in place by poverty.[89] In Chicago, where 40 percent of the population lived in the city's "poverty areas," measles outbreaks between 1967 and 1969 were attributed to the difficulties of delivering vaccine to "a highly mobile, poorly educated, and impoverished population." Epidemiologists also speculated that urban renewal projects—in which tenements had been replaced with high-rise apartments—had facilitated measles' spread, because "children were found to congregate with common babysitters, at recreational facilities, and in other crowded situations."[90] In urban areas, resurgent measles was one of a string of intractable health problems: mental illness, drug abuse, alcoholism, suicide, and new outbreaks of polio and diphtheria, too. CDC officials blamed the latter on low immunization rates, particularly among the youngest children living in urban poverty areas. This time, however, the root explanation was federal budget cuts under the Nixon administration, which were actively protested by a group of federal health employees. "The reason for these epidemics," said a member of the CDC's immunization branch bluntly, "is money."[91]

There was one attractive solution to these problems: a "new and simpler" approach to immunization. Scientists could either demonstrate that it was safe and effective to give children all five recommended vaccines in a single visit to the doctor, or they could come up with a single vaccine that offered combined protection against multiple infections. Combined vaccines would address the problem of doctors who forgot to recommend vaccines and parents who failed to bring children in for required vaccines. They would also reduce the cost of immunizing all children in "a period of relatively strained health resources" and ensure full protection for the "less health-motivated segments" of the population.[92] A combined vaccine against diphtheria, pertussis, and tetanus (as mentioned in the previous chapter) already existed. With three new vaccines against viral infections now available, scientists in government and industry began to press for a way to combine protection against polio, smallpox, measles, mumps, and rubella, too.

But all of these discussions—about the causes of persistent outbreaks and the potential to curb them with simultaneous vaccination—largely sidestepped limitations inherent to vaccines and vaccination. The measles vaccine didn't offer 100 percent protection; no vaccine did. Up to 5 percent of children vaccinated against measles never produced measles antibodies. Another small fraction of children produced antibodies, but the antibodies failed to protect them against the disease. When children were vaccinated too early, antibodies inherited from their mothers interfered with their ability to mount an immune response, rendering the shots ineffective. At the CDC, public health officials suspected that "lapses in technique" in doctors' offices—improper storage, poor record keeping, failure to give vaccines at the recommended age—might be compounding these problems further.[93]

Vaccines had side effects, too; because it was produced in chick cells, measles vaccine was not recommended for children allergic to eggs. In 1969 a CDC physician expressed surprise that not a single serious allergic reaction to the vaccine had been published, even though some were expected. In a letter to Washington, he wrote of plans to corroborate this by examining death certificates. No copies of such certificates appear in CDC files from the time, but one letter describes the death of a two-year-old "Negro Female" who died in an Atlanta hospital four days after receiving the live measles vaccine. The girl had developed a rash after receiving the polio and DPT vaccines the year before. When she was immunized against measles, her left arm swelled, she developed a fever, her parents took her to the hospital, and she died several hours later.[94] No one left note of whether the vaccine played a role in her death, but CDC officials were interested enough to keep the case on file. On the whole, however, the side effects and limitations of vaccines received relatively little scientific attention. The goal of eradication made such issues moot—if a disease could be forced out of existence, and fast, the shortcomings of its vaccine wouldn't matter in the end.

Entering the 1970s, immunization experts and officials took with them lessons learned from their experiences with measles and polio in the 1960s. With polio, fear of disease was universal; with measles, susceptibility to disease was universal. In neither case did all families fully vaccinate their children. The poorer segments of society did not get vaccinated to the same extent as the well-to-do. Federal funds seemed to solve the problem in the case of polio. Federal funds and a campaign that fostered fear of the disease seemed to do the trick in the case of measles. Both campaigns were undertaken to reach the poor in largely urban "slums," but neither campaign was wholly successful. Persistent epidemics left health workers pondering legal and technological solutions to the problem. For the most part, they didn't question whether vaccines weren't a swift means of eradicating disease because of their inherent limitations and because they were a one-size-fits-all solution for a complex and heterogeneous population.

The new funding climate and persistent outbreaks involving preschoolers prompted a shift in focus in the 1970s, away from persuasion and toward compulsion; if families couldn't be convinced to vaccinate their children, they'd be forced to, through laws requiring immunization for enrollment in school.[95] "To reach the single preschool child in the slum is difficult, but to mount a campaign to attain really high levels of immunization in kindergartens and 1st, 2nd grades in school should be relatively easy," said the head of the CDC's immunization branch.[96] The fact that rates of diseases were lower among such children didn't matter; by protecting them, the rest of the community would be protected. As CDC head Sencer had pointed out, "If measles isn't in the schools, it can't be brought home."[97] Such thinking would prompt a push for more uniform and comprehensive school vaccination laws, which would reach an apex at the end of the seventies. At the same time, health officials would grapple with questions about how to use the newer vaccines, which protected against infections that were less serious than smallpox and polio, and most threatening to specific subsets of the population. Who should receive such vaccines, and when? Experience with one of these new vaccines—the vaccine against mumps—would produce a set of answers.

3

HOW SERIOUS IS MUMPS?

When the first effective mumps vaccine was licensed for use, late in 1967, it met with only mild interest from the public, followed at times by confusion.[1] On the rare occasion when the media mentioned the vaccine, they offered uneven portrayals of the infection's gravity and conflicting reports on who should get the vaccine and when.

A year after the vaccine came on the market, medical journalist Faye Marley wrote up an investigation of why the mumps vaccine was so little used. She began by posing a question the vaccine had brought to many people's minds: "How serious is mumps?" The answer she dug up: "It depends on how your doctor views it."[2] The reason so many went without mumps vaccine, Marley concluded, was that doctors themselves didn't agree on whether to give it, when to give it, or whom to give it to. Some wanted to give it to all children, some just to boys, others only to males at puberty and after. Still others saw little need for the vaccine at all, noting that nearly a third of mumps cases produced absolutely no symptoms—but nonetheless bestowed victims with lifelong protection against subsequent bouts of disease.

Within a decade, however, the notion that all children should be vaccinated against mumps early in life came to prevail among both doctors and public health professionals. During the same period, mumps gradually but firmly acquired a new public image, one that downplayed the frequency of its mild and invisible cases in favor of emphasis on the disease's rare but serious complications. Mumps, that is, was transformed from a nuisance widely considered so harmless it was a frequent butt of jokes to an infection defined by its still-unquantified potential to harm the organs, cause deafness, and jeopardize the fertility of men. The first effective mumps vaccine, developed by growing pharmaceutical company Merck, was key in sparking this transformation.

The discovery of the vaccine (and Merck's efforts to promote it) brought attention to mumps at a time when Americans' relationship to infectious disease generally had been transformed by recent triumphs against smallpox globally and polio at home. The near-elimination of those dreaded diseases in the United States prompted some soul-searching on the part of doctors and health professionals, who began to ask how vaccines should be used in this new age. Because of the timing of the mumps vaccine's approval, mumps became an important part of this dialogue. And outbreaks of the disease became an important testing ground for the application of vaccination against what health officials then called the "milder" diseases. Beginning in the late 1960s, that is, the mumps vaccine gave health experts and society at large a new tool with which to make sense of the disease. The new representations of mumps that they fashioned in the process were crucial to ensuring the vaccine's inclusion in the growing arsenal of immunizations that would become mandated for all children by the end of the 1970s.

As historian Charles Rosenberg has shown, society's perception of a particular disease at a given moment in time reflects the contemporary state of scientific knowledge as well as popular values, cultural preoccupations, political priorities, and the like.[3] The very same can be said of society's perception of a vaccine. As this chapter shows, over the course of the mid-twentieth century, mumps represented a threat to U.S. military supremacy, commerce and trade, reproductive capacity, and middle-class security; the mumps *vaccine*, in turn, was configured as a defense against each of these threats and more. The history of the mumps vaccine illustrates how a vaccine can reframe the way that Americans think about a disease and the significance of its prevention. It also vividly illustrates how a vaccine may come to be seen as the remedy for a set of concerns not exclusively medical in nature.

MUMPS IN WARTIME

Between 1963 and 1969, the nation's flourishing pharmaceutical indus-try launched several vaccines against measles, a vaccine against mumps, and a vaccine against rubella in rapid succession. The measles vaccine became the focus of the federally sponsored eradication campaign de-scribed in the previous chapter; the rubella vaccine prevented birth defects and became entwined with the intensifying abortion politics of the time.[4] Both vaccines overshadowed the debut of the vaccine against mumps, a disease of relatively little concern to most Americans in the late 1960s. Mumps was never an object of public dread, as polio had been, and its vaccine was never anxiously awaited, like the Salk polio vaccine had been. Nor was mumps ever singled out for a high-profile immunization campaign or for eradication, as measles had been. All of which made it quite remarkable that, within a few years of its debut, the mumps vaccine would be administered to millions of American children with little fanfare or resistance.

The mumps vaccine first brought to market in 1968 was developed by Maurice Hilleman, then head of Virus and Cell Biology at the burgeon-ing pharmaceutical company Merck. Hilleman was just beginning to earn a reputation as a giant in the field of vaccine development; upon his death in 2005, the *New York Times* would credit him with saving "more lives than any other scientist in the 20th century."[5] Today the histories of mumps vaccine that appear in medical textbooks and the like often begin in 1963, when Hilleman's daughter, six-year-old Jeryl Lynn, came down with a sore throat and swollen glands. A widower who found him-self tending to his daughter's care, Hilleman was suddenly inspired to begin work on a vaccine against mumps—which he began by swabbing Jeryl Lynn's throat. Jeryl Lynn's viral strain was isolated, cultured, and then gradually weakened, or attenuated, in Merck's labs. After field trials throughout Pennsylvania proved the resulting shot effective, the "Jeryl-Lynn strain" vaccine against mumps, also known as Mumpsvax, was ap-proved for use.[6]

But Hilleman was not the first to try or even succeed at developing a vaccine against mumps. Research on a mumps vaccine began in earnest during the 1940s, when the United States' entry into World War II gave military scientists reason to take a close look at the disease. As U.S. engagement in the war began, U.S. Public Health Service researchers began reviewing data and literature on the major communicable infections affecting troops during the First World War. They noted that mumps, though not a significant cause of death, was one of the top reasons troops were sent to the infirmary and absent from duty in that war—often for well over two weeks at a time.[7] Mumps had long been recognized as a common but not "severe" disease of childhood that typically caused fever and swelling of the salivary glands. But when it struck teens and adults, its usually rare complications—including inflammation of the reproductive organs and pancreas—became more frequent and more troublesome. Because of its highly contagious nature, mumps spread rapidly through crowded barracks and training camps. Because of its tendency to inflame the testes, it was second only to venereal disease in disabling recruits. In the interest of national defense, the disease clearly warranted further study. PHS researchers estimated that during World War I, mumps had cost the United States close to 4 million "man days" from duty, contributing to more total days lost from duty than foreign forces saw.[8]

The problem of mumps among soldiers quickly became apparent during the Second World War, too, as the infection once again began to spread through army camps.[9] This time around, however, scientists had new information at hand: scientists in the 1930s had determined that mumps was caused by a virus and that it could, at least theoretically, be prevented through immunization.[10] PHS surgeon Karl Habel noted that while civilians didn't have to worry about mumps, the fact that infection was a serious problem for the armed forces now justified the search for a vaccine.[11] "To the military surgeon, mumps is no passing indisposition of benign course," two Harvard epidemiologists concurred.[12] Tipped off to the problem of mumps by a U.S. Army general and funded by the Office of Scientific Research and Development (OSRD), the source of federal support for military research at the time, a group of Harvard researchers began experiments to promote mumps virus immunity in macaque monkeys in the lab.

Within a few years, the Harvard researchers, led by biologist John

Enders, had developed a diagnostic test using antigens from the monkey's salivary glands, as well as a rudimentary vaccine.[13] In a subsequent set of experiments, conducted both by the Harvard group and by Habel at the National Institute of Health, vaccines containing weakened mumps virus were produced and tested in institutionalized children and plantation laborers in Florida, who had been brought from the West Indies to work on sugar plantations during the war.[14] With men packed ten to a bunkhouse in the camps, mumps was rampant, pulling workers off the fields and sending them to the infirmary for weeks at a time. When PHS scientists injected the men with experimental vaccine, one man in 1,344 went into anaphylactic shock, but he recovered with a shot of adrenaline and "not a single day of work was lost," reported Habel. To the researchers, the vaccine seemed safe and fairly effective — even though some of the vaccinated came down with the mumps.[15] What remained, noted Enders, was for someone to continue experimenting until scientists had a strain infective enough to provoke a complete immune response while weak enough not to cause any signs or symptoms of the disease.[16]

Those experiments would wait for well over a decade. Research on the mumps vaccine, urgent in wartime, became a casualty of shifting national priorities and the vagaries of government funding. As the war faded from memory, polio, a civilian concern, became the nation's number one medical priority. By the end of the 1940s, the Harvard group's research was being supported by the National Foundation for Infantile Paralysis, which was devoted to polio research, and no longer by OSRD.[17] Enders stopped publishing on the mumps virus in 1949 and instead turned his full-time attention to the cultivation of polio virus.[18] Habel, at the NIH, also began studying polio.[19] With polio occupying multiple daily headlines throughout the 1950s, mumps lost its place on the nation's political and scientific agendas.

Although mumps received scant resources in the 1950s, Lederle Laboratories commercialized the partially protective mumps vaccine, which was about 50 percent effective and offered about a year of protection.[20] When the American Medical Association's Council on Drugs reviewed the vaccine in 1957, they didn't see much use for it. The AMA advised against administering the shot to children, noting that in children mumps and its "sequelae," or complications, were "not severe." The

AMA acknowledged the vaccine's potential utility in certain populations of adults and children—namely, military personnel, medical students, orphans, and institutionalized patients—but the fact that such populations would need to be revaccinated every year made the vaccine's deployment impractical.[21] The little professional discussion generated by the vaccine revealed a similar ambivalence. Some observers even came to the disease's defense. Edward Shaw, a physician at the University of California School of Medicine, argued that given the vaccine's temporary protection, "deliberate exposure to the disease in childhood . . . may be desirable": it was the only way to ensure lifelong immunity, he noted, and it came with few risks.[22] The most significant risk, in his view, was that infected children would pass the disease to susceptible adults. But even this concern failed to move experts to urge vaccination. War had made mumps a public health priority for the U.S. government in the 1940s, but the resulting technology (imperfect as it was) generated little interest or enthusiasm in a time of peace, when other health concerns loomed larger.

MUMPS AT HOME

After the war but before the new live virus vaccine was introduced, mumps went back to being what it long had been: an innocuous and sometimes amusing childhood disease. The amusing nature of mumps in the 1950s is evident even in seemingly serious documents from the time. When the New York State health department published a brochure on mumps in 1955, they adopted a light tone and a comical caricature of chipmunk-cheeked "Billy" to describe a brush with the disease. In the Chicago papers, health columnist and Chicago Medical Society president Theodore Van Dellen noted that when struck with mumps, "the victim is likely to be dubbed 'moon-face.'"[23] Such representations of mumps typically minimized the disease's severity. Van Dellen noted that while mumps did have some unpleasant complications—including the one that had garnered so much attention during the war—"the sex gland complication is not always as serious as we have been led to believe."[24] The health department brochure pointed out that "children seldom develop complications," and should therefore not be vaccinated: "Almost always a child is better off having mumps: the case is milder in childhood and gives him life-long immunity."[25]

Such conceptualizations helped shape popular representations of the illness. In press reports from the time, an almost exaggeratedly light-hearted attitude toward mumps prevailed. In Atlanta, papers reported with amusement on the oldest adult to come down with mumps, an Englishwoman who had reached the impressive age of ninety-nine. Chicago papers featured the sad but cute story of the boy whose poodle went missing when mumps prevented him from being able to whistle to call his dog home. In Los Angeles, the daily paper told the funny tale of a young couple forced to exchange marital vows by phone when the groom came down with mumps just before the big day.[26] *Los Angeles Times* readers speculated on whether the word "mumps" was singular or plural, while Chicago *Daily Defender* readers got to laugh at a photo of a fat-cheeked matron and her fat-cheeked cocker spaniel, heads wrapped in matching dressings to soothe their mumps-swollen glands.[27] Did dogs and cats actually get the mumps? In the interest of entertaining readers, newspapers speculated on that as well.[28]

The top reason mumps made headlines throughout the fifties and into the sixties, however, was its propensity to bench professional athletes. Track stars, baseball players, boxers, football stars, and coaches all made the news when struck by mumps.[29] So did Washington Redskins player Clyde Goodnight, whose story revealed a paradox of mumps at midcentury: the disease was widely regarded with casual dismissal and a smirk, even as large enterprises fretted over its potential to cut into profits. When Goodnight came down with a case of mumps in 1950, his coaches giddily planned to announce his infection to the press and then send him into the field to play anyway, where the Pittsburgh Steelers, they gambled, would be sure to leave him open for passes. But the plan was nixed before game time by the Redskins' public relations department, who feared the jubilant Goodnight might run up in the stands after a good play and give fans the mumps. Noted one of the team's publicists: "That's not good business."[30]

When Baltimore Orioles outfielder Frank Robinson came down with the mumps during an away game against the Los Angeles Angels in 1968, however, the tone of the team's response was markedly different. Merck's new Mumpsvax vaccine had recently been licensed for sale, and the Orioles' managers moved quickly to vaccinate the whole team, along with their entire press corps and club officials.[31] The Orioles' use of the new vaccine largely adhered to the guidelines that Surgeon General William Stewart had announced upon the vaccine's approval: it was for preteens, teenagers, and adults who hadn't yet had a case of the mumps.[32] (For the time being, at least, it wasn't recommended for children.) The Angels' management, by contrast, decided not to vaccinate their players — despite their good chances of having come into contact with mumps in the field.

Baseball's lack of consensus on how or whether to use the mumps vaccine was symptomatic of the nation's response as a whole. Cultural ambivalence toward mumps had translated into ambivalence toward the disease's new prophylactic, too. That ambivalence was well-captured in the hit movie *Bullitt*, which came out the same year as the new mumps vaccine. In the film's opening scene, San Francisco cop Frank Bullitt readies himself for the workday ahead as his partner, Don Delgetti, reads the day's headlines aloud. "Mumps vaccine on the market . . . the government authorized yesterday what officials term the first clearly effective vaccine to prevent mumps . . . ," Delgetti begins — until Bullitt sharply cuts him off. "Why don't you just relax and have your orange juice and *shut up*, Delgetti."[33] Bullitt, a sixties icon of machismo and virility, has more important things to worry about than the mumps. So, apparently, did the rest of the country. The *Los Angeles Times* announced the vaccine's approval on page 12, and the *New York Times* buried the story on page 72, as the war in Vietnam and the race to the moon took center stage.[34]

Also ambivalent about the vaccine — or, more accurately, the vaccine's use — were the health professionals grappling with what it meant to have such a tool at their disposal. Just prior to Mumpsvax's approval, the federal Advisory Committee on Immunization Practices at the CDC recommended that the vaccine be administered to any child approaching or in puberty; men who had not yet had the mumps; and children living in institutions, where "epidemic mumps can be particularly disruptive."[35] Almost immediately, groups of medical and scientific professionals began to take issue with various aspects of these national guidelines. For some, the vaccine's unknown duration was troubling: ongoing trials had by then demonstrated just two years of protection. To others, the very nature of the disease against which the shot protected raised philosophical questions about vaccination that had yet to be addressed. The Consumers Union flinched at the recommendation that institutionalized children be vaccinated, arguing that "mere convenience is insufficient justification for preventing the children from getting mumps and thus perhaps escorting them into adulthood without immunity."[36] The editors of the *New England Journal of Medicine* advised against mass application of mumps vaccine, arguing that the "general benignity of mumps" did not justify "the expenditure of large amounts of time, efforts, and funds." The journal's editors also decried the exaggeration of mumps' complications, noting that the risk of damage to the male sex glands and nervous system had been overstated. These facts, coupled with the ever-present risk of hazards attendant with any vaccination program, justified, in their estimation, "conservative" use of the vaccine.[37]

This debate over how to use the mumps vaccine was often coupled with the more generalized reflection that Mumpsvax helped spark over the appropriate use of vaccines in what health experts began referring to as a new era of vaccination. In contrast to polio or smallpox, the eradication of mumps was far from urgent, noted the editors of the prestigious medical journal the *Lancet*. In this "next stage" of vaccination, marked by "prevention of milder virus diseases," they wrote, "a cautious attitude now prevails." If vaccines were to be wielded against diseases that represented only a "minor inconvenience," such as mumps, then such vaccines needed to be effective, completely free of side effects, long-lasting, and must not in any way increase more severe adult forms of childhood infections, they argued.[38] Immunization officials at the CDC acknowledged that with the approval of the mumps vaccine, they had been "forced to chart a course through unknown waters."[39] They agreed that the control of severe illnesses had "shifted the priorities for vaccine development to the remaining milder diseases," but how to prevent these milder infections remained an open question. They delineated but a single criterion justifying a vaccine's use against such a disease: that it pose less of a hazard than its target infection.[40]

To other observers, this was not enough. A vaccine should not only be harmless—it should also produce immunity as well as or better than natural infection, maintained Oklahoma physician Harris Riley.[41] The fact that the mumps vaccine in particular became available before the longevity of its protection was known complicated matters for many weighing in on the professional debate. Perhaps, said Massachusetts health officer Morton Madoff, physicians should be left to decide for themselves how to use such vaccines as "a matter of conscience."[42] His comment revealed a hesitancy to delineate policy that many displayed when faced with the uncharted territory the mumps vaccine had laid bare. It also hinted at an attempt to shift future blame in case mumps vaccination went awry down the line—a possibility that occurred to many observers given the still-unknown duration of the vaccine's protection.

Mumps was not a top public health priority in 1967—in fact, it was not even a reportable disease—but the licensure of Mumpsvax would change the disease's standing over the course of the next decade.[43] When the vaccine was licensed, editors at the *Lancet* noted that there had been little interest in a mumps vaccine until such a vaccine became available.[44] Similarly, a CDC scientist remarked that the vaccine had "stimulated renewed interest in mumps" and had forced scientists to confront how little they knew about the disease's etiology and epidemiology.[45] If the proper application of a vaccine against a mild infection remained unclear, what was clear—to scientists at the CDC at least—was that such ambiguities could be rectified through further study of both the vaccine and the disease. Given a new tool, that is, scientists were determined to figure out how best to use it. In the process of doing so, they would also begin to create new representations of mumps, effectively changing how they and Americans in general would perceive the disease in the future.

A CHANGING DISEASE

Shortly after the mumps vaccine's approval, CDC epidemiologist Adolf Karchmer gave a speech on the infection and its vaccine at an annual immunization conference. In light of the difficulties that health officials and medical associations were facing in trying to determine how best to use the vaccine, Karchmer devoted his talk to a review of existing knowledge on mumps. Aside from the fact that the disease caused few annual deaths, peaked in spring, and affected mostly children, particularly males, there was much scientists didn't know about mumps. They weren't certain about the disease's true prevalence; asymptomatic cases made commonly cited numbers a likely underestimate. There was disagreement over whether the disease occurred in six- to seven-year cycles. Scientists weren't sure whether infection was truly a cause of male impotence and sterility. And they didn't know the precise nature of the virus's effects on the nervous system. Karchmer expressed a concern shared by many: if the vaccine was administered to children and teens, and if it proved to wear off with time, would vaccination create a population of non-immune adults even more susceptible to the disease and its serious complications than the current population? Karchmer and others thus worried—at this early stage, at least—that trying to control mumps not only wouldn't be worth the resources it would require, but that it might also create a bigger public health problem down the road.[46]

To address this concern, CDC scientists took a two-pronged approach to better understanding mumps and the potential for its vaccine. They reinstated mumps surveillance, which had been implemented following World War I but suspended after World War II.[47] They also issued a request to state health departments across the country, asking for help identifying local outbreaks of mumps that they could use to study both the disease and the vaccine.[48] Within a few months, the agency had dispatched teams of epidemiologists to study mumps outbreaks in Campbell and Fleming Counties in Kentucky, the Colin Anderson Center for the "mentally retarded" in West Virginia, and the Fort Custer State Home for the mentally retarded in Michigan.[49]

The Fort Custer State Home in Augusta, Michigan, hadn't had a single mumps outbreak in its ten years of existence when the CDC began to investigate a rash of 105 cases that occurred in late 1967. In pages upon pages of detailed notes, the scientists documented the symptoms (largely low-grade fever and runny noses) as well as the habits and behaviors of the home's children. They noted not only who slept where, who ate with whom, and which playgrounds the children used, but also who was a "toilet sitter," who was a "drippy, drooley, messy eater," who was "spastic," who "puts fingers in mouth," and who had "impressive oral-centered behavior." The index case—the boy who presumably brought the disease into the home—was described as a "gregarious and restless child who spends most of his waking hours darting from one play group to another, is notably untidy and often places his fingers or his thumbs in his mouth." The importance of these behaviors was unproven, remarked the researchers, but they seemed worth noting.[50] Combined with other observations—such as which child left the home, for example, to go on a picnic with his sister—it's clear that the Fort Custer children were viewed as a petri dish of infection threatening the community at large.[51]

Although the researchers' notes explicitly stated that the Fort Custer findings were not necessarily applicable to the general population, they were presented to the 1968 meeting of the American Public Health Association as if they were. The investigation revealed that mumps took about fifteen to eighteen days to incubate, and then lasted between three and six days, causing fever for one or two days. Complications were rare (three boys ages eleven and up suffered swollen testes), and attack rates were highest among the youngest children.[52] The team also concluded that crowding alone was insufficient for mumps to spread; interaction had to be "intimate," involving activities that stimulated the flow and spread of saliva, such as the thumb-sucking and messy eating so common among not only institutionalized children but children of all kinds.[53]

Mumps preferentially strikes children, so it followed that children offered the most convenient population for studying the disease's epidemiology. But in asking a question about children, scientists *ipso facto* obtained an answer—or series of answers—about children. Although mumps had previously been considered a significant health *problem* only among adults, the evidence in favor of immunizing children now began to accumulate. Such evidence came not only from studies like the one at Fort Custer, but also from local reports from across the country. When Bellingham and Whatcom Counties in Washington State made the mumps vaccine available in county and school clinics, for example, few adults and older children sought the shot; instead, five- to nine-year-olds were the most frequently vaccinated.[54] This wasn't necessarily a bad thing, said Washington health officer Phillip Jones, who pointed out that there were two ways to attack a health problem: you could either immunize a susceptible population or protect them from exposure. Immunizing children did both, as it protected children directly and in turn stopped exposure of adults, who usually caught the disease from kids.[55] Immunizing children sidestepped the problem he had noticed in his own county. "It is impractical to think that immunization of adults and teen-agers against mumps will have any significant impact on the total incidence of adult and teen-age mumps. It is very difficult to motivate these people," said Jones. "On the other hand, parents of younger children eagerly seek immunization of these younger children and there are numerous well-established programs for the immunization of children, to which mumps immunization can be added."[56]

Setting aside concerns regarding the dangers of giving children immunity of unknown duration, Jones effectively articulated the general consensus on immunization of his time. The polio immunization drives described in chapters 1 and 2 had helped forge the impression that vaccines were "for children" as opposed to adults.[57] The establishment of routine pediatric care, also discussed in chapter 1, offered a convenient setting for broad administration of vaccines, as well as an audience primed to accept the practice. As a Washington, D.C., health officer remarked, his district found that they could effectively use the smallpox vaccine, which most "mothers" eagerly sought for their children, as "bait" to lure them in for vaccines against other infections.[58] The vaccination of children got an added boost from the news that Russia, the United States' key Cold War opponent and foil in the space race, had by the end of 1967 already vaccinated more than a million of its youngsters against mumps.[59]

The initial hesitation to vaccinate children against mumps was further dismantled by concurrent discourse concerning a separate vaccine, against rubella (then commonly known as German measles). In the mid-1960s, rubella had joined polio and smallpox in the ranks of diseases actively instilling fear in parents, and particularly mothers. Rubella, a viral infection that typically caused rash and a fever, was harmless in children. But when pregnant women caught the infection, it posed a risk of harm to the fetus. A nationwide rubella epidemic in 1963 and 1964 resulted in a reported 30,000 fetal deaths and the birth of more than 20,000 children with severe handicaps.[60] In fact, no sooner had the nation's Advisory Committee on Immunization Practices been formed, in 1964, than its members began to discuss the potential for a pending rubella vaccine to prevent similar outbreaks in the future.[61] But as research on the vaccine progressed, it became apparent that while the shot produced no side effects in children, in women it caused a "rubella-like syndrome" in addition to swollen and painful joints.[62] Combined with the fact that the vaccine's potential to cause birth defects was unknown, and that the vaccination of women planning to become pregnant was perceived as logistically difficult, federal health officials concluded that "the widespread immunization of children would seem to be a safer and more efficient way to control rubella syndrome."[63] Immunization of children against rubella was further justified based on the observation that children were "the major source of virus dissemination in the community."[64] Pregnant women, that is, would be protected from the disease as long as they didn't come into contact with it.[65]

The decision to recommend the mass immunization of children against rubella marked the first time that vaccination was deployed in a manner that offered no direct benefit to the individuals vaccinated, as historian Leslie Reagan has noted. Reagan and, separately, sociologist Jacob Heller have argued that a unique cultural impetus was at play in the adoption of this policy: as an accepted but difficult-to-verify means of obtaining a therapeutic abortion at a time when all other forms of abortion were illegal, rubella infection was linked to the contentious abortion politics of the time.[66] A pregnant woman, that is, could legitimately obtain an otherwise illegal abortion by claiming that she had been exposed to rubella, even if she had no symptoms of the disease. Eliminating rubella from communities through vaccination of children would close this loophole — or so some abortion opponents likely hoped. Eliminating rubella was also one means of addressing the growing epidemic of mental retardation, since the virus was known to cause birth defects and congenital deformities that led children to be either physically disabled or cognitively impaired.[67] Rubella immunization promotion thus built directly upon the broader public's anxieties about abortion, the "crippling" diseases (such as polio), and mental retardation.[68]

In its early years, the promotion of mumps immunization built on some of these same fears. Federal immunization brochures from the

1940s and 1950s occasionally mentioned that mumps could swell the brain or the meninges (the fluid surrounding the brain), but they never mentioned a risk of brain damage. In the late 1960s, however, such insinuations began to appear in reports on the new vaccine. Hilleman's early papers on the mumps vaccine trials opened with the repeated statement that "Mumps is a common childhood disease that may be severely and even permanently crippling when it involves the brain."[69] When Chicago announced Mumps Prevention Day, the city's medical director described mumps as a disease that can "contribute to mental retardation."[70] Though newspaper reporters focused more consistently on the risk that mumps posed to male fertility, many echoed the "news" that mumps could cause permanent damage to the brain. Such reports obscured substantial differentials of risk noted in the scientific literature. For unlike the link between mumps and testicular swelling, the relationship between mumps and brain damage or mental retardation was neither proven nor quantified, even though "benign" swelling of meninges was documented to appear in 15 percent of childhood cases.[71] In a nation just beginning to address the treatment of mentally retarded children as a social (instead of private) problem, however, any opportunity to prevent further potential cases of brain damage, no matter how small, was welcomed by both parents and cost-benefit-calculating municipalities.

The notion that vaccines protected the health (and, therefore, the productivity and utility) of future adult citizens had long been in place by the time the rubella vaccine was licensed in 1969. In addition to fulfilling this role, the rubella vaccine and the mumps vaccine—which, again, was most commonly depicted as a guard against sterility and "damage to the sex glands" in men—were also deployed to ensure the *existence* of future citizens, by protecting the reproductive capacities of the American population. The vaccination of children against both rubella and mumps was thus linked to cultural anxiety over falling fertility in the post–Baby Boom United States. In this context, mumps infection became nearly as much a cause for concern in the American home as it had been in army barracks and worker camps two decades before. This view of the disease was captured in a 1973 episode of the popular television sitcom *The Brady Bunch*, in which panic ensued when young Bobby Brady learned he might have caught the mumps from his girlfriend and put his entire family at risk of infection. "Bobby, for your first kiss, did you have to pick a girl with the mumps?" asked his father, who had made it to adulthood without a case of the disease.[72] This cultural anxiety was also evident in immunization policy discussions. CDC scientists stressed the importance of immunizing against mumps given men's *fears* of mumps-induced impotence and sterility—even as they acknowledged that such complications were "rather poorly documented and thought to occur rarely, if at all."[73]

As the new mumps vaccine was defining its role, the revolution in reproductive technologies, rights, and discourse that extended from the 1960s into the 1970s was reshaping American—particularly middle-class American—attitudes toward children in a manner that had direct bearing on the culture's willingness to accept a growing number of vaccines for children. The year 1967 saw more vaccines under development than ever before.[74] Merck's own investment in vaccine research and promotion exemplified the trend; even as doctors and health officials were debating how to use Mumpsvax, Hilleman's lab was testing a combined vaccine against measles, rubella, and mumps that would ultimately help make the company a giant in the vaccine market.[75] This boom in vaccine commodification coincided with the gradual shrinking of American families that new contraceptive technologies and the changing social role of women (among other factors) had helped engender.

The link between these two trends found expression in shifting attitudes toward the value of children, which were well-captured by *Chicago Tribune* columnist Joan Beck in 1967. Beck predicted that 1967 would be a "vintage year" for babies, for the 1967 baby stood "the best chance in history of being truly wanted" and the "best chance in history to grow up healthier and brighter and to get a better education than his forebears." He'd be healthier—and smarter—thanks in large part to vaccines, which would enable him to "skip" mumps, rubella, and measles, with their attendant potential to "take the edge off a child's intelligence."[76] American children might be fewer in number as well as costly, Beck wrote, but they'd be both deeply desired and ultimately well worth the tremendous investment. This attitude is indicative of the soaring emotional value that children accrued in the last half of the twentieth century.[77] In the 1960s, vaccination advocates appealed directly to the parent of the highly valued child, by emphasizing the importance of vaccinating against diseases that seemed rare or mild, or whose complications seemed even rarer still. Noted one CDC scientist, who extolled the importance of vaccination against such diseases as diphtheria and whooping cough even as they became increasingly rare: "The disease incidence may be one in a thousand, but if that one is your child, the incidence is a hundred percent."[78]

Discourse concerning the "wantedness" of individual children in the post–Baby Boom era reflected a predominantly white middle-class conceptualization of children.[79] As middle-class birth rates continued to fall, reaching a nadir in 1978, vaccines kept company with other commodities—a suburban home, quality schooling, a good college—that shaped the truly wanted child's middle-class upbringing.[80] From the late 1960s through the 1970s, vaccination in general was increasingly represented as both a modern comfort and a convenience of contemporary living. This portrayal dovetailed with the frequent depiction of the mild infections, and mumps in particular, as "nuisances" American no longer needed to "tolerate."[81] No longer did Americans of any age have to suffer the "variety of spots and lumps and whoops" that once plagued American childhood, noted one reporter.[82] Even CDC publications commented on "the luxury and ease of health provided by artificial antigens" of the modern age.[83]

And even though mumps, for one, was not a serious disease, remarked one magazine writer, the vaccination was there "for those who want to be spared even the slight discomfort of a case."[84] Mumps vaccination in fact epitomized the realization of ease of modern living through vaccination. Because it kept kids home from school and parents home from work, "it is inconvenient, to say the least, to have mumps," noted a Massachusetts health official. "Why should we tolerate it any longer?"[85] Merck aimed to capitalize on this view with ads it ran in the seventies: "To help avoid the discomfort, the inconvenience—and the possibility of complications: Mumpsvax," read the ad copy.[86] Vaccines against infections such as mumps might not be perceived as absolutely necessary, but the physical and material comfort they provided could not be undervalued.

MUMPS POST-VACCINE

Despite mild apathy and confusion about mumps immunization in its early years, by 1972 Merck had sold more than 11 million doses of Mumpsvax.[87] By 1974, 40 percent of U.S. children had been vaccinated against the disease.[88] There was at least one very practical reason for the growing uptake of the mumps vaccine. In 1971 Merck released two combined vaccines, in which a single shot offered protection against multiple infections. One of the company's combined vaccines protected against measles and rubella; the other against measles, rubella, and mumps. Company representatives were quick to point out that the measles, mumps, and rubella, or MMR, vaccine reduced total vaccination costs as well as the number of doctor visits needed to completely immunize children.[89] The vaccine thus offered a tidy solution to the low overall rates of vaccination among poor, mobile, and urban populations, a reality that became increasingly apparent as immunization rates among middle-class children continued to rise and outbreaks of preventable diseases continued to plague the nation's cities in the 1970s.

Merck's combined vaccine was the endpoint of corporate market-share and profitability concerns; the company had a product to sell and figured out the best way to sell it. But the combined vaccine also directly impacted evolving immunization policy, which in turn helped sell Mumpsvax. In its 1972 update on mumps vaccination, the CDC's Advisory Committee on Immunization Practices (ACIP) broadened its recommendation to state that anyone over the age of one *could* be vaccinated—but that mumps vaccination should never take priority over "more essential ongoing community health activities."[90] (The committee's final report declined to specify what these might be.) Combined vaccines, they noted, were still being tested. In 1977, however, the committee reaffirmed the stance that mumps was a low-priority disease but added that "large-scale production" of combined vaccines "have made mumps vaccination a practical component of routine immunization activities." As a result, they wrote, "Live mumps virus vaccine is recommended for all children at any age after 12 months."[91] The combined vaccine enabled mumps to piggyback on acceptance of the vaccines against measles and rubella and overrode for good any questions about the necessity of universal protection against mumps.[92] It also secured a profitable position for Merck, which in the years to come would view MMR as its "work horse" vaccine.[93]

But even as mumps immunization rates gradually climbed, overall rates of immunization stagnated or fell in the seventies. As health officials continued their quest to encourage broad use of all of the childhood vaccines, they attributed poor uptake to permissive school laws, public apathy toward childhood disease prevention, and maternal ignorance about the severity of childhood diseases. These observations drove efforts to prove the value of school laws and the economic benefits of vaccination. They also drove efforts to inform the public, and mothers in particular, about the dangers of childhood disease. (For more on this, see chapters 4 and 5.) As a result, as the seventies progressed, *all* of the vaccine-preventable diseases were increasingly portrayed as a monolithic category whose members were uniformly threatening to the health of American families. In 1973 the Orange County, California, health department adopted a particularly intimidating motto to convey this idea, dubbing a parents' failure to immunize his or her children against diphtheria, whooping cough, tetanus, polio, measles, rubella, and mumps "the Seven Deadly Health Sins."[94]

The practice of portraying the vaccine-preventable diseases in this way culminated with an unprecedented nationwide childhood immunization effort spearheaded by President Jimmy Carter's administration in 1977. In that initiative, described in detail in the following chapter, campaign materials issued by local health departments and the flurry of accompanying media reports all contained the same message: *any* disease that could be prevented with a vaccine was dangerous, if not deadly, to children. In such accounts, any difference in risk or severity among infections was depicted as negligible. Measles was a disease that caused brain damage and deafness, mumps a disease that caused deafness, polio a disease that caused paralysis or death.[95] Media portrayals of the time reflected this now-homogenized view of all of the vaccine-preventable infections. A 1978 article in *Good Housekeeping* told the stories of seven-year-old Joey, who became deaf after a bout of mumps; five-month-old

Marcy, who died of whooping cough; and twelve-year-old June, who suffered measles-induced brain damage. Wrote the author, "If this sounds scary, it's meant to."[96]

The rhetorical transformation of mumps into a serious disease of children was neither deliberately calculated nor entirely smooth. Hints that such a shift was imminent were evident in the late 1960s, when the prospect of a new mumps vaccine had prompted a few observers to argue that the infection's occasional severity made it a disease worth not just preventing, but eradicating.[97] Between 1968 and 1978, scientific journal articles, press reports, and advice books for parents and doctors contained uneven portrayals of the disease's severity. On the whole, however, mumps' image in this time period morphed from that of childhood nuisance to that of deadly crippler. In 1968 the *Washington Post* referred to mumps as a "relatively harmless childhood disease," the *Los Angeles Sentinel* called it a "mild childhood disease," and the *New York Times* reported that "serious complications in young children are unusual."[98] Ten years later, all references to mumps as a mild disease vanished from popular portrayals, with the exception of reports that sought to illustrate the danger of perceiving mumps in this way: Measles, mumps and the like are not just "part of growing up," reported the *New York Times* in 1978, ". . . those 'minor' ills can cripple and kill, too."[99]

Mentions of the disease's severity in children versus adults also disappeared over the course of the 1970s. Mumps "is an acute viral disease . . . which can lead to orchitis [testicular swelling] and meningoencephalitis," reported the *Atlanta Daily World* in 1978, making no reference to the actual risk of complications nor any distinction between the risk of sequelae in adults versus children, or males versus females.[100] Portrayals of the disease were now consistently daunting: "Mumps frequently leads to inflammation of the covering of the brain (meningitis) or, more rarely, of the brain itself (encephalitis). As many as one in every seven to nine children with mumps may show signs of these complications (but actual cases are not as common). Occasionally, permanent damage such as deafness results," reported *Good Housekeeping*.[101] Such portrayals (if perplexing) leveled the field between mumps and the rest of the vaccine-preventable diseases, enabling the once mild and chucklesome infection to keep close company with long-dreaded diphtheria, smallpox, and polio.

Depictions of vaccine-preventable diseases in the immunization promotion materials from the late seventies often directly and unquestioningly parroted the information coming from government sources.[102] While CDC immunization recommendations from the early seventies had referred to mumps-induced meningeal swelling and deafness as "rare" and mumps-induced sterility as "very rare," a 1978 Department of Health, Education, and Welfare brochure informed the public simply, "Mumps can cause deafness, diabetes, and brain damage. It can make boys sterile."[103] The mention of diabetes is a clear signal that government health officials had decided that any risk of disease complication was worth wielding in the crusade to encourage vaccination. For at that time, evidence of a link between mumps and diabetes was hypothetical, based on isolated case reports and the infection's potential to cause pancreatic swelling. The link was too tenuous to be mentioned in any of the ACIP's published recommendations on mumps immunization in the 1970s, and it in fact remains unconfirmed to this day. But health officials appeared to have bet that Americans would agree that an infection that reportedly caused diabetes, deafness, and brain damage was certainly as worthy of prevention as an infection that caused paralysis and death.

And so they did. The Carter campaign discussed in the next chapter would help take immunization rates to new heights by the very end of the seventies. By the early eighties, 97 percent of American children would be vaccinated against mumps.[104] And by 1985, incidence of the disease would fall to just under 3,000 cases—an all-time low and a 98.1 percent drop from mumps levels back in 1968.[105]

Before research on a mumps vaccine began in World War II, mumps had long been categorized as a common childhood disease. Resources to control the disease were corralled only when the infection posed a threat to the nation's security, by infecting grown men engaged in military or economically productive pursuits. From the beginning of the twentieth century to the end, efforts to thwart mumps' spread through crowded settings, whether barracks or work camps or mental institutions or school districts, belied a concern for the impact of epidemic mumps on the social order of the state. Vaccines against mumps—from Lederle's half-effective vaccine to Merck's MMR—were thus tools of governance that served the national interests of security and economic efficiency as much as they served the medical needs of individuals.[106]

That Merck's new live-virus mumps vaccine was ultimately sold and bought as an integral part of childhood care is testament to the many sociocultural meanings that mumps and its vaccine acquired from 1967 on. Before 1967, mumps was a threat to U.S. supremacy because it harmed soldiers and laborers; after its vaccine became available, it represented a threat to U.S. supremacy because it harmed the nation's ability to produce future soldiers and laborers. Before 1967, mumps was an uncomfortable if generally innocuous part of childhood, as captured in a 1962 Jimmy Dean song about a daughter growing older: "It's a rapid journey and you'll travel light, leaving behind you measles, mumps, freckles, bumps, bubble gum and me."[107] After 1967, mumps vaccine, especially once given as MMR, gradually replaced mumps itself as the uncomfortable if generally innocuous part of childhood. And playful cultural references to the disease became relics of another time.

The fate of mumps and its vaccine foreshadowed a similar fate for infections that would be targeted by vaccines of the future, including those against hepatitis B, chicken pox, and HPV—all discussed in later chapters. Federal health officials and pharmaceutical giant Merck played undeniably active roles in securing a place for mumps in the roster of childhood vaccines, but mumps vaccination was accepted by Americans in this period because of what the disease itself came to signify. Mumps may not have been a public health priority when Merck's vaccine was introduced, but the vaccine warranted the disease special attention from scientists, health officials, and lay Americans alike. Once in the spotlight, mumps infection was framed by the cultural preoccupations of the day, including access to technological conveniences, the growing emotional valuation of middle-class children, and predominantly middle-class concerns about reproduction and mental retardation. As the seventies progressed, mumps vaccine and the childhood vaccines generally would get an added boost from domestic health policy priorities in President Jimmy Carter's White House—which offered its own spin on the childhood vaccination expansion efforts inaugurated during the Kennedy era.

PART II

4

CARTER'S CHILDHOOD
IMMUNIZATION INITIATIVE

Mothers and fathers browsing the latest issue of *Parents* magazine in November 1977 came across an article from an unusual contributor: U.S. Health, Education, and Welfare Secretary Joseph Califano. Califano told readers that President Jimmy Carter had made a commitment to protect American children from preventable disease, but that this commitment could be fulfilled only if parents took the steps necessary to get their children vaccinated. The secretary urged parents not to allow diseases that were once "deadly, daily facts of life" to resurge through "apathy or ignorance." And then he stressed a separate concern. "The cost to each family, and the nation as a whole, of fully immunizing our nation's young is negligible compared to the cost, in dollars and human suffering, when children are attacked by diseases such as polio, tetanus, whooping cough and diphtheria," he wrote. "All it takes to prevent these diseases is a few simple and inexpensive shots for every child."[1]

Califano's assertion—that shots were cheap and the diseases they prevented pricy—was no mere selling point. It was the very reason that the Carter administration had recently launched a nationwide initiative to vaccinate all American children against seven preventable infections. Carter's Childhood Immunization Initiative was announced three months after the new president took office in 1977, as the nation's struggling economy, oil crisis, and rising health care costs made frequent headlines. In the explicit interest of saving health care dollars, over the next eighteen months the president's immunization program urged states to strengthen, update, and enforce childhood vaccination laws and identify and crack down on children lacking vaccines against any of the preventable infections. Califano was a key player in the initiative, personally recruiting governors, corporate leaders, local PTAs, and celebrities to the cause. By 1980, when Carter left office, upwards of 96 percent of all children entering school were vaccinated against measles, rubella, polio, diphtheria, pertussis, and tetanus, marking the highest rates of vaccine coverage the country had ever seen.[2]

The Carter immunization initiative captured the essence of everything Jimmy Carter believed about government and its appropriate place in the American health care system. It was not a traditional big government endeavor, reliant instead upon the cooperation of local governments, the private sector, and individual citizens, and designed to give individuals a sense of ownership and investment in the effort. The initiative rallied countless volunteers to carry out the government's vision in towns and cities across the country, and parents heeded the message to immunize their children promptly and thoroughly. The program cited causes and consequences of under-immunization that stood in stark contrast to those cited by the Kennedy administration's immunization program. Parental ignorance (not poverty) was the reason so many parents didn't immunize their children, and soaring health care costs were the result of a nation too complacent to prevent disease before it struck. The program's emphasis on personal responsibility, cost-efficiency, and prevention was mirrored in the administration's later proposal for national health insurance. That proposal failed; the Childhood Immunization Initiative succeeded. But even as it took child immunization levels to new heights, the Carter program failed to account for changing social currents and unresolved dilemmas that were just beginning to beleaguer the nation's vaccine enterprise.

HEALTH SPENDING AND THE CALL FOR REFORM

Jimmy Carter ran for president in 1976 as a resolute Washington outsider who pledged to return transparency, fairness, and efficiency to the federal government in the post-Watergate era.[3] The 1976 campaign took place in the midst of a deep economic recession, high unemployment, and a staggering rise in fuel prices—all of which colored Carter's approach to setting his domestic policy agenda.[4] In his inaugural address, the newly elected president called on the nation to take on "moral duties" and work together with government "in the spirit of individual sacrifice for the common good."[5] In office, he routinely urged the American public to do their part to address the nation's woes. At the height of the nation's crippling fuel crisis, in one memorable example, the president asked Americans to carpool or use public transportation, obey the speed limit, and turn down their thermostats to help cut back on total national energy consumption.[6]

Carter's administration applied that same spirit to steadily mounting health care costs, which the president had also pledged to rein in during the campaign. Earlier in the decade, President Richard Nixon had declared American health care to be in a state of "crisis"; politicians and constituents across the political spectrum agreed that legislation was necessary to make access to care more equitable and to curb skyrocketing health costs. By the late seventies, however, substantial legislative reform had yet to materialize, and sky-high hospital costs and health care inflation were routine front-page news. Health care costs were rising faster than the cost of living: 10 percent a year, or 3 percent faster than the rise in costs of other goods and services. By 1977 health care spending constituted more than 8 percent of gross national product (enviable by today's standards, but alarming at the time), and individual health care expenses ate up more than 10 percent of the average household's income.[7]

The message of prevention was central to the administration's stance on health care. Too many Americans were reckless with their health, ignoring simple steps to stave off illness. "We tend to forget that our improved health has come more from preventing disease than from treating it once it strikes," noted Carter. "Our fascination with the more glamorous 'pound of cure' has tended to dazzle us into ignoring the even more effective 'ounce of prevention.'"[8] In crafting the administration's message on health inflation in early 1977, Carter's chief domestic policy adviser Stu Eizenstat emphasized personal responsibility for prevention. "There is no health inflation policy we can develop that would do more good than sensible living by the American people," he stressed. "The American people should reduce their intake of alcohol and tobacco; they should get more exercise and drive more safely; they should have their children immunized; and they should seek early care for pregnancy, hypertension, and other conditions," he added, citing approaches that had become popular in the field of public health over the past decade.[9] Government, he assured, would help Americans save health care dollars, but responsibility also lay with individual citizens to protect their own health.

On the matter of how much government should help solve the growing problem of health care inflation, Carter and many of his closest advisers were at odds with more liberal Democrats. Carter had made a campaign pledge to create a national health insurance plan—but he did so begrudgingly, to secure an endorsement from organized labor.[10] In office, Carter put hospital-cost containment at the top of his domestic policy agenda and stalled on introducing overall health reform. His predilection was for an incremental approach to reform that built on public-private partnerships; both preferences pitted him against Democrats who wanted a quick and comprehensive overhaul of the health care system. Carter envisioned a federal-state-private plan that emphasized disease prevention, cost containment, and efficiency, and that called on patients to share costs.[11] The proposal he finally unveiled in the summer of 1979 was to provide catastrophic insurance to all Americans, require employers to provide health coverage to their workers, and create a new federal program to insure the aged, the poor, and the remaining uninsured.[12]

Health care costs were a widely acknowledged domestic policy priority—and had been since the beginning of the seventies—but the president's proposed solution found itself too politically embattled to succeed. Denounced by powerful congressional Democrats as too conservative and overshadowed by other domestic and foreign crises, the proposal went nowhere in Congress. But it wasn't the administration's only solution to runaway health care costs. Early in his presidency, Carter had championed a plan to ratchet up childhood immunization. With its small price tag, guaranteed return on investment, and a constituency that couldn't object, childhood immunization promised to be a politically expedient approach to scaling back on the nation's health spending. As one immunization program proponent put it, "Supporting immunization was like supporting the cause of motherhood and apple pie—you can't go wrong."[13]

LAUNCHING THE CHILDHOOD IMMUNIZATION INITIATIVE

On April 7, 1977, Carter announced an unprecedented "high-visibility" two-year initiative to promote immunization in the United States. The administration's program had two main goals: to immunize 90 percent of all children against seven preventable infections by October 1979, and to establish a permanent system to ensure the full and timely immunization of the 3 million children born each year. The initiative's list of targeted diseases included diphtheria, pertussis, tetanus, polio, measles, rubella, and, of course, mumps. "Our goal is to reduce as much as humanly possible the numbers of youngsters without medical protection against many major childhood afflictions," Carter told his cabinet.[14]

The announcement marked the nation's observance of World Health Day, an annual event launched by the World Health Organization (WHO) two decades earlier to mark the anniversary of the organization's founding and to bring attention to pressing global health concerns. The president's initiative aligned with the WHO's theme for the year: "Immunize and Protect All Children."[15] The six diseases targeted by the WHO for prevention through immunization in developing nations reflected a slightly different set of priorities than in the United States: globally, there would be no push to immunize against mumps or rubella, but children would be immunized against tuberculosis, in addition to diphtheria, tetanus, polio, pertussis, and measles. For Americans, Carter noted, the global theme was "a reminder of our responsibilities to our children" and a reminder to "take pride in the dedicated research which has produced safe and effective vaccines against polio, diphtheria, measles, and other childhood diseases."[16]

With apartheid in South Africa and the atrocities of the Khmer Rouge in Cambodia and dictatorial regimes in Argentina and Chile a constant theme of current events, Carter had placed the defense of global human rights at the center of his foreign policy agenda. His advisers noted that the immunization initiative extended Carter's "worldwide campaign for human rights . . . to the children of America."[17] But the initiative's connection to global health priorities and Carter's own human rights agenda was mostly just a convenient rhetorical point. Carter didn't embrace an inalienable right to health; as Peter Bourne, his special adviser on health, noted, the president "did not see health care as every citizen's right, which the government had an obligation to meet."[18] Indeed, Carter was ambivalent toward government and medicine, two of the pillars that upheld a fully vaccinated population. Carter himself railed against "waste, excess, and profits" in medicine.[19] And his administration's staunchest immunization supporters criticized both physicians and the CDC. (As immunization champion Senator Dale Bumpers once put it, the CDC "has never been high on my credibility list."[20]) But the fact remained that vaccines themselves were available, proven effective, cost just cents per dose, and yielded significant savings in the long run. Vaccines also existed in abundance—American families just weren't making use of them. The fundamental belief underlying Carter's immunization initiative was not that government needed to oversee the distribution of health care commodities, vaccines included, as a matter of social equity or justice. Rather, the immunization of children was an efficient use of resources, an incremental and cost-beneficial solution to a much larger, much more intractable problem.

The immunization of children was also, arguably, urgent. The Vaccination Assistance Act renewal had expired at the end of the sixties, and in the early seventies, under President Richard Nixon, states had instead received block grants from the federal government to spend on public health programming generally.[21] Many states chose to use the funds for projects unrelated to immunization, and across the country the number of children getting vaccinated against polio, diphtheria, tetanus, and pertussis had declined by the time Carter was inaugurated. The fraction of children immunized against measles, by contrast, had increased— among children from upper- and middle-income families. But by 1976, just 42 percent of children from *lower*-income families were immunized against the disease.[22] As a result, a growing number of measles epidemics had erupted in cities across the United States in 1975 and 1976, and health officials were predicting more outbreaks (along with out-breaks of rubella and mumps) for 1977.[23]

Parents who had vaccinated their children against measles may have done so for a number of reasons: in response to the CDC's eradication campaign, because they saw the many ads vaccine-maker Merck placed in women's and parenting magazines, or because new laws required them to.[24] At the end of the sixties, the Kennedy Foundation, in its on-going efforts to prevent mental retardation, had pushed states to adopt laws requiring measles vaccine for school enrollment.[25] In the context of growing epidemiological evidence showing that school mandates were effective in curbing epidemics, many states did.[26] By 1974, forty states had adopted laws requiring measles (and often other) vaccines for school.[27] When Carter took office, however, eleven states still had no measles vaccination laws and five states had no school vaccination laws at all.[28] In total, some 20 million of the nation's 52 million children under fourteen weren't fully protected against measles or the other vaccine-preventable infections.[29] To the Carter administration, such statistics signaled a clear opportunity for improvement. Schools, and the nation, "can and must do a great deal more" to promote immunization and halt the spread of infectious disease, Califano declared.[30]

BUMPERS'S BLUEPRINT AND THE SWINE FLU MACHINE

Health care cost concerns, global events, falling immunization rates, and persistent epidemics all suggested that the timing was right for a renewed federal effort to promote immunization. The initiative came together quickly, however, in part because two friends of the Carters—Betty and Dale Bumpers—provided the blueprint for getting such an effort off the ground with relatively little federal investment, and in part because a recent series of events had put the necessary infrastructure in place.

In the early 1970s, Betty Bumpers—wife of Arkansas senator and former governor Dale Bumpers and a friend of the Carters—had led a massive immunization campaign to fully vaccinate that state's children. With a budget of just $100,000 and the cooperation of county clinics, community groups, the National Guard, and more than 10,000 volunteers, the campaign immunized nearly 300,000 children in a single weekend in 1973. (Local celebrity "Miss Arkansas" also plugged the campaign across the state.) The campaign moved the state's immunization levels from forty-eighth to second in the country.[31] It also translated into huge savings: it cost the state upwards of $13,000 a year to educate each blind or deaf pupil, but immunizing every child against infections that caused such disabilities had involved a one-time cost of just $5 per child, a state health officer informed then-Governor Bumpers. Just after Carter took office, the Bumpers reminded the president and First Lady in detail of their success and urged them to try something similar on a national scale.[32]

Following the Bumpers model, Carter's Childhood Immunization Initiative was billed as a low-cost program with the promise of extraordinarily high returns—critical positioning in the midst of the deep economic recession that Carter had inherited from his predecessor, Gerald Ford. The Carter plan increased spending on childhood immunization 50 percent over the previous administration, but proponents promised it was an expenditure that would yield dramatic savings in the long run.[33] In press conferences, releases, brochures, and speeches announcing the initiative, administration officials tallied the historical costs of not immunizing children. The rubella epidemic of 1964 had cost the United States $1.5 billion. Every child disabled by preventable disease cost states and the federal government a total of $900,000 in direct and indirect costs.[34] By contrast, mass immunization against measles was estimated to have saved the country $1.3 billion in health care spending between 1963 and 1974.[35] "By immunizing children . . . we save millions of dollars that would otherwise have to be used on hospital costs and long-term care for those who are seriously afflicted," argued Califano.[36] For $6 per child—the cost of a complete round of childhood vaccines in 1977—the nation might even see its health care spending decline by $5 to $10 billion per year, Senator Bumpers predicted.[37]

The economic climate that gave these calculations such weight was not all that Carter had inherited from Ford. A year earlier Ford, flanked by polio vaccine developers Jonas Salk and Albert Sabin, had announced a campaign to immunize an unprecedented 200 million Americans—95 percent of the entire population—against a prophesied epidemic of swine flu.[38] The disease had struck five hundred soldiers at Fort Dix, New Jersey, early that year, and when scientists took a closer look at the offending virus, they found it frighteningly similar to the strain that had killed half a million Americans in the deadly flu pandemic of 1918.[39] The plan to vaccinate nearly the entire U.S. population against the virus had its detractors; some called the comparison to the 1918 flu "specious" and argued that the chances of an epidemic were overblown.[40] On the recommendation of CDC head David Sencer, Ford forged ahead with the plan nonetheless in October 1976. But by December—after 40 million Americans had been vaccinated—the epidemic had failed to materialize. And early reports of illness and death following vaccination were confirmed to be something serious: the swine flu vaccine had caused dozens of cases of Guillain-Barré syndrome, a rare neurological disorder. The $135 million project was abruptly brought to a halt. The news was a public relations disaster for the Ford administration, and ultimately cost Sencer and a number of health officials their jobs.[41]

The swine flu "affair" was widely mocked in the press and led to confusion, doubt, and consternation among the public and many public health professionals. Given "this whole swine flu business . . . I am wondering now whether all vaccinations are really safe and effective," wrote a *Chicago Tribune* reader.[42] "There will be massive resistance to the next public health campaign, perhaps a much wiser and more necessary one," predicted a former CDC officer stationed in Hawaii.[43] In a triple strike for public health officials that year, the swine flu debacle coincided with two unforeseen outbreaks, one an unexplainable, widely reported "mystery disease" (later named Legionnaires' disease) that broke out in a Philadelphia convention center, the other an outbreak of flu caused by another strain of the virus, which health officials had inadequately prepared for. "The inability of federal health officials to explain the deaths of the people who attended the American Legion convention in Philadelphia last summer, the hullabaloo over a swine flu epidemic that has not yet materialized, and the irony of not being able to get the A/Victoria flu vaccine without receiving the swine flu shot obviously have left the public shaken," Harris survey pollsters concluded in early 1977, just after Carter had taken office. "One must now assume that if another health crisis arises and efforts are made to immunize large numbers of people, public skepticism can be expected to be high."[44]

Given that public skepticism, a massive federal immunization program could easily have been a tough sell in early 1977. But the Carter initiative's advocates skillfully used the swine flu affair to their advantage. The massive amount earmarked for the swine flu campaign positioned the administration to demonstrate that they could do much more — namely, protect millions of children from multiple known threats — with far less money. Califano and Bumpers also pointed out that the Childhood Immunization Initiative offered an opportunity to "turn the swine flu machine" — the distribution networks and state health offices mobilized to distribute and administer swine flu vaccine — into "a positive effort on behalf of children."[45] Both the CDC and the National League for Nursing had organized huge networks of health workers and volunteers in 1976. For the Carter administration, the two pools of mobilized volunteers amounted to a tremendous ready resource.[46] So did unspent swine flu vaccination funds, which Congress transferred to the CDC's immunization division for the express purpose of immunizing children.[47]

Crafting a nationwide initiative that lacked the imprint of big government also helped the Carter administration distance itself from the previous administration's disgraced campaign. "No nationwide campaign comparable to last year's swine flu vaccination program is planned," administration officials told the press of the president's new initiative — in a calculated exaggeration of the truth.[48] In keeping with the Bumpers model, Carter's immunization initiative was frugal: $57 million was appropriated for the program over three years, totaling less than half the budget for swine flu immunizations alone in 1976.[49] Such a slim budget prevented Congress from looking like "a bunch of fools" the way the failed swine flu program had made them look, said Indiana Senator Birch Bayh.[50] The initiative would also be decentralized. As Bumpers pressed his fellow senators to approve added funding to launch the childhood vaccination program in the spring of 1977, he stressed that any such initiative should be guided by federal oversight but carried out by state health workers and legions of volunteers. The Carter administration had every intention of hewing to this vision. The campaign, said Califano on the day it was announced, would primarily "be waged with the will of our citizens, not the dollars of our treasury."[51]

CALIFANO AND THE CAMPAIGN

As the initiative got under way, that will was mustered in large part by massive outreach and groundwork on the part of Health, Education, and Welfare officials, especially Joseph Califano. Califano, who had helped author Great Society legislations as chief domestic adviser to President Lyndon Johnson, had actively sought the post of Secretary of Health, Education, and Welfare in Carter's cabinet, eager to be part of a renewed push for social programs in the first Democratic administration in eight years.[52] Indeed, the media, and many Democrats, saw Califano's appointment as presaging "a new dawn of social concern" and a return to the "fond yesterdays" of the John F. Kennedy and Lyndon B. Johnson years, when civil rights legislation was passed, the War on Poverty declared, and Medicare and Medicaid established to provide health care to the elderly and indigent.[53] Califano assumed the HEW helm with high energy and visibility, prompting one reporter to remark that "he took on his job as though he intended to carry out every Carter campaign promise before Congress even went home for summer vacation."[54] When Carter and his wife, Rosalynn, asked Califano to spearhead a nationwide initiative to immunize children, the secretary took on the task with his trademark zeal.

On announcing the initiative, Carter had explained that HEW's role would be to "stimulate immunization action," while the CDC gave states grants to purchase vaccines and run clinics, and the National League of Nursing (with Betty Bumpers as chair) mobilized and coordinated volunteers.[55] When it came time to "stimulate action," Califano was creative and forthright. He called and wrote directly to governors, labor leaders, newspaper and magazine publishers, heads of Fortune 500 companies, and television executives across the country, asking—and convincing—them to inform their constituents, employees, readers, and viewers about the importance of vaccination.[56] The secretary's personal pleas led to a television spot and posters featuring two characters from the block-buster hit *Star Wars*, R2-D2 and C-3PO, asking, "Parents of Earth: Are Your Children Fully Immunized?"[57] Popular children's show host Captain Kangaroo filmed a spot encouraging immunization, the National Football League made the campaign the focus of its public service announcements, and widely read advice columnist Dear Abby implored her followers to protect their children with vaccines. At Califano's direct urging, governors of the states without any immunization laws agreed to put such laws on the books, and states with existing laws took steps to close loopholes and step up enforcement.[58]

Califano also reached out directly to doctors—even as he held them partly culpable for recent declines in immunization rates. "Too many doctors fail to ensure that youngsters in their care are fully immunized," he asserted.[59] To the immunization initiative's proponents in Washington, doctors' negligence was an important reason why immunization rates weren't low just among the poor. A full 7 million of the 20 million inadequately immunized children were "in the higher-income levels," noted Senator Bayh.[60] "It is striking," Califano agreed, "that two-thirds of the 20 million children who don't have complete immunization are in families that use private health care providers, rather than public agencies."[61] For that reason, the program would make a strong and concerted effort to reach out to "better-off families," administration officials told the press.[62] Two key messages were embedded in such statements. Low immunizations among "better-off families" meant that the public sector was not at fault; government, not the cause of the problem, could therefore serve as solution. Moreover, if well-off families were not fully vaccinating their children, there were two parties to be held responsible: physicians and parents themselves.

In his article for *Parents* magazine, Califano had called immunization "the job of every parent."[63] But educational materials developed by his agency to support the initiative were far less diplomatic, attributing low immunization rates to "oversight," "misunderstanding," and "parental negligence."[64] In an editorial written for the *Washington Post*, Rosalynn Carter lamented the "laissez-faire attitude of many parents" when it came to immunizing their children.[65] Some members of the public—parents among them—agreed with such assessments. "Place the blame where it belongs: on the neglectful, careless, thoughtless parent," wrote a *Los Angeles Times* reader in a letter to the editor.[66] The administration's emphasis on parental failure made sense alongside the mantra of personal responsibility that the Carters stressed with respect to addressing domestic policy concerns. "Massive government programs" didn't cure social ills, said Rosalynn Carter in a joint interview with the president; people had to "assume responsibility for the problems around them."[67] Of course, not everyone agreed that careless parents were the reason every child wasn't immunized. Some who heard this message wanted to make sure the administration knew that parents and communities could accomplish only so much without help from government sources, precisely because poverty was still an important obstacle to the immunization of children. "We work hard to maintain a 97% immunization level in a rural county," an Ohio nurse wrote to the First Lady, "but can do it only through free clinics."[68]

To truly keep costs low and the imprint of big government faint, the administration sought an immunization campaign waged, like Arkansas's, at the community level—with the help and dedication of people like the concerned nurse from Ohio. "If we can tap the deep well-spring of American idealism and draw upon America's notable tradition of voluntarism, then this immunization campaign can stand as a bright example, not of government helping people, but of people helping themselves," Califano told a gathering of doctors and health officials.[69] His comment—and his vision—invoked the polio immunization campaigns of the 1950s and 1960s, which drew on widespread popular investment and the help of tens of thousands of volunteers. But the administration's efforts were also calculated to invoke a more recent trend. The 1970s were a decade in many ways defined by community organizing. Grassroots social movements fought for women's rights, gay rights, disability rights, peace, and a cleaner environment. The immunization initiative sought to capitalize on this social current, deliberately—if incongruously—attempting to stimulate a grassroots movement from the top down.

In letters and brochures sent out to local PTAs, women's clubs, chapters of the Red Cross, the National League for Nursing, and other community groups, Califano's agency called on communities to form Immunization Action Committees. HEW literature spelled out precisely how to wage a "community-based" campaign: "set goals, recommend policies and procedures, initiate local action . . . and make sure that things get done." "Call to Action" guides suggested meeting with newspaper editors, holding fund-raisers, giving presentations to school groups and medical societies, and visiting schools and doctors' offices to review records, identify children in need of vaccines, and directly contact their parents. But the agency warned would-be organizers to consult their local health department first: "It is vital that the health professionals be aware of your plans," the brochures stressed.[70] HEW wanted a grassroots campaign—albeit one that conformed to a mandate from above.

The nationwide push for immunization received an added boost from school-based measles outbreaks that made more headlines across the United States in 1977.[71] Local governments had already begun to strengthen and enforce vague or outdated laws requiring children to be vaccinated for school entry, but that spring the immunization initiative helped fuel a nationwide crackdown on unimmunized children. Los Angeles County schools, responding to an outbreak of 1,416 measles cases, announced that as many as 50,000 unimmunized children would be barred from school if not vaccinated by May 2.[72] On the morning of May 3, in an unprecedented move, the district turned away more than 23,000 still-unvaccinated students when they showed up for class.[73] Other districts soon followed the California county's lead. Parents in Chicago were told to keep their kids home from school if they weren't vaccinated by fall; Washington, D.C., area schools suspended thousands of students for showing up without their shots; even small-town Watertown, New York, suspended fifty-six students, including five kindergartners, for failing to rack up all of their required vaccines.[74] Throughout 1977, media outlets across the United States cooperated with the federal initiative's mandate: when not reporting on measles outbreaks or suspensions, newspapers publicized the dates and times of public immunization clinics, at which children were vaccinated not just against measles, but all seven targeted infections.

Eighteen months into the campaign, Califano and Rosalynn Carter addressed a conference held to thank the state health and education officials and volunteers who had spearheaded local immunization efforts.

Califano announced that thanks to the initiative, twenty-three states had updated and strengthened their immunization laws, adding new vaccines, such as those against rubella and mumps, to existing laws, for example, or expanding laws to apply to children in all grades in public and private schools.[75] Thirty-one states had pledged to begin enforcing laws requiring children to be immunized before entering school, and thirty-seven had undertaken review of all school records to identify children who had "slipped through the net." To state immunization rates rapidly approaching 90 percent, Califano attributed a 53 percent decline in measles cases, a 21 percent drop in mumps, a 13 percent decline in diphtheria, and an 11 percent drop in rubella.[76]

But the program's success was also measured, critically, in dollars and attributed to widespread cooperation across all sectors of society. As Califano put it:

We doubt, in the 70s, that the Postal Service can deliver a letter in five days, and our doubt is too often justified. So it gives me great pleasure . . . to talk about a government program that . . . is achieving what it set out to do, a program that has drawn the best of the voluntary sector of the states, of the cities, of our health systems, of our school systems, and of our federal system. . . . For an investment of only 3 million Federal dollars in polio vaccine over the past year, we saved the American public $262 million. . . . For $2.1 million spent by the government for rubella immunizations this past year, we believe we've saved $16 million in health care and other costs that would have been required had we not acted.[77]

The secretary sounded a clear message: Vaccines were cheap, they prevented disease, and by taking responsibility and working in partnership with government and the private sector, the American public could tackle communicable diseases and the toll they took on the nation's economy.

The Carter initiative was successful in immunizing unprecedented numbers of children against an unprecedented number of diseases. But it failed to address or even acknowledge fissures in the nation's immunization infrastructure, many of which had been highlighted by the botched swine flu campaign. At a conference held in November 1976—the same month Carter was elected—vaccination experts met at the National Institutes of Health to discuss pressing issues in immunization policy. They discussed the problem of obtaining meaningful informed consent in programs designed to vaccinate millions. They debated the need for a national program to compensate individuals who suffered serious vaccine side effects. And they considered the growing public demand for a greater say in social policy decisions, including decisions on vaccination policy. An emerging patients' rights movement had led to louder and louder demands for patient autonomy and informed consent to all medical procedures. And a new consumer movement was demanding transparency and protection from abuses in the health care system writ large. Noted one of the meeting's participants, "Consumer groups feel that the time is long past when scientists can meet in private and make important social policy decisions without involvement by the public."[78]

At a conference held in April 1977—the very same week Carter announced his administration's immunization initiative—vaccine scientists, doctors, manufacturers, and policy makers grappled yet again with the array of problems facing national vaccination efforts. They bemoaned the dearth of a reliable, continued source of financial support for vaccine research and development. They discussed the need for a federal policy to protect vaccine makers from lawsuits when vaccines caused harm. And they again debated how best to inform the public of both the importance of vaccinations and the risks they carried.[79]

As policy makers and ethicists struggled with the meaning of informed consent to medical procedures required by law—as vaccination now often was—health officials struggled with vaccine supply problems and manufacturer liability concerns. For several years, the only cases of paralytic polio occurring in the United States had been caused by Sabin's live oral polio vaccine, although estimates of just how often the vaccine resulted in a case of polio varied widely.[80] A series of polio vaccine injury lawsuits had driven a significant number of vaccine makers out of the market, putting the government in the position of assuming both liability and the responsibility for informing the public of vaccine risks. With climbing vaccination rates, the number of adverse events would climb too—and not just for polio. Every vaccine came with inherent risks, some more serious than others. As a 1977 HEW brochure pointed out, vaccine side effects could be as mild as a sore arm or as serious as encephalitis or brain damage.[81] Absent any agreement on how those harmed by vaccines should be compensated, the issue of liability remained—and might even be aggravated by the administration's new initiative. "There is no question but that the achievement of the initiative is complicated by persisting problems associated with liability for injury resulting from immunization," noted the CDC's new director, William Foege.[82] His comment foretold a set of problems to come.[83]

The Carter initiative was one piece of a much broader agenda to promote the immunization of children in the 1970s. Health officials at all levels, drug manufacturers, pediatricians, and parents themselves all played a part in making a long list of vaccines the norm for every child by the end of the decade. At the very same time, the major social movements of the decade—not just the patients' rights and consumer movements, but also the women's movement and environmentalism—prompted some Americans to look more closely at vaccines, what was in them, who was promoting them, and why. The swine flu campaign gave Americans one reason to question both the safety of vaccines and the wisdom of government vaccination programs. Instead of providing answers to these questions, however, the Carter initiative gave Americans a different type of government vaccination program. The program vaccinated children against real threats, but it did little to address unresolved questions about vaccines and the policies guiding their use. Even as it brought immunization levels to impressive new heights, the initiative failed to adequately account for profound societal shifts that would ultimately revive the kind of vigorous vaccination skepticism not seen since the Progressive Era.

5

A MOTHER'S RESPONSIBILITY

Late in 1978, Rosalynn Carter and Joseph Califano thanked a crowd of health workers and volunteers for their hard work to date on the Carter administration's national immunization initiative. As they urged the group to keep working toward the goal of immunizing 90 percent of all children, they stressed the need to reach out to parents and inform them of the importance of vaccination. But they emphasized the need to reach one type of parent in particular: mothers. Califano told conference attendees they'd be discussing new plans for reaching the mothers of the 3 million children born each year, to ensure that they received the message that vaccines were vital for their children's health. The First Lady followed up by telling the crowd, "Mothers need to know the crucial importance of shots early in their children's lives."[1]

As described in the previous chapter, the Carter initiative frequently held parents responsible for immunization rates that fell short of public health goals. But in political, popular, and scientific conversations about vaccination at the time, "parental" responsibility typically meant "maternal" responsibility. Vaccination campaigns from the fifties through the seventies had routinely emphasized maternal responsibility for obtaining needed vaccines for children. The Carter campaign likewise targeted mothers; it also relied to a great extent on women volunteers to reach out to these mothers. But as the seventies wore on, the centrality of maternal engagement to national immunization goals increasingly came into conflict with the tenets of second-wave feminism. A small but growing number of Americans—many of them mothers—began to question the safety of vaccines and the sagacity of vaccination requirements, echoing as they did so the rhetoric and ideas of feminism and the related women's health movement. In short, as federal involvement in vaccination continued to expand and as the number of vaccines recommended for children grew, the rise of second-wave feminism gave vaccination skeptics a framework for criticizing vaccines and the ways they were used.

This chapter unravels two related stories that are crucial to understanding the rise of popular resistance to vaccination in the last decades of the twentieth century. It shows, first, how ideas about motherhood and the changing social, civic, and economic roles of women informed vaccination promotion efforts in the new era of vaccination. It also uncovers the connection between the women's movement of the seventies and the tide of vaccine skepticism that swelled in the early eighties. Feminists and women's health activists raised awareness about the risks of widely prescribed drugs, popularized the idea that the male-dominated medical profession mistreated women and withheld critical information, and encouraged women to take control of their own health. Vaccine critics, too, came to doubt medical and government assurances of vaccine safety. They argued that information about vaccine risks had been withheld from them by a paternalistic medical profession. And many argued that, as mothers, they knew better than doctors about their children's health, a corollary to health feminists' claim that only a woman could truly understand her own body and experiences of health and illness. Such doubts lay at the heart of the organized vaccine-safety movement that emerged in the 1980s.[2]

Americans who questioned vaccines in the 1970s and 1980s were not necessarily self-described feminists, nor were health feminists necessarily skeptical of vaccines. Even so, vaccine critiques from the time bear the unmistakable imprint of health feminist ideas.[3] Feminism, however, wasn't alone responsible for Americans' growing doubts about vaccines and vaccination. The seventies were marked by a general crisis of authority, as the public grew increasingly skeptical of government and industry, and social movements coalesced around a range of issues from consumer protection to rights for the disabled.[4] Second-wave feminism itself was a product of the rise of the New Left, whose members fought for civil rights, free speech, an end to the Vietnam War, and a more participatory democracy. The New Left also spawned a new environmentalism, whose influence on Americans' vaccination ideas is examined in the following chapter. The influence of these social movements on American vaccine beliefs highlights a paradox in the modern history of vaccination. A liberal view of government expanded federal involvement in vaccination and the use of vaccines, on the one hand; on the other, a leftist view of politics and society produced ideas suited to critiquing vaccination and the growing use of vaccines.

Vaccination resistance began to flare up in the 1970s not just because the movements of the New Left, including feminism, offered a new set of tools with which to critique vaccines, but because they spoke specifically to problems with the nation's inherently gendered approach to vaccination promotion. Modern vaccination recommendations generally built on socially determined expectations of women as child-bearers, members of the nation's workforce, mothers, and the primary caretakers of children. In the modern era of vaccination, that is, policies and practices both implicitly and explicitly recognized and reinforced socially constructed gender norms. As this chapter shows, this pattern gave shape to the vaccination resistance that mounted as the final decades of the twentieth century began.

A MOTHER'S RESPONSIBILITY

At the "grassroots" level, Carter's Childhood Immunization Initiative was carried out largely by women and mothers: members of women's clubs, nursing leagues, and Parent-Teacher Associations who volunteered to reach out to other mothers and urge them to vaccinate their children. The Carter campaign's dependence on women volunteers was, by then, a well-established tradition of immunization promotion. The March of Dimes (formerly the National Foundation for Infantile Paralysis) had relied heavily on countless women volunteers to raise money for polio treatment and vaccine research and, later, to help carry out vaccine field trials. These legions of volunteers imprinted upon American memory the legendary image of mothers marching en masse through communities, collection cans in hand. They were women on hospital boards and in Parent-Teacher Associations with, as author Jane Smith put it, "both the time and the passion to work against childhood disease." They also possessed a culturally informed sense that, as mothers, participating in such causes was their civic duty.[5]

The Carter campaign relied on volunteers in a manner that closely resembled the early days of the March of Dimes—as well as the Arkansas immunization campaign of the early 1970s. In that campaign, described in the previous chapter, Arkansas First Lady Betty Bumpers and beauty queen Miss Arkansas rallied mothers across the state to spread the word about immunization from door to door and to volunteer at vaccination clinics.[6] The campaign made direct appeals to women's sense of duty and potential for fulfillment as mothers. "Protect These Treasured Moments," stated campaign materials that featured a sentimental illustration of a young mother seated in a rocking chair, her son and daughter nestled at her sides. In the years preceding the Carter campaign, this type of entreaty—to a mother's sense of unique responsibility and love for her children—was popular not just in Arkansas, but across the nation. "Every mother who loves her children will get them vaccinated both against rubella and against ordinary measles," wrote nationally syndicated medical columnist Walter Alvarez in 1972.[7] Modern motherhood, as historian Rebecca Jo Plant has argued, may have been an increasingly private affair, but American mothers in the seventies were still culturally pressed upon to assume a sense of lifelong and exclusive responsibility for their children's well-being.[8]

A mother's perceived duty to vaccinate her children cut in two different directions. For health officials and politicians promoting vaccination in the seventies, mothers were often viewed as a ready resource already dedicated to the cause of protecting their children. On the other hand, when children went unvaccinated, mothers were often held culpable and labeled thoughtless, uneducated, and irresponsible.[9] When measles outbreaks erupted across the country in the late sixties, many in the medical and public health community found fault with mothers: mothers who failed to bring their children to clinics, mothers who failed to realize the vaccine was available, and mothers who failed to recognize the new measles vaccine's importance.[10] Mothers were chastised for mistaking measles for simple colds and for treating it as a "mild" infection.[11] When measles erupted in Texarkana, a city straddling the Texas-Arkansas border, Alvarez blamed it on "unwise" mothers "too poor" or "too ignorant" to vaccinate their children.[12] Even when the fault for low vaccination rates was distributed across multiple parties, the responsibility ultimately rested with mothers: "The unnecessary case of diphtheria, measles, or poliomyelitis may be the responsibility of the state legislature that neglected to appropriate the needed funds, the health officer who did not implement the program, the medical society that opposed community clinics . . . or the mother who didn't bother to take her baby for immunization," noted one group of health officials.[13]

It followed, then, that as states attempted to combat outbreaks using the set of new vaccines at hand by the early 1970s, they turned again and again to vaccination promotion efforts that specifically targeted mothers. As Washington, D.C., attempted to stem a resurgence of measles in 1970, health officials there implemented a plan to mail immunization reminder notices to mothers three months after their child's hospital birth.[14] New York City health officials worked with local hospitals to identify, at birth, mothers without pediatricians, so they could later be visited by local health station representatives and encouraged to bring in their children for free vaccines.[15] Hospitals were an important gateway to reaching mothers, one CDC official pointed out, because hospitals had birth records of mothers and their children and could easily reach out to those mothers during the first year of their newborn's life.[16] When measles struck New Jersey in 1974, state health officials there asked doctors to cull their files for patients in need of immunizations—and then call their mothers. "We want those mothers to get their kids to their doctor as soon as possible," said the state's assistant health commissioner.[17]

Such plans justified the decision to reach out to mothers as a matter of convenience; after all, mothers, not fathers, gave birth to children in hospitals. Women generally and mothers in particular have long been viewed as a gateway to improved children's health, for both biological and cultural reasons that have bound children more closely to them than to their fathers. But in the 1960s and 1970s, efforts to encourage mothers to vaccinate their children—either out of a sense of duty or shame—were embedded within larger conversations about the social and economic roles of women. As medical professionals and health officials debated, beginning in the late 1960s, whether children should be universally vaccinated against measles, rubella, and mumps, a particular economic argument in favor of requiring vaccines for children (one distinct from that embraced by the Carter initiative) gained currency. Namely, the new viral vaccine targets may have been "mild," but their prevention via vaccination offered an unprecedented convenience for families with two wage earners. When a child comes down with mumps, argued a Washington State health official, "[a] working mother may have to stay home to care for him and more often than not, two to three weeks later, mumps develops in the susceptible siblings and adults . . . with another week or two of family disability." The new class of vaccines, however, made the potential loss of income associated with such disability "preventable and unnecessary."[18] Moreover, not only could vaccination protect a woman's economically productive hours; it could also make—or break—her career. In a 1973 column, Alvarez promoted rubella vaccination by telling the tale of a "very intelligent woman whose promising career as a university professor was stopped" because she caught rubella while pregnant. Because of her infection, her child was born deaf, and her career hopes were dashed as she devoted her time to her child instead of her work—a fate the vaccine could have prevented.[19]

Alvarez's column was part of a national push to promote rubella immunization, an effort that, as historian Leslie Reagan has shown, reinforced the idea that vaccination was first and foremost a mother's responsibility. Health officials promoted the vaccination of children against that infection not to protect children themselves but to protect their mothers, as noted in chapter 3, because rubella posed the greatest risk of harm to developing fetuses. In the wake of the nation's 1963–64 rubella epidemic, expectant and potential mothers eagerly embraced the new vaccine for their children. Campaign efforts also appealed directly to children, emphasizing their responsibility—girls as the nation's future mothers, boys as future protectors and family men—for rubella prevention. This approach, as Reagan argues, made rubella prevention a "gendered civic responsibility" that fit into and reinforced existing gender norms.[20] The rubella vaccine may have been warmly received, but it was precisely this type of reinforcement that would later help give rise to creeping doubts about vaccine recommendations.

In the meantime, stories like the one told by Alvarez carried multiple meanings. They further concretized the idea that the vaccination of children was the exclusive province of mothers. They also suggested that vaccines against the "milder" diseases could effectively prevent children from interfering with women's economic, professional, or personal goals. That women prioritized such goals reflected changing demographic and social realities: the continued rise in the number of women in the workforce, and shifts in the status and longevity of women in the workplace.[21] Despite these shifts, however, women nonetheless retained primary responsibility for their children's care and medical needs, including vaccination.

But not all working mothers were aspiring university professors, like the mother described by Alvarez. Epidemiological studies conducted in the seventies continued to indicate that vaccination rates were particularly low among poor inner-city residents of color (a reality the Carter campaign often minimized). For mothers in these communities, class and race compounded the effects of gender in the eyes of health professionals and politicians, who attributed disease outbreaks in "ghettos" to "poor mothers" "struggling to get up the rent money," or "waiting until their children entered school for free . . . inoculations."[22] When white middle-class mothers failed to vaccinate their children, by contrast, vaccination experts and bureaucrats attributed the oversight to age and naïveté, not race or income. "Today's mothers . . . don't remember the polio epidemics of the 1940s and 50s, the pictures of children in iron lungs or the mass closing of swimming pools in mid-summer," a CDC official told the *Washington Post*. The front-page article in which the official was quoted featured a large photo of Karen Pfeffer, a white twenty-two-year-old mother whose daughter contracted a near-fatal case of whooping cough. "Whooping cough? Who's ever heard of whooping cough?" said Pfeffer. "I just didn't realize how serious it could be."[23]

If mothers—poor or rich, young or old—were the target of vaccination campaigns in the 1970s, nonworking mothers were sometimes seen as the key to reaching them. The frugal Carter campaign was deeply dependent on the services provided by women in voluntary groups from

Alaska to Florida.[24] By the time the Carters entered the White House, however, second-wave feminists had spent several years chipping away at the notion that volunteerism should be the universally accepted domain of women. In the early 1970s, the National Organization of Women had taken an official position against what they called the exploitative nature of volunteer work.[25] Rosalynn Carter—whose high-profile involvement in political affairs and equal partnership with her husband were favorite subjects of news outlets, even as feminists criticized her for lacking an identity separate from her husband—nonetheless championed the cause of volunteerism while in the White House.[26]

When one reporter asked her if it wasn't "denigrating" to ask women to engage in important work without pay, however, Carter acknowledged that it wasn't a widely popular cause. "Voluntarism has a little bit of a bad connotation," she explained. "I've been trying to say 'public initiative' or 'public responsibility.'"[27] Carter's support for voluntarism was just one example of how her political choices sometimes rested uneasily in the shifting landscape of women's social roles. As First Lady, she declined to wear her motherhood on her sleeve, repeatedly turning down invitations to chair both the Childhood Immunization Initiative and the International Year of the Child—even as some of her female constituents saw her as uniquely qualified to support such causes. "Mrs. Carter, Please use your influence as a concerned mother and as an intelligent participant in national planning to reinstate money in the budget for vaccines," one mother pleaded in a letter to the White House.[28]

Similarly moral conceptions of motherhood, which held motherhood as the basis for female civic engagement, also informed letters that mothers wrote to advice columnist Ann Landers on the subject of vaccines in the 1970s. "Heartsick Mother," whose son suffered permanent hearing loss after a bout of measles, wrote to ask that "thoughtless, irresponsible" mothers see to it that their children got vaccinated. "I am sending my letter to Ann Landers," she wrote, "because this problem is bigger than our own two children. It involves all children everywhere."[29] "Mom Who Cares" wrote to ask, "Why do mothers and fathers who claim they love their children neglect to have them vaccinated . . . ? Don't they realize they can get these shots free at the county or city health centers?"[30] Such writers wielded their identity as mothers to legitimize the civic act of chastising other parents for neglecting the care of their children and—given the communicable nature of vaccine-preventable diseases—the community at large.

Testimony from mothers who chose not to vaccinate their children in the late 1960s and 1970s is harder to find, but it is clear that some mothers made this choice deliberately. In the late 1970s, letters to Landers began to hint at a sense of doubt regarding the need for across-the-board immunizations against all childhood infections. A mother in Baton Rouge described a disagreement with her sister-in-law over whether it was better for children to get the childhood diseases themselves rather than the vaccines, so that they would have lifelong immunity.[31] A mother in Champaign, Illinois, described an argument with her sister over the need for polio vaccine. "I have not heard of a child getting polio for several years," she wrote. "Why go to the trouble if there is no danger?"[32] As the Carter campaign got under way, vaccination was an increasingly visible topic not only in the advice columns, but in a variety of women's and parenting magazines, including *Good Housekeeping, Redbook,* and *Ladies' Home Journal.* Most of these magazines parroted federal pro-immunization messages, but their coverage hinted at reader doubts and fears. "Misguidedly, some of us fear that vaccines are dangerous; but the minimal risk must be weighed against the much greater benefit," stated *Harper's Bazaar.*[33] "Parents frequently ask whether it's really necessary to immunize their children against measles, rubella, mumps, and poliomyelitis, as well as against diphtheria, whooping cough, and tetanus . . . the answer is an unequivocal yes," reported *Parents.*[34] Such articles often played up the dangers of vaccine-preventable diseases, sidestepping readers' specific vaccine worries in the process. But the very presence of reader doubts and fears indicated an information gap, which a growing number of mothers began to question.

QUESTIONING AUTHORITY

In the late 1960s, a social movement focused on women's health emerged as a component of the broader second-wave feminist movement. Health feminists argued that women's health-related knowledge belonged to women, and they fought to wrest control of women's health issues from the predominantly male medical profession.[35] Women's health activists founded clinics, held self-examination workshops, conducted abortions, and wrote books for women on health issues directly relevant to them. They also exposed the negative effects of specific drugs commonly prescribed to women, helping to bring national attention to the serious risks associated with oral contraceptives, estrogen taken for menopause, and diethylstilbestrol (DES) taken during pregnancy. The revelation of these medical hazards drove feminists' further demands for informed consent in medical decision-making and increased access to information about medical treatments.[36] Feminists weren't the only ones criticizing the medical profession and drug industry; concurrent anti-medicalization and consumer rights movements leveled charges too. Health feminists' critiques were also felt beyond the arena of women's reproductive health. As historian Susan Speaker has shown, a general disillusionment with the prescribing practices of doctors and growing doubt about the safety of commonly prescribed drugs influenced women's turn against widely prescribed minor tranquilizers (such as Valium) in the 1970s. For women, what was "wrong" with the industry of medicine generally was that physicians, who were mostly male, "refused to listen to or believe female patients, withheld knowledge or lied to them, overcharged them, [or] performed unnecessary procedures."[37]

The reach of this general disillusionment with professional medicine began to spread to vaccines at the tail end of the 1970s. Vaccine worries were evident, for instance, in the 1978 child-rearing guide *Ourselves and Our Children*, a follow-up volume to *Our Bodies Ourselves*, the groundbreaking lay manual to women's health first published by the Boston Women's Health Book Collective in 1971. The feminist authors of *Ourselves and Our Children* acknowledged the "controversy surrounding the medical risks of immunization"—and then went on to lament the expiration of the Vaccination Assistance Act and argue that government should do more to make vaccines available to those who wanted them.[38] Feminists at the influential collective appeared to have been somewhat divided on the subject of vaccines. Readers who wanted more information on the "controversy" or "medical risks" so briefly mentioned had to look elsewhere.

Perhaps no venue was more open to discussing vaccine risks than *Mothering*, a magazine devoted to "natural family living," founded in Colorado in 1976. As mainstream parenting magazines urged mothers to vaccinate their children (often at the direct behest of the Carter campaign), *Mothering* printed samples of the skeptical reader letters it was receiving on the topic.[39] *Mothering*'s readers were likely all drawn to the magazine's back-to-nature ethic, but they came from all corners of the country, and on the issue of immunization they were sharply divided. Readers chimed in on the matter from California, Indiana, Ohio, Maryland, Texas, Montana, Alaska, Massachusetts, Washington, New York, Hawaii, and beyond. Their letters represented the views of staunch immunization advocates (including pediatricians and general practitioners), lifelong anti-vaccinationists (including natural hygienists and homeopaths), and parents working to sort through it all to make an informed decision for their own children.[40] Some seemed convinced by pro-vaccination arguments; others held off in worried doubt.

As more mainstream women's magazines urged mothers to vaccinate their children, *Mothering* alone advised mothers to "be cautious with vaccines." The magazine's writers warned those allergic to eggs and chickens to avoid the measles vaccine. They informed readers that vaccinating a child against polio could cause the disease in other family members. And they listed encephalitis and death as possible side effects of the pertussis vaccine. These warnings, noted the editors, were taken directly from vaccine package inserts, which mothers themselves should ask to see before having their children immunized.[41] Other articles encouraged readers to become informed consumers by doing their own research on the subject beforehand—advice directly informed by the women's health movement. *Mothering* editor Peggy O'Mara captured the movement's influence when she later told readers that she began questioning vaccination while pregnant with her first child, in 1973: "Because I was accustomed to making personal healthcare decisions, it seemed like the obvious thing to do," she wrote.[42] Her own questioning mirrored that of her readers, who from the late 1970s through the early 1980s sent more letters on vaccination than any other topic (save circumcision).

In the decade after *Mothering*'s readers first took up the issue of immunization, two key exposés alerted the broader public to the occasionally devastating side effects of the pertussis vaccine in particular. Following in the tradition of the 1969 book *The Doctor's Case Against the Pill*, a rousing health feminist text that accused experts of concealing known hazards of the birth control pill, the 1982 NBC investigative report *DPT: Vaccine Roulette* and the 1985 book *DPT: A Shot in the Dark* lambasted scientists and physicians for producing and promoting a vaccine known to cause convulsions, paralysis, and deaths. Reports of the pertussis vaccine's risks had been publicized in the UK, Sweden, and Japan in the 1970s.[43] But before *DPT: Vaccine Roulette* aired in the Washington, D.C., area in April 1982, discussion of the vaccine's risks in the United States had been largely confined to scientific journals. The broadcast, subsequently excerpted nationwide on the *Today* show, showed extensive footage of mentally and physically disabled American children whose handicaps were attributed, by parents and doctors, to the pertussis component of the combined DPT vaccine. The report informed parents that 1 in 7,000 children suffered serious adverse effects related to the vaccine, including high fevers, inconsolable crying, seizures, brain damage, and death.[44] Said reporter-producer Lea Thompson, "The medical establishment" had been "aggressive in promoting . . . the most unstable, least reliable vaccine we give our children."[45]

Doctors and scientists were swift and harsh in their response. They called the report imbalanced, distorted, and inaccurate, and accused Thompson of misinterpreting the science and committing "journalistic malpractice."[46] In the nationwide panic that ensued, physicians fielded endless calls from concerned parents, whom they often labeled "hysterical." Thousands of parents called the D.C. television station to report that they believed their own children had been harmed by the vaccine.[47] Station representatives put some of the parents in touch with one another, and a handful of them — Kathi Williams, Barbara Loe Fisher, Jane Dooley, Donna Middlehurst, and Middlehurst's husband, Jeffrey Schwartz — banded together to form an advocacy group they dubbed Dissatisfied Parents Together, or DPT for short. The following month, Williams and Marge Grant, one of the mothers who had appeared in *DPT: Vaccine Roulette*, testified before a Senate subcommittee. The May 1982 hearing had originally been scheduled to address federal immunization funding cuts and strategies for reaching children who remained unvaccinated in the wake of the Carter-era campaign. Instead, the hearing, called by Senator Paula Hawkins of Florida (whose own son had contracted polio from the polio vaccine), featured extensive testimony by parents of vaccine-injured children, health officials, and other parties on the risks of vaccination.[48]

DPT: Vaccine Roulette and its fallout—including the congressional hearings, the formation of the parents group, and the publication of DPT: A Shot in the Dark—reveal that women's gendered experiences shaped popular responses to the news of pertussis vaccine risks. Thompson, a "consumer reporter" for WTOP-TV who received an award for her reporting from the American Academy of University Women in 1978, did not focus exclusively on women's issues, but she did indicate that her reporting was at times directly shaped by her experiences as a woman and mother.[49] Her report on asbestos-lined hair dryers led to a recall of 12.5 million hair dryers, and her report on nutritive deficiencies in baby formulas, which she pursued following her own child's birth, helped bring about a federal law enforcing routine formula testing.[50] In DPT: Vaccine Roulette, she interviewed male doctors and health officials who denied the pertussis vaccine's risks, and she intercut these with interviews and footage of mothers struggling to care for their severely handicapped children. In several shots, the mothers were seated alongside their husbands, but in each case, the mother was the main spokesperson for her child and the expert on her child's condition.

By giving voice and credence to their personal experiences and observations, the broadcast's format elevated mothers to the level of scientific experts on the subject of children's vaccine reactions. As historian Wendy Kline has shown, health feminists themselves had cultivated the notion that knowledge gained through personal experience was just as valuable or even more valuable than medical knowledge.[51] The content of Thompson's interviews echoed this and other themes of health feminism. Mothers of vaccine-damaged children complained that their doctors hadn't listened to them. Dissident doctors testified that the vaccine was no longer necessary. And government scientists suggested that federal agencies had ignored and suppressed data implicating the vaccine in causing harm.[52] Throughout the broadcast, the mothers who spoke on-screen delivered a common narrative, one that followed the format of earlier health feminist revelations. Mothers like Emily Yankovich and Evelyn Gaugert described their children's frightening reactions following vaccination, their doctors' denial that the two were connected, and their later discovery of what they knew all along: despite their doctors' dismissals, their children's symptoms were in fact vaccine related.[53]

This very same narrative appeared in *DPT: A Shot in the Dark*, the 1985 pertussis vaccine exposé coauthored by DPT co-founder Fisher and independent scholar Harris Coulter. The book interwove personal accounts of vaccine injuries with detailed exposition of decades of scientific research on the pertussis vaccine. In it, a Montana mother described her doctor's dismissive reaction to her concerns about her son Mark's inflamed leg and incessant piercing cry after his pertussis shot: "He said, 'Don't worry. Just give him Tylenol and he'll be fine.' So I didn't call him back again, because I thought, well, this is the way it is supposed to be."[54] Over the next three months, Mark stopped eating, developed allergies, and weighed only twenty pounds by the age of two—prompting his mother to fight against giving Mark any more pertussis shots.

Mark survived; Richie, the son of a nurse named Janet Ciotoli, died the day after his first DPT shot. Ciotoli described her battle with the doctor and coroner over the attribution of her son's death to sudden infant death syndrome and not the vaccine. Her identity as a mother legitimized not only this struggle, but the larger one she vowed to take on. "These doctors and officials in the government, who keep talking about the benefits and risks of this vaccine, better take fair warning. My baby may be just another statistic to them, but he was my child, and there is nothing more powerful than a mother's fight for her child," she said. "I will fight no matter what I have to do and no matter how long it takes to keep this from happening to other babies."[55] Ciotoli was one of many female vaccine critics who emphasized their maternal identities as the basis for their vaccine-safety advocacy. The approach allied these women with a long-standing tradition of "maternalist" activism in the United States and beyond.[56]

Before Congress and in *DPT: A Shot in the Dark*, Ciotoli described herself as an educated, professional woman who took her doctor's medical advice at face value, only to learn that this quiescence had cost her son's life. Her story is one of several in *DPT: A Shot in the Dark* that link the book to a series of popular books published in the late 1970s that chastised organized medicine for its intimidation and mistreatment of women, especially mothers and mothers-to-be. Books such as Gena Corea's *The Hidden Malpractice* and Suzanne Arms's *Immaculate Deception* argued that the medical establishment had instilled a sense of powerlessness in women, subjecting them to unnecessary, overmedicalized procedures that harmed them and their babies.[57] Women, they contended, were administered unnecessary sedatives and subjected to procedures, such as pubic-hair shaving and fetal monitoring, without their consent. They were "frightened into believing" that anesthesia and other drugs were crucial for childbirth, and that birth, "once a natural process," must take place in the hospital, among strangers.[58] Vaccine critics echoed these very same themes. "I, like so many mothers, lacked the information necessary to even ask intelligent questions. . . . Instead I trusted the experts," said Gerri Cohn, whose daughter Traci suffered brain damage subsequent to her DPT vaccine.[59]

The authors of the aforementioned volumes focused on the process of reproduction and usually left off shortly after childbirth. They promoted breastfeeding over formula but ventured no further into child rearing; as a result, they rarely, if ever, touched on immunization. But they were closely related to a separate but contemporaneous body of work that broadly accused the medical profession of a range of transgressions and that *did* specifically criticize mass vaccination as a disease-prevention strategy. In his 1976 book *Medical Nemesis*, historian and philosopher Ivan Illich argued that factors other than "medical progress"—including water and sewage treatment, better nutrition, and sociopolitical equality—were primarily responsible for improvements in health, and that professional medicine did not deserve the live-saving reputation it was so universally and exclusively accorded. To Illich, the medical profession could duly accrue only partial credit for the defeat of smallpox through vaccination; in his analysis, the importance of mass vaccination as a medical intervention had been dramatically overstated. Deaths due to diphtheria, whooping cough, and measles, he pointed out, had declined 90 percent prior to widespread immunization.[60]

Illich, in turn, was often cited by physician-turned-popular author Robert Mendelsohn, who became an outspoken and widely quoted critic of vaccines in the early 1980s. Mendelsohn—who wrote in his 1981 book *Male Practice* that women were the "primary victims" of "medical and surgical overkill"—listed vaccines as one of several controversial and risky practices and procedures that women were coerced into accepting for their newborns.[61] In his 1979 book *Confessions of a Medical Heretic*, he questioned the need for vaccines against mumps, measles, and rubella, because the diseases weren't nearly as severe as smallpox, tetanus, and diphtheria. He pointed to evidence that the diphtheria vaccine was sometimes ineffective, and he described the controversy over the safety of pertussis vaccination that was, at that point, still brewing only within the profession.[62] One didn't have to read Illich or Mendelsohn too closely to come away with the idea that mass vaccination was both risky and an example of medical overkill.

DPT: Vaccine Roulette, *DPT: A Shot in the Dark*, and the media coverage they prompted transmitted this notion of medical overkill to a national audience, tying it to a critique of the pertussis vaccine. The exposés pointed out that whooping cough rarely caused children to die in the modern era, and that (borrowing Illich's point) the disease had declined significantly prior to widespread vaccination. Both Sweden and West Germany had abandoned the vaccine over concerns about its side effects, *DPT: Vaccine Roulette* and *DPT: A Shot in the Dark* reported, and neither country had suffered epidemics as a result.[63] Medical overkill was just one idea that the widespread attack on pertussis vaccination borrowed from works that had criticized medicine in the seventies. Critiques of the vaccine also reflected a mistrust of doctors, government officials, and the drug industry, and implored Americans to challenge these figures and institutions of authority.

In *DPT: A Shot in the Dark*, women referred to as "a mother on the West Coast," "a mother in Massachusetts," "Sharon's mother," "Marie's mother," and "Patrick's mother" were just a handful of the mothers who recounted asking their doctors about their children's high-pitched screaming, high fevers, and muscular spasms following vaccination, only to be told not to worry. In each mother's story, the child developed a seizure disorder or brain damage; a few died. Their accounts stressed the need for mothers to question their doctors' opinions; they also suggest a dramatic loss of faith in medicine, expressed in sometimes starkly gendered terms. "We are so conditioned to the idea that our doctor's word is to be trusted without question that we don't think for ourselves. I am a nurse. I watched my son die that day, and I didn't even know what was happening until it was all over," said Janet in *DPT: A Shot in the Dark*. "If this had not happened to my baby . . . I would still be taking my doctor's word as the word of God, like most mothers do."[64] Ellen, who described questioning her doctors about her daughter Sherry's post–DPT shot brain damage, recalled being "officially labeled a 'troublemaker' and 'hysterical mother' in Sherry's medical records." Her outrage at the doctors' paternalism was unmistakable: "They can be so damn patronizing," she said. "You know, pat the little mother on the head and tell her to calm down."[65]

Other vaccine-critical mothers blamed not just doctors but also the government and drug industry, alluding as they did so to a large-scale cover-up of the dangers of the by then widely administered vaccines.[66] "It appears to me that the manufactures [*sic*] and/or certain government agencies are intentionally withholding vital information," said Wendy Scholl, who testified before Congress in 1983 about her daughter Stacy's measles vaccine–induced paralysis, learning disabilities, and seizures.[67] Senator Hawkins shared this perception of deliberate dissemblance when she asked federal vaccine officials, "What symptoms or warning signals should the parents look for from the adverse reaction from the vaccine, which I believe is the secret that has been held from them?"[68] The sense of a conspiracy was only heightened when officials defended the practice of administering vaccines without informing parents of potential risks, as one FDA official appeared to do in *DPT: Vaccine Roulette*: "If we told parents there was a risk of brain damage," he said, "there's no question what their response would be."[69]

The benevolent paternalism belied by the official's comment was proof that if parents wanted objective information on medical risks, they would have to demand it or seek it out for themselves. The women's health movement had adopted "informed medical consumerism" as a core principle, and indeed women who spoke out against vaccines in the late 1970s and early 1980s followed in this tradition.[70] In *Mothering* magazine, a Berkeley, California, health educator described her search for information on vaccine risks in the medical literature. "What is known about vaccines is a whole other story from what is told," she concluded. "Health care consumers should insist on reading the package inserts which come with vaccines."[71] At the end of *DPT: Vaccine Roulette*, a prominent vaccine scientist stated on-screen that convulsions were not a contraindication against DPT vaccination. The camera then cut to reporter Lea Thompson, who read directly from an American Academy of Pediatrics' warning against giving the shot to children with a history of convulsions.[72] When the coroner refused to attribute her son's death to DPT vaccination, Janet, in *DPT: A Shot in the Dark*, recounted returning to him with a copy of *The Physician's Desk Reference*, in which her son's precise condition was described.[73] Informed medical consumerism was also a core message of *DPT: A Shot in the Dark*, which concluded with the following admonishment: "The time has come to educate ourselves about vaccines."[74]

FRAMING THEIR DEMANDS

The effect of the feminist and women's health movements was such that over the course of the 1970s, women were more and more likely to receive information from their doctors regarding the risks and benefits of their own medical treatments. In turn, as historian Elizabeth Watkins has argued, women were increasingly *expected* to participate in their own medical decisions.[75] In the late seventies and early eighties, women who expressed concern about vaccine safety applied this approach to their children, noting that, after all, their children were "part of" them.[76] Moreover, they argued, they, more than anyone, had to live with the consequences of vaccinating their children. In *DPT: Vaccine Roulette*, Gail Browne described how her son's disabilities had led her and her husband to abandon hopes of another child as they struggled to pay for his extensive care.[77] Testifying before Congress, Wendy Scholl told of an endless quest for providers and financial aid for her disabled daughter's care, made worse when her husband lost his job and their new insurer wouldn't cover Stacy's condition.[78] Mother after mother in *DPT: A Shot in the Dark* described how her life had been fully restructured to accommodate her vaccine-injured child's costly and all-consuming needs.

Whereas feminists and women's health activists demanded a form of social justice, however, vaccine activists demanded political justice. DPT, as a group, acknowledged the importance of vaccines and the dangers of vaccine-preventable diseases. Instead, they criticized the risk-benefit calculus cited by public health officials, who pointed out that the vaccine might cause a few dozen cases of brain damage, but that the alternative, whooping cough, would cause thousands of deaths each year. "No parent should be put in the untenable position of having to choose between a bad vaccine and a bad disease," argued DPT founder Barbara Fisher.[79] To the parents of vaccine-injured children, it was unjust that they alone should suffer the high cost of achieving better health for the nation as a whole. "Did these children, like soldiers, give their lives so that others might live?" asked mother Gerri Cohn at a Maryland hearing.[80] Because they believed the answer was yes, DPT demanded that government take the lead in providing safer vaccines, more information for parents, support for better studies of adverse reactions, and justice, in the form of remuneration, for the families of vaccine-injured children.

That parents were able to view vaccines as a threat to their children's health in this period relates to epidemiological and demographic shifts that had occurred over the previous decades. Pertussis had diminished to just a couple thousand cases per year. Many vaccine-worried parents nonetheless expressed simultaneous trepidation about both pertussis and its vaccine. But to most, the threat of vaccine injury loomed larger. "I live with the fear that they might get whooping cough. It's scary. But until they come up with a purer vaccine, I will have to live with these fears," said the mother of Cindy, a little girl who developed neurological symptoms following her first DPT dose.[81] The chances of contracting pertussis still outweighed the chances of a vaccine injury; government statistics indicated that collapse or convulsions occurred once in every 1,750 shots, and brain damage once in every 100,000 to 172,000 shots.[82] But such statistics held no sway with the parents of vaccine-injured children and their friends, neighbors, and relatives. As Marge Grant explained in her Senate testimony: "I can tell you most assuredly, WHEN IT HAPPENS TO YOUR OWN CHILD, THERE ARE 'NO BENEFITS' AND THE RISKS ARE 100 PERCENT!"[83] (It was certainly no accident that vaccine critics borrowed the very maxim that health officials had used two decades before to encourage the adoption of mass vaccination against the milder diseases.[84])

The dispute between parents and health officials over the appropriate risk-benefit calculation for justifying mass vaccination took place not only in the context of diminishing pertussis disease rates, but also in the context of diminishing birth rates, particularly among white middle-class American women, who comprised the bulk of the vaccine-safety movement's members. The value of the individual child to the American family continued its climb and was epitomized by the emergence of a national obsession with the protection of children. With the advent of the child protection movement at the very end of the seventies, anti-smoking and anti-drug campaigns focused on the sanctity of children, and citizens mobilized against a host of perceived social threats to children, including not just drugs, but also mass murderers, sexual deviants, cultists, homosexuals, child pornographers, drunk drivers, and child abusers.[85] The child protection movement itself was also, to an extent, an outgrowth of feminism; it was feminists who brought the issue of child abuse to public light, and rape crisis centers founded by feminists that revealed the extent of sexual crimes committed against children.[86] But aspects of the child protection movement also stemmed from conservative reactions to the previous decades' advancement of a liberal social agenda—an anti-gay campaign of the time thus took the name "Save Our Children" to frame itself as a movement to protect youths from a host of vices.[87]

The organized vaccine-safety movement that emerged contemporaneously rightly fits within this larger child protection movement, and its underlying political ideologies were similarly mixed. The vaccine-safety movement coalesced in the early 1980s, when a pronounced shift toward political conservativism had taken place across the nation, signaled by the 1980 election of Ronald Reagan and his promises to slash big government. In this context, some outspoken vaccine critics decried the Carter-era expansion of vaccine laws as an undue encroachment of government upon personal freedoms. As a solution, these critics fought to undo the Great Society–type laws that had made vaccines mandatory for their children in the first place. Partnering with other Wisconsin parents, for instance, Marge Grant founded the Research Committee of Citizens for Free Choice in Immunization, which advocated dismantling all state vaccine mandates, and which effectively lobbied Wisconsin legislators to amend a philosophical exemption clause to that state's vaccine laws.[88] In Pennsylvania, parents pressured state officials to remove pertussis completely from the list of vaccines required for school.[89] In Idaho, parents lobbied for and achieved the same.[90]

That vaccine resisters in the late 1970s and early 1980s saw the newly invigorated vaccine laws as an undue expansion of government is exemplified by the frequency with which, in the context of heightened Cold War tensions, they compared the laws to those of the Soviet Union and Eastern European nations. When Maryland began enforcing its law requiring vaccines for school entry, Barbara Syska's son was expelled for lacking vaccines, and Syska, in response, filed suit against the board of education. Syska, who had immigrated from Poland, told reporters, "I'm a refugee from a communist country. There the good of the largest number of people is important, not the individual. I came here where the individual is supposed to have a say."[91] The comparison of U.S. vaccine laws to the practices of oppressive regimes was soon a common refrain among vaccine critics. In her Senate testimony, mother Isabelle Gelletich, whose son suffered brain damage following DPT vaccination, called the cover-up of vaccination risks "an American Holocaust." "I wonder," she wrote, "are my son and I the survivors of a modern day Auschwitz, both of us left crippled and maimed by apathy and deceit?"[92] Wrote Fisher and Coulter in DPT: A Shot in the Dark, it is only in "totalitarian societies where powerful bureaucrats routinely decide what is best for the rest of the population."[93]

As pervasive as this ideology was, it did not shape the demands that took center stage within the nascent vaccine-safety movement. DPT, which rapidly became a prominent national organization, initially demanded more government as a solution to the problem of unsafe vaccines. In Maryland in 1983, DPT members drafted model legislation to require doctors to keep records of and report vaccine reactions to the state.[94] Over the next few years, organization members leaned on members of Congress, health officials with the FDA and CDC, members of the American Public Health Association, representatives of the American Academy of Pediatrics, and others to drum up support for a federal bill to establish greater government oversight of vaccine safety and a new compensation system for vaccine-injured children.[95] The plan succeeded. In late 1986, Reagan signed the National Childhood Vaccine Injury Act into law.[96] In addition to establishing a vaccine tax that would provide funds for the families of vaccine-injured children, the act required doctors to record and report vaccine reactions to federal authorities, and it required the Department of Health and Human Services (DHHS; the new name for the Department of Health, Education, and Welfare as of 1979) to develop and disseminate informational materials on vaccine benefits *and* risks for parents.[97]

DPT members initially celebrated their 1986 victory. But in the months and years that followed, they expressed growing frustration with government as a solution to the problem of vaccine safety. DPT members bemoaned DHHS delays in the production of informational pamphlets for parents and Congress's sluggishness in appropriating funds for the new compensation program. They wrangled with federal health officials over the determination of vaccine-injury-related deaths. And they expressed outrage over a New Jersey law that restricted parents' right to sue vaccine makers.[98] "The federal government, organized medicine, and the pharmaceutical industry are closing ranks, determined to prevent the growing number of educated parents from exercising their right to make informed vaccination decisions for their children," wrote Fisher in late 1987.[99] With co-founders Kathi Williams and Fisher at the helm of the organization, DPT spent the last years of the decade continuing to lobby for a safer pertussis vaccine. They also helped families navigate the new federal compensation system, and collected and disseminated materials on the risks of vaccination, state-by-state vaccine-related rights and obligations, and the latest research on safer pertussis vaccine alternatives. With public education increasingly central to the organization's mission, in 1989 they founded the National Vaccine Information Center (NVIC), a side project initially devoted to producing informational publications and conferences. Within another two years, DPT assumed NVIC as its new name as it expanded its focus beyond the pertussis vaccine and adopted a much broader mission.[100]

DPT/NVIC's commitment to the democratization of vaccine knowledge mimicked a primary tactic of the women's health movement. But health feminism's influence on the vaccine-safety movement also ran deeper than this. Fisher, whose son Chris suffered a convulsion and encephalitis following his fourth DPT shot, often described her dedication to the cause in terms that made reference to both her gender and her awakening to the fallibility of professional medicine. "I was an educated woman," she said. "But, when it came to medicine, I was clueless about vaccines. . . . To know that I participated in what happened to my son because I did not become informed and because I trusted medical doctors without question is a difficult thing to live with, even now."[101] It was a lesson she aimed to inculcate among DPT's constituency from early on: "Mothers, who are primarily responsible for taking children to the doctor and holding them while vaccinations are given, must stop being intimidated by physicians," she wrote in the organization's first newsletter. "We must educate ourselves about vaccines, start asking questions and demanding answers."[102]

Fisher embodied the influence of both feminist and maternalist ideas on vaccine reception and activism at the end of the twentieth century. But she and her organization also captured the challenges inherent in attempts to categorize vaccine critics, either by politics, gender, class, or geography. The founding members of DPT held professions ranging from lawyer to cosmetologist. They were men and women, Republicans and Democrats. Fisher described herself as both a Republican and an original subscriber to *Ms.* magazine, someone who looked to Gloria Steinem and Bella Abzug as role models.[103] The social and political diversity of DPT's members echoed the geographic and ideological diversity of *Mothering*'s readers (whose debates on whether women who worked did a disservice to their children and families were almost as contentious as their debates on immunization). This diversity, combined with shifting political winds, in turn shaped the confluence of political ideologies that informed vaccine critics' ultimately varied demands.

For despite its liberal inheritance, the vaccine-safety movement increasingly sounded distinctly libertarian complaints regarding the nation's vaccine enterprise. In the early 1990s, DPT/NVIC's faith in government as a means of both disseminating information on vaccine risks and guaranteeing safe vaccines became subsumed by their view of government as an obstacle to parents' rights to make educated vaccine decisions for their children. "Our original goal was to get a safer pertussis vaccine for American babies," Fisher said, but with time, "we understood our fight was part of a larger fight for freedom of choice in health care."[104] Libertarianism and allied values have been important factors shaping vaccination resistance throughout American history. But in this particular historical moment, a complex set of ideologies informed popular vaccination criticism. From the 1970s into the 1980s, vaccine critics perceived a wrongful concealment of important information on vaccine risks and an abuse of social and political power embodied in the practice of vaccination. These perceptions allied vaccine critics' nascent cause with a leftist political ideology. Some, like Marge Grant, saw a smaller government role in vaccine promotion as an important solution to these problems. To DPT's founders, in contrast, the solution lay with an expanded government role in overseeing vaccine safety, ensuring dissemination of information on vaccine risks, and helping families injured by government-mandated vaccines. As progress toward these goals was frustrated, however, the organization reframed its stated demands, bringing them more in line with classic libertarian objections to vaccination. But this development would not undo the complex ideological inheritance that gave rise to their initial critiques and demands in the first place.

The organized vaccine-safety movement that was spearheaded by Dissatisfied Parents Together was entirely distinct from the women's health movement, but its origins nonetheless bear the imprint of feminism. Women who spoke out against vaccines in the early 1980s made clear that, like health feminists, they felt patronized and oppressed by the medical profession. Like women's health activists, they also argued that the profession's tight control over information on drug risks prevented them from making informed health care decisions—in this case, for their children. The earlier movement produced, in effect, what Ellen in *DPT: A Shot in the Dark* referred to as two broad categories of mothers: "those mothers who blindly accept a pediatrician's every word and can be easily reassured or controlled; and those mothers who question a diagnosis, ask for more information, and cannot be easily controlled."[105]

The idea that feminism resulted in "two types of mothers" was—if an oversimplification of matters—directly relevant to the movement that began to loudly criticize vaccination policies in the 1980s. It was relevant precisely because doctors and health officials had long viewed mothers as primarily responsible for children's vaccination status, and because mothers, too, often saw themselves in this light. It was relevant because vaccine-related discourse had reflected changing conceptions of women's social roles, both as mothers and as citizens. And it was relevant because both the women's health and anti-medicalization movements had equipped women specifically with a framework for demanding information on vaccines and responding to vaccine risks as they came to light.

The nation's history of vaccination promotion based on gendered assumptions combined with the emergence of the women's health and related movements of the seventies to give shape and content to vaccine critiques that gained visibility and credence in the eighties. A set of collective gendered experiences was just one factor that shaped popular responses to vaccines in this period. Concern for the protection of highly valued children and shifting political winds also loomed large. So did the lingering influence of other key social movements of the seventies, including environmentalism, which gave vaccine-hesitant Americans yet another framework for criticizing vaccines and vaccination.

6

TAMPERING WITH NATURE

In 1985 the mother of a girl named Heather wrote to pediatrician and newspaper columnist Robert Mendelsohn for advice on which vaccines to give her daughter. She had vaccinated Heather against polio, planned to vaccinate her against tetanus, was dead set against the pertussis vaccine, but wasn't sure what to do about diphtheria. She summed up her and her husband's vaccine worries in a single sentence: "We have been afraid to give them to Heather because we are concerned that they contain dreadful toxic things, that they would not contribute to her health and might cause harm to her immune system."[1]

Heather's mother was one of a growing number of parents who, like the parents in the previous chapter, increasingly vocalized concerns about vaccines from the late seventies into the eighties, a period in which the environmental movement burgeoned, environmental politics grew ever more prominent, and environmentalist metaphors and worldviews became mainstream. In this context, environmentalist rhetoric (like health feminist rhetoric) seeped into lay vaccine critiques, giving parents who were even just a little bit squeamish about vaccines—like Heather's mom—a new vocabulary with which to describe their hesitations. Those who began questioning vaccines in this period absorbed key theories and ideas from the environmental zeitgeist. Like the environmentalists who fought to ban the pesticide DDT a decade before, they became concerned about the possible relationship between the widespread use of new technologies and the rising prevalence of chronic and newly emergent diseases, and they began to see vaccines as modern technologies with unknown but potentially devastating long-term consequences for human health.

The anxiety expressed by Heather's mom best classifies her as a vaccine skeptic. She wasn't categorically opposed to vaccines; after all, she had vaccinated her daughter against polio and tetanus and was open to vaccinating her against diphtheria, too. But the environmental movement that informed her own vaccination hesitations also helped cultivate a distinctly modern form of anti-vaccinationism. In contrast with vaccine skeptics, anti-vaccinationists drew no distinctions between vaccines; they rejected them all out of an age-old belief in the wisdom and beneficence of nature. Vaccination, to them, was harmful because it represented an "artificial" way of developing immunity. This artifice, they feared, prompted otherwise benign nature to unleash even worse diseases in revenge. Like more circumspect vaccine skeptics, they saw vaccines as modern technologies with inadequately studied long-term effects. They also worried specifically about the chemical components of vaccines.

This last, distinctly modern, concern marked an important departure from the natural philosophical objections of earlier anti-vaccinationists, whose fears centered on the biological matter in vaccines. Anti-vaccinationists and vaccine skeptics alike in the late twentieth century saw vaccines as troublingly similar to pesticides, cigarettes, artificial sweeteners, and mercury. These chemical products and substances—and many more just like them—had been widely used, the public long assured of their safety, before the long-term dangers of their use came to light. Parents in particular wondered if the widespread use of vaccines represented a similar technological hubris that put the well-being of their children at risk. In the seventies and early eighties, trepidation about the unknown long-term consequences of vaccination was quite vague: parents like Heather's mom worried that vaccines were "toxic" but weren't quite sure what ill effects they might have. In the later eighties and nineties, however, these generalized anxieties evolved into well-defined fears of specific harms, such as autism and learning disabilities, and specific chemical vaccine components, such as thimerosal and aluminum.

Vaccination worries inspired and invited by the environmental movement originally found expression in obscure magazines and newsletters, self-published pamphlets and books, and volumes by small presses. In the 1980s, with the founding of the vaccine-safety group Dissatisfied Parents Together and the publication of *DPT: A Shot in the Dark*, this type of thinking became much more visible. In the 1990s and 2000s, it became more visible yet. Environmental organizations, it bears noting, were not directly responsible for this trend (by the late 1990s, some groups called for mercury to be removed from vaccines, but by then vaccine critics had been expressing anxiety about the "toxicity" of vaccination for close to two decades). Moreover, this trend had relatively little impact on vaccination uptake, which would climb to new heights in the late 1990s and 2000s. But this trend is worth examining nonetheless for the light it sheds on growing vaccination anxieties in the last decades of the twentieth century. The imprint of environmentalist thinking has been visible in the theories and complaints of a range of vaccine critics for the last several decades. For the environmental movement that took off in the 1970s had a lasting effect on the way some Americans came to think about the environment, risk, and disease, with profound implications for the way some came to view vaccines.

THE POISONED NEEDLE: THE ANTI-VACCINATIONIST LEGACY

Vaccine anxieties in the late twentieth century were not a new phenomenon. Nineteenth- and early twentieth-century anti-vaccinationists had decried the smallpox vaccine's potential to "poison" the blood by transmitting either the disease itself, other diseases, or animal matter with unknown consequences for health. Anti-vaccinationists' fears were not unfounded. At a time when vaccination might involve scratching lymph from previously vaccinated individuals into a laceration in the arm under less-than-sanitary conditions, syphilis or other infections were sometimes passed along as well.[2] The practice of vaccinating with human or calf lymph also chafed against the ideals of many nineteenth- and early twentieth-century adherents of nature cures, who saw health as deriving from proper hygiene, diet, and environmental conditions; to some, disease was a necessary means of ridding the body of impurities acquired by eating meat, drinking alcohol, or engaging in other unscrupulous behaviors. In Victorian England, for instance, where anti-vaccination activity was particularly robust, anti-vaccinationism found firm devotees among medical botanists, hydropathists, hygienists, and other alternative medical practitioners and their followers.[3]

As in Victorian England, in Victorian and subsequently in the Progressive Era United States, many homeopaths, botanical physicians, and hydropaths threw their weight behind the anti-vaccinationist cause.[4] In the United States, as in England, anti-vaccination agitation was as much a rejection of dominant medical ideology as it was a struggle over the limits of state power; prominent anti-vaccinationists often denounced compulsory immunization as a form of medical oppression akin to the religious and political oppression they believed their government was meant to protect them from.[5] Organized anti-vaccinationist activity grounded in such leanings was particularly robust in the Progressive Era United States.[6] But with changes in the nation's public health priorities and in organized medicine, and with the deaths, in close succession, of anti-vaccination leaders Charles Higgins and Lora Little, anti-vaccination activity had faded considerably by the 1930s.[7]

It by no means disappeared, however, as a few devoted writers, including outspoken social critic and activist Annie Riley Hale, author of *The Medical Voodoo,* continued to attack vaccination through the 1930s and 1940s; to Hale, vaccination was a form of tyranny propped up by false science and capitalism.[8] In the 1950s, California chiropractor and naturopath R. G. Wilborn founded Health Research, a small press that began reprinting nineteenth- and early twentieth-century works on teetotalism, fasting, natural hygiene, and other nature cures, several of which rejected vaccination as part and parcel of an overall rejection of orthodox medicine. Wilborn's enterprise also sought out original works by contemporary alternative medicine adherents, and in 1957 the press published California naturopath Eleanor McBean's *The Poisoned Needle.*[9]

McBean's book, a harangue against both orthodox medicine and the practice of vaccination, synthesized more than a century of commentary on the moral transgressions and physical hazards posed by vaccination. While she cited the work of a few contemporaries, including American natural hygienist Herbert Shelton and British anti-vaccinationist Lily Loat, the bulk of her volume revisited the arguments of Victorian and Progressive Era philosophers, scientists, healers, and anti-vaccinationists, including British philosopher John Stuart Mill, British naturalist Alfred Russel Wallace, and American hydropath and health reformer Russell Trall. Quoting from this array of sources, McBean argued that vaccination poisoned the blood with animal proteins, was based on the "false premise" of germ theory, and served only to gild the coffers of profit-hungry doctors. She decried compulsory vaccination as an act of "medical oppression" and a form of "enslavement" practiced only by the most "backward" of states. She also denounced vaccination—compulsory or not—as a direct affront to the laws of nature.

This particular premise of McBean's had long been held by anti-vaccinationists and so-called medical irregulars of all types. Wallace, best known for articulating the theory of natural selection, saw vaccination as "an attempt to cheat outraged nature" at its own necessary endeavors.[10] Quoting Wallace, Trall, Shelton, anti-vaccinationist physician John W. Hodge, and others, McBean articulated a philosophy that saw nature and its human inhabitants, in their untouched states, as existing in perfect equilibrium. In humans, poor nutrition and the consumption of processed foods disrupted the body's natural equipoise; it was these habits, not germs, that resulted in disease. Disease, wrote McBean, was the body's way of cleansing itself of "excess poisons, waste matter, obstructions, and incompatible food." "DISEASE IS NOT SOMETHING TO BE CURED; IT IS A CURE," she emphatically wrote.[11] McBean quoted extensively from nineteenth-century French biologist Antoine Béchamp, who had proposed that germs did not cause disease but were rather the result of disease, drawn to diseased tissue to consume it and return it to nature. Combining Béchamp's premise with Wallace's theory of evolution, McBean argued, as other post–germ theory anti-vaccinationists had before her, that germs were "useful wherever they are found in nature."[12] By fighting germs and not the true causes of disease, McBean wrote, "modern medical methods"—including first and foremost vaccination—"delay and frustrate the unexcelled healing efforts of nature."[13]

Modern medical methods (understood as inherently "unnatural") were not the only threat to health; McBean documented a litany of modern commodities and habits that destroyed the well-being of both humans and their environment. Like the century's worth of natural healers who came before her, she emphasized the central importance of a diet of whole, unprocessed foods to good health. McBean's list of modern "poisons" included not only canned, refined, and otherwise processed foods, but also food additives and preservatives, Coca-Cola, tobacco, and chemical fertilizers and insecticide sprays. Writing just before Rachel Carson would begin work on *Silent Spring*, McBean denounced the use of DDT and blamed it and other sprays for a host of modern ills, including cancer, heart disease, and polio. In her view, mass vaccination was a calculated distraction from "foodless foods" and "poison sprays," the true causes of allegedly vaccine-preventable diseases, such as polio.

Like Carson, McBean (referencing a recently emergent literature on organic agriculture) drew an ecological view of health, in which insecticides caused harm not only to birds, butterflies, and bees, but also to humans, because of their dependence on the health of the ecosystem. Insecticide sprays poisoned food directly, McBean wrote, and were washed from crops into soil, where they killed earthworms and other organisms vital for healthy soil, which was vital for producing nutritious crops to fortify humans. The sprays caused even further damage by contaminating water supplies that both humans and animals relied upon. To McBean, widespread pesticide applications, which citizens were powerless to avoid, were, like vaccination, crimes committed by government acting in the interest of powerful corporations with no regard for human health. She wrote:

This staggering *increase* in a *preventable disease* is a grave reflection upon our present system of living with its *popularized blood pollution* practices by way of *vaccination* campaigns and *mass poisoning* as a result of government enforced *spraying of fruits and vegetables with deadly lead arsenate* and other poisons. The power politics of the drug and chemical companies have also influenced legislation to set aside vast sums of the taxpayer's money with which to buy their poison chemicals. . . . The people are told that these practices are beneficial but facts disclaim these statements.[14]

Just as Carson would highlight the association between rising cancer rates and pesticide use, McBean made much of the parallel increases in cancer prevalence and the deployment of vaccines and insecticides. She also drew a parallel between these two categories of modern hazards and a third: atomic radiation. In McBean's view, atomic radiation was the only modern poison that caused more harm to human health than vaccination did.

McBean's book was a response to the nation's massive polio vaccination efforts in the 1950s. It's difficult to say how widely it was read, but when the first print run of 5,000 copies sold out, Health Research printed a second run of 5,000 in 1959. After that, the book went out of print for close to two decades, but the press brought back the title in 1974.[15] By then, of course, polio and smallpox had all but disappeared from the United States, but vaccination efforts targeting other infections were on an upswing across the country. In McBean's hometown of Los Angeles alone, there was a rush of official immunization activity in the early seventies. A countywide campaign began promoting rubella immunization in 1970. A new law requiring that all children be immunized against diphtheria, tetanus, pertussis, polio, and measles before starting school sent parents rushing to the doctor in 1972. And in 1973, the first polio case in a decade had health officials warning that the disease was poised to return in epidemic proportions unless vaccination rates improved.[16] Health Research may simply have run out of first-edition copies of McBean's book by 1974, but the press's decision to reprint it may also have been a response to increased attention to the subject of immunization, coupled with increasing popular cynicism regarding authority. McBean's anti-professional stance, her assertion of the value of American rights and freedoms, and her claim that justice had too long been the exclusive province of the rich and powerful must have resonated with readers who identified with the New Left movements that emerged in the sixties and swelled in the seventies. Her litany of environmental concerns undoubtedly did as well.

By 1974, the year *The Poisoned Needle* was reissued, *Silent Spring* had helped foster a new environmental movement, which had popularized concerns about radiation, heavy metals, pesticides, and other chemicals, and had prompted passage of a series of federal laws to protect the quality of air, water, and other natural resources.[17] The values that McBean wove together in *The Poisoned Needle*—the preciousness of freedom and nature combined with a mistrust of government and industrialists—anticipated popular attitudes of many 1970s social activists by the better part of a generation. Indeed, as the 1970s progressed, McBean, then in her seventies, appeared to have found a new audience for her thoughts and renewed energy for her cause. She published three more anti-vaccine books between 1977 and 1980, and commentaries and articles by her inspired followers began to appear as well.[18] McBean's books may never have garnered a tremendous readership, but renewed interest in *The Poisoned Needle* in the 1970s was significant in that the book carried over a set of anti-vaccinationist ideas from the first half of the century (and earlier) to the latter half. The book thus served as a bridge between the anti-vaccinationism that faded in the 1930s and the renewed vaccine skepticism that began to gain momentum in the last decades of the century.

NEW VACCINE FEARS

The variety of vaccine skepticism that became increasingly prevalent at the end of the century inherited several key ideas from McBean, and, in turn, from the natural healers and anti-vaccinationists whose work had inspired her own. The predominant vaccine concerns expressed by parents, doctors, and others in self-published and small-press books and pamphlets were grounded in the ideas that nature, even in the form of disease, was purposeful if not benevolent; that health derived not from artificial immunity, but from balance and harmony with the natural order; and that vaccines were akin to environmental hazards inasmuch as they were products of industry with uncertain and potentially harmful long-term consequences. These vaccine beliefs mirrored some of the defining characteristics of the environmental movement of the 1960s and 1970s: the rise of popular ecology, concern about the health consequences of environmental choices, and the prevailing sense of a ubiquitously toxic environment.[19]

They also reflected environmentalist anxieties about everyday consumer product exposures, which became increasingly prevalent in the 1970s.[20] From the 1960s through the 1970s, one episode after another had driven home the notion that unseen dangers were widespread in the consumer environment. Cigarettes, lead, asbestos blankets, red food dye, and more were all proven harmful one after the other, well after Americans had been long assured of their safety (if not their benefits). The accumulated revelations that such products could cause cancer, brain damage, and even death gave consumers legitimate cause to question the safety of consumer products generally—and the assurances that government and industry had taken sufficient measures to protect them from harm. The country's experience with cigarettes in particular was a lesson not only in the power of industry obfuscation, but also the scientific challenges of proving a cause-and-effect relationship between a single chemical exposure (such as tobacco smoke) and a health outcome (such as lung cancer) that did not arise until very much later in life.[21] Cigarettes, lead, asbestos, and radioactive materials couldn't be proven unsafe until they were long and widely used, which in effect left an array of consumer products open to the critique that they, too, might someday be proven hazardous, but only well after much damage had been done.[22] Indeed, it was this very argument—about the challenge of uncovering long-term health effects—that environmentalists famously used to win a ban on DDT in 1972.[23]

By the late seventies, environmentalists' focus on the uncertain and unknowable long-term consequences of consumer-product interactions provided a framework for critics to question vaccines by highlighting the scientific uncertainties inherent in their use. Writing in 1978, Maine chiropractor Daniel Lander worried that "in reality, no one knows for sure how effective or safe immunization really is and it is unlikely that we will ever know."[24] Across the country in Oregon, childbirth educator Cynthia Cournoyer echoed this notion in a self-published pamphlet: When you vaccinate your child, "you cannot be *sure* you are not also administering a serious side effect," she wrote, "much is left unknown. There is no conclusive evidence vaccines are completely safe."[25] Harvard- and New York University–trained physician-turned-homeopath Richard Moskowitz pointed out that not only were there uncertainties inherent in vaccine use; no attempt had been made to uncover the long-term implications of their use. "The fact is that we do not know and have never even attempted to discover what actually becomes of these foreign substances once they are inside the human body," he wrote in 1984.[26]

These vaccine critics and others often adopted an environmentalist lexicon to explain why these uncertainties were worth worrying about. For many vaccine skeptics, for instance, the image of a dangerously polluted environment served as a powerful metaphor for the contemporary condition of the human body. These critiques held that the human body was a microcosm that faced the same onslaught of toxins that threatened the well-being of the natural macrocosm; this idea mirrored environmentalist laments about nature's fragility and its loss of purity.[27] McBean and her followers, in particular, embraced the idea that no environment, ambient or bodily, could be purified via the addition of ever more pollutants. "Certainly a city or other area cannot be immunized from pollution by introducing more contaminating substances into it," wrote one of McBean's admirers in the pages of *Mothering* magazine. "Could a thinking public be so brainwashed as to believe that the addition of more smog to their city would possibly have the effect of purifying it?"[28]

The pollution metaphor also evoked another hallmark of environmentalism: the hubris and shortsightedness of science and industry, often contrasted with the inherent wisdom of nature. Influential environmentalists such as Barry Commoner popularized in the 1960s and 1970s the idea that society had established a pattern of committing to new technologies—nuclear weapons, fertilizers, insecticides, detergents, and automobiles among them—before the consequences of mass deployment were completely understood.[29] In the aftermath of several widely publicized drug scares (such as those involving thalidomide and DES), and following the revelation that the 1976 swine flu inoculation campaign had done more harm than good, lessons initially relevant to the environmental arena began to color perception of mass vaccination as well. Wrote a *Boston Globe* reporter in response to the swine flu fiasco: "It was as if Mother Nature were warning us against arrogance: there are many things in a world full of biological hazards that we don't understand, don't even have the tools to understand."[30] Moskowitz, among others, applied the lessons of environmental disasters to vaccination more broadly: "We have been taught to accept vaccination as a sacrament of our . . . participation in the unrestricted growth of scientific and industrial technology, utterly heedless of the long-term consequences to the health of our own species, let alone to the balance of nature as a whole."[31]

In envisioning the potential for undesirable long-term consequences, many vaccine skeptics invoked the environmentalist metaphor of the chemical time bomb. In *How to Raise a Healthy Child . . . in Spite of Your Doctor*, Robert Mendelsohn suggested that vaccines might be a "medical time bomb" simply because "no one knows the long-term consequences of injecting foreign proteins into the body of your child."[32] To Marian Tompson, a co-founder of the breastfeeding advocacy group La Leche League, the consequences were clear. In her view, vaccination was creating a generation of weak, defenseless beings. "Instead of taking personal responsibility for our body's immunological system, we try to handle everything with a vaccine, insulting our bodies and creating a sicker, more endangered species. We are, literally, walking time bombs!" she wrote in 1982.[33] Tompson believed, as many other vaccine critics did, that the artificial nature of vaccination was compromising children's natural defenses. But neither Mendelsohn nor Tompson pointed specifically to the explosive potential of any particular component of vaccines; to both, the vaccines in their entirety were the explosive materials encapsulated in human bodies.

Tompson's worry—that vaccination was weakening the species—was related to a broader set of perceptions regarding the artificial nature of vaccines, the superiority of natural immunity, and the importance of balance to the pursuit of health. While a number of people began to question vaccines in the late 1970s and early 1980s, a much smaller subset rejected vaccines outright. Many of these staunch anti-vaccinationists rejected—as McBean had—the idea that germs caused disease. In this view, vaccines were unnecessary because sickness resulted not from microorganisms but from an imbalance between "a person's inner environment and the external world," as McBean follower Leonard Jacobs put it.[34] Jacobs and like-minded vaccine skeptics argued that immunity was "the natural ability to maintain balance with the environment" and could be obtained through breastfeeding, a balanced diet, and exercise.[35] (Others added such elements as "relaxation" and a "positive attitude" to the list.[36]) This ecological view of health was not the exclusive purview of germ-theory nihilists, however. Some, like New York pediatrician Victor LaCerva, shared the belief that health derived primarily from internal balance and balance with one's environment, while asserting that viruses and bacteria did cause disease in bodies "out of balance."[37]

But what, precisely, did "balance" signify for people with vaccine doubts? For many, including LaCerva, it signified an approach to health that emphasized lifestyle choices, including the decisions to breastfeed, exercise, avoid processed foods, and get adequate rest, an approach that gained increased traction in the 1970s.[38] It also signified the acknowledgment that humans were part of an ecosystem, their own well-being dependent on the well-being of the larger environment in which they lived. It was this perception that prompted Albuquerque physician Sue Brown to write that while she worried about vaccine safety, she believed nuclear war and "an uninhabitable planet" were more important threats to children's health.[39] Balance also referred to a quasi-religious belief in the precise, perfectly calibrated interplay of natural systems, from the molecular to the macro, that was designed to promote health but was so intricate and complex that it was beyond the comprehension of humans. "There is a wisdom within the body," wrote Maine chiropractor Lander, illustrating this point of view. "The human body has the most complex organic machinery in the world. It produces all the chemicals one will ever need to be healthy. . . . The wisdom that created our bodies is far superior to the finite mind of all scientists in the world."[40]

Lander and other vaccine critics deliberately distinguished between what they saw as "artificial" immunization—that is, vaccination—and natural immunity, which derived not from a pharmaceutical product but from lifestyle choices.[41] Indeed, it was the artificial nature of vaccines that most often came under attack from critics from the 1970s onward. This revived a complaint common among anti-vaccinationists of the nineteenth and early twentieth centuries; McBean herself also frequently decried the "artificial" nature of vaccines. Both old and new nature-based vaccine critiques held that vaccines were artificial because they represented a contrived way of encountering disease, an "unnatural way of handling foreign material."[42] But in the brand of vaccine skepticism that arose in the 1970s, critics looked for scientific evidence (as environmentalists often did) to prove their position. "Did you know," wrote La Leche League's Tompson, "that when immunity to a disease is acquired naturally, the possibility of reinfection is only 3.2 percent? If the immunity comes from a vaccination, the chance of reinfection is 80 percent."[43] Others argued that because vaccines conferred a lesser degree of protection from natural infection, they didn't produce "true immunity," and thus the term "immunization" was a misnomer.[44]

That vaccines were perceived as artificial—and that their artificial nature was something to be abhorred—was implied by the widespread adoption of the term "toxic" to describe them. McBean and other earlier anti-vaccinationists had routinely referred to vaccines as "poisons"; writing in the 1970s, McBean continued to employ the term. But in the 1970s and 1980s, the word had less traction than it once did, and vaccine critics broadly turned instead to "toxin" and "toxic," both of which were becoming popularly adopted with the diffusion of contemporary environmentalist thinking.[45] Concerns about the toxic environment were ubiquitous in the early phases of the environmental movement, and in the 1970s these morphed into concerns about how chemical substances wreaked toxic effects on human bodies.[46] Pesticides, artificial sweeteners, food dyes, drugs, and other consumer products were all subject to popular attack and government regulation for their potential or demonstrated toxic effects on humans. Vaccines, to some, seemed a logical addition to this list. Some adapted this idea to age-old natural healing philosophies: "All vaccines, like drugs, are toxic. None render the body healthy, but rather more toxic," wrote one natural hygienist in defense of her anti-vaccinationism.[47] Others employed it to articulate their newfound vaccine worries, as did one *Mothering* reader who referred to vaccines as "standardly accepted injectable toxins."[48] In a new edition of her self-published pamphlet, Cournoyer told readers that "Manufacturers of vaccines admit they are highly toxic and by their very nature, cannot be made safe." Cournoyer went on to enumerate the "toxic ingredients" in vaccines. In addition to "horse blood, dog kidney tissue . . . and other decomposing proteins," she listed:

Phenol—(Carbolic acid) a deadly poison.

Formaldehyde—A known cancer causing agent which is commonly used to embalm corpses.

Mercury—A toxic heavy metal that is not easily eliminated from the body.

Alum—A preservative.

Aluminum phosphate—Used in deodorants. Toxic.

Acetone—A solvent used in fingernail polish remover. Very volatile.

Glycerin—A tri-atomic alcohol extracted from natural fats which are putrified and decomposed. Some toxic effects of glycerine are kidney, liver, lung damage, diuresis, pronounced local tissue damage, gastro-intestinal damage and death.

Aluminum and Oil Adjuvants—Carcinogenic (cancer-producing) in laboratory mice.[49]

This focus on the chemical components of vaccines marked a distinct departure from the criticism of Victorian and Progressive Era anti-vaccinationists, who had also decried vaccination's artificial nature and likened the process to pollution. From the 1970s on, however, these specific complaints encompassed a new meaning: they evoked not just biological contamination with "horse blood" and "cow pus," a complaint that originated with earlier anti-vaccinationists, but the irreversible chemical pollution of bodies, a distinctly modern concern.[50] Vaccine critics pointed out that "most parents who are trying to feed their children properly would not let them eat a food which contained any of the many ingredients in immunizations."[51] They employed the modifier "toxic" to describe the chemicals they found listed on vaccine package inserts, and by pointing out that some of the chemicals were carcinogenic, they implicated vaccines in the ever-growing epidemic of cancer. By pointing out that vaccines included heavy metals, they implicated immunization in other epidemics as well. To Cournoyer and other critics, the presence of such compounds in vaccines was proof that the products were under-regulated and thus unsafe. "When cancer causing elements are found in foods, they are either banned (remember cyclamates?) or an obvious warning label appears on the package (saccharin, cigarettes)," she wrote. Her conclusion: "There seems to be a double standard for vaccines!"[52]

CHRONIC DISEASE FEARS

In the eyes of vaccine critics, both aspects of the artificial nature of vaccines—their unique interface with the immune system and their contents—implicated vaccines in an ever-adjusting list of seemingly modern epidemics. The suspicion that vaccines might be responsible for the emergence or increasing prevalence of new (or perceived-as-new) diseases was long held by anti-vaccinationists. Lora Little linked vaccination to cancer in her 1905 anti-vaccine tract; Annie Riley Hale did the same in 1935.[53] In 1957 and 1974 (her book was reprinted a third time, in 1993), McBean argued that the "900 percent increase" in cancer deaths in the first half of the century was brought on by vaccination.[54] McBean (quoting Hale) maintained that modern medicine's greatest achievement had been to "swap smallpox for cancer and typhoid fever for diabetes and insanity."[55] The suspicion that vaccination had resulted in an unfavorable trade-off was oft-repeated by vaccine critics from the 1970s on, who argued that Americans had "traded mumps and measles for cancer and leukemia"—or, put another way, for "the far less curable epidemic of chronic diseases of the present."[56]

For much of the twentieth century, anti-vaccinationists and vaccine skeptics were unanimous in their suspicion of a link between vaccination and cancer. But around 1980, fears that vaccines were responsible for climbing cancer rates gave way to fears that vaccines were responsible for other epidemics, namely, epidemics of autoimmune diseases, learning disabilities, and childhood behavioral disorders. These anxieties were linked to the belief that vaccination constituted an unnatural bodily intrusion, and they shifted in response to other cultural anxieties. At the 1975 Asilomar Conference on Recombinant DNA, which followed an explosion of findings on genetic disorders and great advances in (and trepidations about) genetic engineering, some scientists speculated that vaccines might be introducing disease-inducing genetic material into the body. Vaccine skeptics quickly picked up on and circulated such hypotheses. In a 1976 issue of his newsletter *The People's Doctor*, Mendelsohn described a theory proposed by a Rutgers University geneticist, which in Mendelsohn's interpretation suggested that vaccines were seeding humans with genetic material that could form "pro-viruses" in the body; under the right conditions, he warned, the pro-viruses might lead to "rheumatoid arthritis, multiple sclerosis, lupus erythematosus, Parkinson's disease and perhaps cancer."[57]

Eight years later—following, not coincidentally, the discovery of HIV—Mendelsohn wrote of the "growing suspicion" among scientists that immunization "against relatively harmless childhood diseases" might be responsible for the recent rise in autoimmune diseases.[58] With the emergence of AIDS, preoccupations with the potential link between vaccines and cancer were largely supplanted by concerns that vaccines were responsible for epidemic levels of immune dysfunction. Some feared that AIDS itself might have been caused by the vaccines against polio and hepatitis B.[59] Others believed that AIDS and the "overall immunologic weakening of our children" were brought on by the generally artificial nature of vaccination.[60] In 1985 Pennsylvania physician Harold Buttram (who long had unorthodox medical views and ties to anti-vaccinationists) articulated a theory linking vaccines to the recent epidemic of immune disease. Because injections permitted the material in vaccines to bypass mucous membranes and instead directly stimulate a set of antibodies that were not the body's usual first line of defense, they constituted "immunologic shock treatment" and depleted the immune system's resources. Buttram and a co-author proposed that this artificial route to immunity (coupled with exposure to environmental pollution and processed foods) was responsible for causing an "AIDS-like state" in children that was manifesting as an overwhelming rise in allergic disorders.[61] One didn't need to be a physician or a staunch anti-vaccinationist to draw a line between vaccines, their purportedly artificial nature, and the epidemic that came to define the 1980s. As one parent wrote to Mendelsohn in 1988, "The thought of injecting toxins (of fairly dubious origins) into my children, who have never known any illness more serious than an occasional cold, is absurd. . . . In this era of malfunctions of the immune system—cancer and AIDS specifically— our country would be better off spending its research money on learning about immune functions."[62]

As times changed, Buttram's theories evolved, and in the early 1990s he argued that the most worrisome trend in the health of American children was the rise of behavioral disorders, including hyperactivity and learning disorders. Those very same conditions stood out to Barbara Loe Fisher and Harris Coulter as they compiled *DPT: A Shot in the Dark*.[63] After unearthing evidence from the medical literature linking pertussis vaccination to convulsions, seizures, encephalitis, and permanent brain damage, Coulter and Fisher proposed that the vaccine might also be responsible for a far greater number of cases of "minimal brain dysfunction, or learning disabilities," including hyperactivity, dyslexia, and autism. Citing an idea popularized by the environmental movement, they argued that scientists had likely missed this connection because of the lag time between exposure and effect. To strengthen their case, they turned to the bellwether of the environmental movement, Rachel Carson's *Silent Spring*.

In her 1962 book, Carson delineated the devastation that pesticides had wrought on wildlife and then argued that scientists had no way of knowing whether the very same exposures weren't endangering in subtle ways the health of humankind as well. Coulter and Fisher seized on this latter point, quoting Carson at the start of a chapter they titled "Long Term Damage": "We are accustomed to look for the gross and immediate effect and to ignore all else. Unless this appears promptly, we deny the existence of hazard. Even research men suffer from the handicap of inadequate methods of detecting the beginnings of injury."[64] *A Shot in the Dark* described the allergies, deafness, and behavioral disorders in children who had appeared to suffer acute reactions upon vaccination as infants. It documented the "phenomenal" increase in learning disabilities and hyperactivities over the previous three decades, the same period in which pertussis vaccination came into widespread use. To Coulter and Fisher, Carson's words suggested that no matter the length of time that had elapsed between the shot and the onset of such conditions in a child, a connection between the two could not be ruled out.

Coulter and Fisher also noted that infantile autism was first documented by doctors just a few years after the pertussis vaccine was widely deployed in the United States. They quoted one mother who described the autistic-type behaviors in her son Richard, who suffered encephalitis following his third DPT shot and could assemble puzzles but not put on his own shoes. Richard's mother was "convinced" of a connection between autism and her son's pertussis vaccine, and Coulter and Fisher concluded that this was one of several possible vaccine-related harms that deserved greater scrutiny from the scientific community.[65] Autism was also one of just a few harms that Coulter, an independent medical historian who had written extensively on alternative health, went on to investigate himself. For a subsequent book, he interviewed sixty parents of vaccine-damaged children in addition to the hundred that he and Fisher had interviewed for *A Shot in the Dark*. From these, he devised a theory that linked widespread vaccination to what he described as epidemic sociopathic behavior in American society.

Coulter's 1990 book, *Vaccination, Social Violence, and Criminality: The Medical Assault on the American Brain*, proposed that vaccines had inflicted such widespread damage on the brains of children that they were responsible for the explosion of psychiatric disorders among American children since the 1950s, the social "turmoil" of the 1960s, increases in crime, alcohol and drug abuse, and even the pathological behavior of serial killer and rapist Ted Bundy. Coulter gave special attention to autism, attributing increasing prevalence of the condition to vaccine-induced encephalitis, a phenomenon well-documented in the medical literature (though Coulter proposed that it occurred far more commonly than documented). Coulter pointed out that the first medical descriptions of autism in the 1940s immediately followed the widespread use of pertussis vaccination. He also argued that the reason autism had been limited to offspring of upper-class parents prior to the 1970s was because they were the only ones who could afford to vaccinate their children. Only after the 1970s, when new policies made vaccination widely available to all socioeconomic classes, did autism spread more widely in his view.[66]

Coulter's 1990 volume never achieved the readership or media attention that his book with Fisher enjoyed. But it's worth noting because it joined a wave of books on immunization hazards that came out after *DPT: A Shot in the Dark*, a trend that signaled an expansion of vaccine skepticism. However, whereas *DPT: A Shot in the Dark* focused its attention on a single vaccine, and held out hope that science would provide a safer vaccine in time, many of the authors of the subsequently published vaccine critiques were far less sanguine about modern medicine and believed that safer routes to immunity existed. Books such as Walene James's *Immunization: The Reality Behind the Myth*, Randall Neustaedter's *The Immunization Decision*, Neil Miller's *Vaccines: Are They Really Safe and Effective?*, Jamie Murphy's *What Every Parent Should Know about Childhood Immunization*, and Viera Scheibner's *Vaccination* decried the artificial nature of vaccines, the lack of research on long-term consequences, and vaccines' potential role in fostering the spread of chronic diseases.[67]

On the whole, these books reflected the growing popularity of natural and alternative healing methods, which expanded in the late eighties and early nineties.[68] Many of their authors subscribed to nature-based healing philosophies, and environmental metaphors and ideas filled their work. Murphy, a Massachusetts-based herbalist, equated vaccination to the pollution caused when oil drums were buried underground or industrial solvents dumped into open waters.[69] Neustaedter, a San Francisco homeopath and acupuncturist, worried about "subtle and long-term damaging effects on the immune system and nervous system."[70] New Mexico journalist and father Miller argued against tampering with the "delicate structure of the human organism."[71] Scheibner, an Australian physician whose vaccine-skeptical book was published in the United States, argued that doctors and parents needed to "start respecting nature and recognize infectious diseases for the value they bring to children."[72] She was one of many who argued that even infectious diseases serve a purpose, and that nature gives all creatures everything they need to survive.

To these and other writers, recent history offered proof that their vaccination objections were justified. A recent resurgence of measles, which struck vaccinated as well as unvaccinated children in the late 1980s, was a troubling sign of nature's obstinacy. The revelation that pertussis vaccine could cause brain damage was proof that vaccines could be proven harmful after many decades of use. And the cases of Guillain-Barré syndrome that occurred in people vaccinated against the swine flu back in 1976 offered a similar example of harm following assurances of safety on a compressed time scale.

MERCURY MADNESS

DPT: A Shot in the Dark nudged vaccine skepticism into the national spotlight and proved that there was an audience for information on vaccination alternatives—or at least for more information on vaccines and side effects than was generally provided in doctors' offices. But most authors of subsequent vaccine-critical books struggled to produce further evidence of harm. Several pointed to Coulter's theory, repeating the observation that the climbing prevalence of both autism and hyperactivity correlated with increased vaccination. And more and more critics enumerated the chemicals in vaccines and hunted down details on their toxicity.

Murphy, in the 1994 book *What Every Parent Should Know about Childhood Immunization*, wrote that vaccines contained a "witch's brew" of chemicals, including known carcinogens (such as formaldehyde), mercury (in the form of thimerosal), aluminum, and formalin. He described aluminum poisoning in factory workers and the Environmental Protection Agency's decision to classify formaldehyde as "hazardous waste."[73] Thimerosal, he pointed out, was a mercury derivative, but no tests had been done to see how much of the metal remained in the body after vaccination. Bits of evidence from other studies, however, did provide cause for concern. Murphy described a report in the *British Medical Journal* that concluded there was a "theoretical risk" of harm to patients receiving immunoglobulin injections, which contained thimerosal. And he pointed out that the mercury in once widely used teething powders had been traced—after seven decades of use—to a condition called pink disease, a form of mercury poisoning, in children. Other vaccine critics noted that mercury was toxic to the kidneys and central and peripheral nervous systems and associated with tremors, dementia, and memory loss—symptoms that closely resembled those seen in children who had reportedly reacted to the DPT, pertussis, and the new Hib (*Haemophilus influenza* type b) vaccines, all of which contained thimerosal.[74]

The increasing suspicion that the chemical components of vaccines—especially thimerosal—might be toxic to children's developing nervous systems took place against a backdrop of growing national concern about mercury exposure. Studies in the 1980s had revealed dangerously high levels of mercury in fish—including some of the highest levels ever recorded in the United States.[75] The findings, reported the *New York Times*, bore out environmentalists' warnings about pollutants in the food supply and were especially troubling in light of growing fish consumption among increasingly health-conscious Americans.[76] As federal agencies examined the health and environmental effects of mercury emissions in the environment and debated the amount of mercury that individuals could safely consume, state governments and community groups went ahead with steps to reduce exposure to the metal.[77]

States issued warnings against consuming fish from local waters and passed regulations to monitor disposal of mercury-containing appliances; a few states even banned children's flashing sneakers, which contained mercury switches.[78] Advocacy groups began campaigns to phase out the use of mercury-containing thermometers and warned pregnant women and children to limit canned tuna consumption.[79] News reports on mercury throughout the 1990s emphasized that even "tiny" amounts of the metal were toxic, especially to the fetus, and that exposure could cause subtle but permanent damage. Many articles also quoted officials who compared mercury to lead and cigarettes, the hazards of which were long ignored and then suppressed. In a July 1999 article that referred to "mercury madness," *Mothering* notified readers about widespread warnings against fish consumption and the dangers of mercury thermometers, adding, "It only takes a drop of the toxin to contaminate a whole lake—or a child."[80]

The article in *Mothering* made no reference to mercury in vaccines, but it likely would have had it gone to press just one month later. In July 1999, the Public Health Service and the American Academy of Pediatrics (PHS/AAP) issued a joint statement asking vaccine manufacturers to "eliminate or reduce as expeditiously as possible the mercury content of their vaccines." The statement indicated that because of new vaccines and new vaccine recommendations (by now, the list of childhood immunizations had grown to include vaccines against hepatitis B, chicken pox, and Hib), children in their first six months were at risk of getting a cumulative level of mercury that exceeded a federal guideline on methyl mercury, a chemical relative of the mercury compound in thimerosal. The risk of harm from this exposure was unknown but certainly smaller than the risk of infectious disease, the statement went on, but it was nonetheless worth addressing "because any potential risk is of concern."[81]

The determination that there was any risk from the mercury in vaccines came as a result of an amendment to the Food and Drug Administration Modernization Act of 1997 that required the FDA to evaluate the mercury contents of drugs and biologic products.[82] It also followed the 1997 publication of the *Mercury Study Report to Congress*, the Environmental Protection Agency's examination of the health and environmental effects of mercury emissions undertaken in response to the Clean Air Act Amendments of 1990.[83] During—and after—the production of the *Mercury Study Report*, the safe exposure level was the subject of intense political debate. Industry cited results of a study finding no adverse effects among mercury-exposed children in fishing communities in the Seychelles, while public health and EPA officials cited a study of mercury-exposed children in a fishing community in the Faroe Islands, whose motor function, language, and memory were all diminished compared to non-exposed children.[84]

The 1999 PHS/AAP statement on vaccines set off its own debate over the health effects of mercury exposure, although this time the disagreement split the public health community. Pediatrician and outspoken vaccine proponent Paul Offit called the statement a "flawed" policy for elevating "a theoretical risk above an actual risk."[85] Veteran vaccine researcher Stanley Plotkin called it a "public health disaster" for delaying the vaccination of infants against hepatitis B.[86] In an editorial defending the call to remove thimerosal from vaccines, Johns Hopkins University epidemiologist and federal vaccine adviser Neal Halsey equated the debate over mercury to that over lead, "where sequential studies over many years provided evidence for subtle effects with progressively lower exposures."[87] To Halsey, the theoretical risk of harm from thimerosal had to be balanced against the public's tolerance of that risk, which in his estimation was growing ever more limited. Policies to limit risk also needed to appear consistent. As he later explained to a *New York Times* reporter, his own position on the presence of thimerosal in vaccines crystallized while he was canoeing on a lake in Maine, where he came across a sign reading "protect your children—release your catch." It was a problem, he told the reporter, for the government to tell parents not to feed their children fish contaminated with mercury but to ask them to be injected with the same substance.[88]

Leslie Ball and Robert Ball, the FDA scientists who conducted the agency's risk assessment on thimerosal, also defended the recommendation to remove the preservative from vaccines, both for its feasibility in the interest of limiting total human exposure to mercury and as a way of maintaining public confidence in vaccines.[89] Indeed, at the end of the century, it was hard to ignore the fact that lay confidence in vaccines was flagging. CDC officials reported that the national immunization hotline had been receiving increasing numbers of calls about vaccine safety.[90] The new vaccines added to the childhood vaccination schedule in the late 1980s and 1990s had evoked some consternation among parents and drew increased scrutiny from the press, which throughout the 1990s ran reports on potential vaccine harms, news of recalls, and investigations into the machinations of the "billion-dollar" vaccine industry. This pileup of negative, end-of-the-century vaccine news — discussed in chapter 9 — often seemed to confirm skeptics' long-held fears and convictions. When, for example, the *Lancet* published a well-publicized report on a potential link between measles vaccination and a form of autism in 1998, Barbara Loe Fisher received the news with equanimity. After all, she pointed out, the link between autism and vaccines was "first reported in *DPT: A Shot in the Dark* fifteen years ago" and had been "simmering" for over a decade.[91] To skeptics, such "news" about vaccines was nothing new.

A CRISIS OF FAITH

According to historian Thomas Dunlap, environmentalism, like religion, "invokes the sacred," "refuses to choose between intellect and emotion," and "gives moral weight to the apparently trivial decisions of daily life."[92] Environmentalism and religious belief systems both grapple with questions of human existence and conceptualize life and the universe as a complex, quasi-mysterious whole whose intricate workings are beyond the comprehension of humanity. Over three decades of resistance, vaccine skeptics relied on a quasi-religious environmental belief system to make sense of vaccines and their encounter with human bodies and the environment. In doing so, they conceptualized not just the body but the immune system specifically as sacred and intricately complex beyond human comprehension.

As vaccine policies were strengthened in the 1970s, and as the vaccine schedule was expanded in the 1980s and 1990s, nature offered hesitant parents answers to the increasingly fraught questions of whether, and against which diseases, they should vaccinate their children. Nature also helped resolve the broader question of how to care for children in a complicated and incomprehensible modern world. The notion that everything in nature — viruses, bacteria, and diseases included — served an unknowable but crucial purpose was more comforting than the prospect of gambling with haphazard and ever-changing scientific knowledge. Not infrequently, vaccine skeptics attributed the sanctity of everything in nature to the workings of a higher power. As the mother of a vaccine-damaged son explained in *DPT: A Shot in the Dark*:

I thought I was being a good parent to give him that shot. If I had known about the risks, if I had been given an option, I might have taken my chances with the natural disease. . . . I was so happy when he was born. He was so beautiful, with ten toes and ten fingers. God gave me a perfect child, and man, with his own ways, damaged God's perfect work.[93]

Many parents who eschewed vaccines expressed a desire, like the mother above, to leave (or to have left) their child's fate in the hands of divine wisdom, be it nature's or God's. In the 2000s, this desire manifested in the revival of the chicken-pox party: "A little playing, some conversation and some passing of the pox to the next family," as one Virginia mother put it.[94] The development of a vaccine against chicken pox in the 1980s was followed in the 1990s by the adoption of state laws requiring proof of vaccination or immunity for children entering day care or school. But media reports shared the news that parents in Britain were opting for "disease parties" over the new vaccine.[95] In the United States, Internet discussion boards soon buzzed with talk of the parties, and in 2004 *Mothering* published advice on establishing immunity the "natural" way. Varicella, the chicken pox virus, "is communicated easily through saliva," the authors wrote, so they suggested having children at a "Pox Party" share whistles.[96] Parents who chimed in online suggested passing around kazoos, popsicles, lollipops, or M&M's while the gathered children did art projects or played games.

Over the next few years, popularity of the parties swelled, achieving a cult-like following and considerable press, particularly in upper-middle-class enclaves from coast to coast. When interviewed by the media, mothers who hosted or took their children to Pox Parties said they wished to give their children natural, lifelong immunity that they, in turn, could pass on to their own children someday. Often they described their thinking in religious terms. "I believe in the body's ability to build immunity and heal itself," said a Brooklyn mom and registered nurse who hosted a chicken pox party to spread her son's infection.[97] "I'm of the belief system that by putting a synthetic, or a dead virus, into a body as an inoculation creates an ongoing response to trying to build up immunity to that, and that ultimately it creates a situation where you have an overworked immune system," said a California mom.[98]

Parents who saw ultimate doom in acceptance of the chicken pox vaccine saw not only a progressive, cumulative assault on natural immunity in the making but also (because the duration of the shot's protection was unknown) a series of man-made disasters on the horizon: a generation of women with no maternal antibodies to pass on to their own children; an epidemic of shingles, the more serious manifestation of chicken pox virus infection in adolescents and adults; and a generation of daughters who could be susceptible to chicken pox during their own pregnancies, putting two generations removed at risk of birth defects. These fears echoed vaccine scientists' own initial questions about the vaccine when it was first developed. But vaccine critics amplified these hypothetical scenarios until they became predictions of unavoidable disaster in which, like the devoutly religious, they firmly believed, and which guided their decisions and actions.[99]

Writing in the 1980s, historian Roderick Nash argued that American environmental ethics had come to be dominated by the ideas that nature has both "intrinsic value" and "the right to exist," the latter an idea that placed environmentalism (an albeit "radical" movement) "squarely in the mainstream of American liberalism."[100] With the expansion of both the immunization schedule and immunization programs in the 1990s—a subject described in greater detail in chapters 7 and 8—vaccine-safety groups increasingly advocated for the freedom to make health care decisions for their children, arguing that this was a right of parents upon which the state should not intrude. This demand directly echoed the requests of anti-vaccinationists of a century before, and it fit neatly within the tradition of American emphasis on the importance of liberty and equality. At the same time, however, many vaccine critics, including those who resisted the new chicken pox vaccine, argued for the importance of letting childhood diseases run their course; they argued, that is, for the inherent value of these creations of nature. The belief that everything in nature has a purpose may not have been as common among vaccine skeptics as was the belief in a parent's right to choose her or his child's health care, but both beliefs have direct correlates in American environmentalism as well as American political ideology as a whole.

Even if every last germ was not of value, vaccine critics who derived their critiques from either environmentalist rhetoric or natural healing philosophies found nature intrinsically valuable in other ways. Many of the same critics who likened vaccinated human bodies to polluted lands or waters often referred to the body as "terrain" or "soil" and believed that only the most natural inputs could truly promote health. In this view, the body, like a pristine wilderness, stood its best chance of survival in an untouched state, sustained on pure water and whole, unprocessed foods—the life-sustaining fruits of nature, that is, not the products of humanity. Those who rejected vaccines because they followed natural healing practices generally ascribed to nature a wisdom and beneficence—and sometimes even a religious authority—perceived by centuries of nature-cure adherents before them.[101] To cross nature would weaken the body and indeed the species, either by eroding "natural" immunity or prompting the transmutation of existing diseases into more horrific maladies (a possibility that McBean dubbed a "Frankenstein monster of immense proportions"). Fighting nature thus smacked of blasphemy and hubris, as Moskowitz indicated when he argued that vaccination was a misguided "attempt to beat nature at her own game, to eliminate a problem that cannot be eliminated, i.e., the susceptibility to disease itself."[102]

AN EXHORTATION

Prompted in good part by the advent of the environmental movement, a prevailing philosophy of vaccine harm took shape and persisted in largely unchanged form across the final decades of the twentieth century. This philosophy was based on an ecological view of health; it regarded vaccines or their components as an artificial and therefore harmful intrusion upon the body, and it held that vaccines were responsible for chronic diseases that had reached epidemic proportions—even if science was as yet unable to prove a link between the two. In the 1970s, this philosophy was informed by generalized consumer worries about a toxic environment and the bodily effects of widespread pollution. In the 1980s, it sought to explain the spread of immune disorders, including AIDS, and growing rates of neurological and behavioral disorders in children. In the 1990s, the admission of an unknown risk of harm from the mercury compound in vaccines confirmed environmentalism-inspired charges that man-made technologies were all too often adopted before their harms were fully calculated or understood. And in the 2000s, vaccine rejection once again accompanied a wholesale rejection of technology in favor of the unfathomable wisdom of nature.

The environmentalism that emerged in the 1960s and evolved through the 2000s popularized this ecological view of health, fostered popular concern over toxic materials seen and unseen, highlighted the long-term consequences of hastily adopted technologies, and cultivated popular reverence for nature—in some cases in all of its forms, right down to the germ. Moreover, the environmental movement buttressed and popularized the idea that a lack of evidence of harm was no proof of safety, a notion that the country's experiences with radiation, lead, cigarettes, DDT, and ultimately thimerosal appeared to prove. Finally, building on these very experiences, environmentalism (like health feminism) established an exhortation to question the assurances of those in authority that new technologies were fully vetted and safe. Coulter and

Fisher illustrated this line of thinking when they quoted Rachel Carson at the very end of *DPT: A Shot in the Dark*:

Just as we have polluted our environment with man-made chemicals, we may well be polluting ourselves with a myriad of man-made vaccines in our quest to eradicate all disease and infection from the earth. In her exploration of the ways that Americans have polluted the air, water, and earth with synthetic chemicals, Rachel Carson concluded in *Silent Spring* that "The choice, after all, is ours to make. If, having endured much, we have at least asserted our 'right to know,' and if, knowing, we have concluded that we are being asked to take senseless and frightening risks, then we should no longer accept the counsel of those who tell us we must fill our world with poisonous chemicals; we should look about and see what other course is open to us."[103]

Indeed, it was this "other course" that vaccine skeptics sought, whether they were hygienists or homeopaths or merely parents worried about the long-term consequences of their children's toxic exposures. To all, environmental thinking justified the search for alternatives to immunity via vaccination. And to some, the search became more urgent as the number of vaccines required for children and federal involvement in vaccination continued to expand.

PART III

7

CLINTON'S VACCINES FOR CHILDREN PROGRAM

The kids at the White House Easter Egg Roll in April 1993 may have looked happy and healthy to most observers. But to President Bill Clinton, every last one of them was being held "hostage" by a Senate that had been withholding vaccines — not just since Clinton took office three months earlier, but for the last twelve years. During that time, vaccine prices had skyrocketed, families had been bankrupted by illnesses that could have been prevented with simple shots, and the country's immunization rate had sunk to third worst in the Western Hemisphere. That was the story Clinton told on the morning of the Easter Egg Roll, when he and Secretary of Health and Human Services Donna Shalala announced a comprehensive new program to hoist child immunization rates. "When I go out there on the lawn and I think about those kids picking up Easter Eggs, I want to be able to think about them all being immunized, and all those children coming along behind them being immunized," said the president. "It is time that we broke the gridlock and stopped making excuses for not doing anything."[1]

Clinton and Shalala promised a new national program that would lower vaccine prices, extend public clinic hours, reach out to more parents, set up a national tracking system, and be "a model of what we can do again to make this government work." The program was announced early in the new president's first term — Clinton mentioned it in his first State of the Union Speech — and the administration acted quickly to move it through Congress.[2] Along the way, Clinton and his cabinet cited pressing reasons why Congress should act fast: a recent wave of measles epidemics, a resurgence of new and old infectious diseases, the dramatic cost-effectiveness of vaccines, and a drug industry that would continue to price parents and localities out of the vaccine market unless the federal government intervened.

On the surface, the Clinton immunization program sounded remarkably similar to the Carter plan of the late 1970s. To both administrations, it was in the federal government's interest to promote immunization because vaccines were powerful cost-saving tools that could rein in health care spending and help nudge the economy out of the doldrums. The details of the Clinton program and the philosophy behind it, however, echoed Kennedy more than Carter. The Clinton program focused on preschoolers, not schoolchildren. It portrayed vaccines, and health care generally, as every American's right, not their duty. And it held government and industry—not parents—responsible for unvaccinated children. Like Carter, however, Clinton envisioned his childhood immunization initiative as a proof-of-concept. Through it, government would provide free vaccines to all children, and this prototype of a single-payer program would pave the way for broader health reform, with government taking on a bigger role than ever before in the provision and payment of health care—or so the administration hoped.

The Vaccines for Children Program ultimately launched by Clinton was more modest than that initial vision, though it did take federal involvement in childhood immunization to unprecedented heights. It made vaccines a federal entitlement for certain children, and it gave the CDC the authority to determine the scope of that entitlement.[3] But it was established with little heed to the popular concerns about vaccine risks detailed in the previous two chapters. In fact, to garner a broad base of support for a new federal immunization program, the administration cultivated American anger over an entirely different vaccine issue: high costs. To the Clinton administration, vaccines were not objects of a paternalistic medical profession, nor were they toxic time bombs being forced on the American public. They were, instead, safe and effective prophylactics being held out of reach by profit-hungry drug companies—a situation that only the federal government could effectively address. Vaccines for Children proved successful: within a decade, the program purchased 41 percent of all vaccines administered to children (Kennedy's Vaccination Assistance Act purchased an additional 11 percent).[4] It also gave states incentives to expand their list of compulsory vaccines for children—which would inadvertently help set the stage for more popular vaccination resistance as the millennium drew to a close.

ERADICATION TO EPIDEMIC

Back in the 1970s, the Carter administration's immunization initiative had such a pronounced effect on vaccination rates and disease incidence that in the fall of 1978—on the heels of the World Health Organization's announcement that it had wiped smallpox from the face of the earth—Joseph Califano announced yet another goal: the eradication of measles from the United States.[5] (Unsurprisingly, he made no reference to the country's failed attempt of a decade before.) The Carter administration's measles eradication effort combined disease surveillance and response measures with immunization promotion activities, which included enforcing school-entry laws and making the measles vaccine available to all citizens, free of charge, at local health departments. By 1980, 96 percent of all children entering school for the first time were immunized against measles. In 1981, the year Carter left office, there were just over 2,600 measles cases, an all-time low that led public health officials to wax optimistic about the prospect of eradication within a year.[6]

Measles continued to decline, to just under 1,500 cases in 1983—the lowest number of cases ever reported—but the trend was short-lived.[7] By the end of the eighties, measles had joined a list of resurgent infectious diseases that included tuberculosis, rubella, syphilis, and gonorrhea, among others. Beginning in 1989, a series of measles outbreaks struck cities across the United States, causing a total of 18,000 cases that year and over 27,000 the next. By the end of 1991, more than 50,000 children had caught the disease, more than 11,000 had been hospitalized, and over 150 had died.[8] The epidemic disproportionately struck minorities in inner cities, and in particular the very young: in 1990 roughly half the measles cases recorded across the nation occurred in children under the age of five.[9] The outbreaks disheartened health officials, who in early 1989—just prior to the start of the epidemic—had revised federal vaccination guidelines to recommend two measles shots before the age of fifteen months. The revision had stemmed from observations that throughout the eighties, persistent measles cases had occurred among infants and schoolchildren who had received just one dose of vaccine.[10]

To some in public health, the measles epidemic of 1989–91 indicated that revised recommendations alone were insufficient to stop the spread of the disease. When a federal vaccine advisory panel studied the outbreaks, their resulting "Measles White Paper" found that although measles immunization levels were close to 100 percent by the time children entered school, the epidemic was driven by a widespread failure to immunize children at the appropriate time—namely, before the age of two. A large number of factors contributed to this failure. The country lacked a nationally coordinated system to promote, monitor, and ensure immunization. Doctors failed to encourage immunization at routine appointments. There were too few public clinics and personnel to serve those in need. Clinics had inconvenient clinic hours, and health plans had rules that forced children to have physical exams, doctor referrals, or be enrolled in well-baby programs before they could receive any vaccinations. The measles epidemic was not just the result of poorly timed immunizations, but of major problems with the delivery of health care across the nation.[11] "We . . . think there is some acute deterioration of the health care system, and we fear that this"—the recent measles epidemic—"may be a warning flag for other problems to come," said the head of the National Vaccine Program Office.[12] Epidemiologist D. A. Henderson, who had led the global smallpox-eradication campaign, was more blunt: "This isn't a measles problem. It's a systems problem," he told the Washington Post.[13]

The measles epidemic granted health officials a prime opportunity to highlight fissures in the nation's health care system, and many took advantage of the outbreaks to blame the Republican administrations of the eighties for shortchanging public health.[14] Under President Ronald Reagan, funds for children's immunization had been leveled or cut, even as the prices of vaccines rose. In 1983 CDC officials complained that they'd be able to vaccinate only half as many children with federal funds as they had in 1981.[15] But by the early nineties, health officials were not alone in pointing out the cracks in the nation's health care systems, both public and private. As in the years leading up to Carter's election, the perilous state of health care was frequent headline news. In 1991 *U.S. News & World Report* called American health care "scandalous" for its failure to safeguard health and protect families from economic ruin, despite billions spent on insurance and services.[16] *Time* declared the nation's health care in "critical condition," citing prices in defiance of supply and demand as well as millions of employed uninsured.[17] The magazine declared health care the new "litmus test of American politics," and indeed the issue topped the list of American voters' major concerns during the 1992 election season, helping to sweep Democrat Bill Clinton into office late that year.

CLINTON'S COMPREHENSIVE CHILD IMMUNIZATION ACT

Like Carter, Clinton took office in the midst of a severe economic downturn and a federal budget crisis—factors that prompted Carter to comment that Clinton was "inheriting more problems than any other president in my memory."[18] Clinton, like Carter before him, identified the nation's broken health care system as a critical factor driving the country's economic woes. Since the 1970s, millions had been added to the ranks of those lacking health insurance, and health care spending had continued to climb as a fraction of family income and gross national product. By early 1993, more than 37 million Americans had no health insurance, health care spending totaled 12.3 percent of gross national product, and the inflation of health care prices was still far outpacing inflation of other goods and services.[19] On the campaign trail, Clinton promised a comprehensive overhaul that would ensure health coverage for every American and rein in runaway health care spending once and for all.

When Clinton took office in January 1993, he acknowledged that health reform was going to be a formidable task, but it was one that, according to the polls, the majority of Americans strongly supported: one survey revealed that 75 percent of Americans were willing to pay higher taxes in order to improve access to health care.[20] Although Clinton had initially promised a health care reform plan within his first hundred days in office, a week after his inauguration he announced that his health care task force, headed by First Lady Hillary Clinton, would unveil a plan in May.[21] In the meantime, he announced before a joint session of Congress in February, his administration would face its top priority: the economic recession. He announced plans for immediate measures to tackle the budget deficit and jump-start the economy by creating jobs, cutting back on government "waste" and spending, and switching "the balance in the budget from consumption to investment."[22]

Clinton's speech also emphasized the centrality of health reform to any plan for an economic turnaround. He listed the toll that health care costs would take by the new millennium, when they were projected to swallow up 20 percent of the average American's income and account for half of the growth in the nation's deficit. "All of our efforts to strengthen the economy will fail unless we also take this year, not next year . . . bold steps to reform our health care system," he told the nation. "Reducing health care costs can liberate literally hundreds of billions of dollars for new investment in growth and jobs." The nation faced an imperative for reform and recovery, and both would begin, he continued, with an immediate investment in the health of children. "Each day we delay really making a commitment to our children carries a dear cost," he said. "Half of the 2-year-olds in this country today don't receive the immunizations they need against deadly diseases. Our plan will provide them for every eligible child. And we know now that we will save $10 later for every $1 we spend by eliminating preventable childhood diseases. That's a good investment no matter how you measure it."[23] That dollar figure had been widely cited by health officials during the measles epidemic, but in Clinton's speech, it carried a deeper meaning: improving children's health would set the country on a path to greater wealth.

The seeds for what would become Clinton's Comprehensive Child Immunization Act had germinated during the 1992 campaign, but they were initially sown by the persistent measles outbreaks of the eighties. In the aftermath of the 1989–91 measles epidemic, states had informed the CDC that in order to stay abreast of immunization targets in the future, they needed not just a steady vaccine supply (during the measles epidemic, some clinics had run dry following unanticipated demand for shots), but also thousands of outreach workers and nurses to educate the public and administer vaccines. Clinton campaign strategists took note of this figure and placed immunization on a list of short-term public health stimulus measures to potentially move on in a new administration. An immunization initiative offered an opportunity for jobs creation, in their view, in addition to future cost savings. Strategists and advisers also saw such an initiative as a potentially cheap and politically popular move. Under the previous administration of George H. W. Bush, they noted, childhood immunization legislation was authorized "without dissent."[24]

If children's immunization was politically palatable, it was in part because children's health and welfare generally had climbed the country's political agenda. Children's welfare advocates noted that in the eighties, the nation had entered a new era of "renewed national consciousness concerning the problems of children," which were manifold: the number of children who were homeless, living in poverty, or living with just one parent had been steadily climbing. The infant mortality rate had stagnated, more and more children were living with HIV, and fewer and fewer children had health insurance. Children were also more likely to be poor than any other group in the country.[25] One pollster informed the new administration that a whopping 86 percent of Americans said they would pay more taxes to guarantee basic health care for children.[26] Clinton's advisers, in response, urged the president to put children's health (and education) at the top of the domestic policy agenda. Not only did children's issues "look good politically"; they were also highly consistent with the Clinton campaign's "pro-family, pro-middle class themes," they noted.[27]

In the days leading up to Clinton's February speech before Congress, Secretary of Health Shalala outlined five key components of a national child immunization program. The initiative would need to lower the cost of vaccines, educate the public, increase opportunities for people to get vaccinated, fund research in new and improved vaccines, and establish a computerized tracking system to monitor children's vaccination status nationwide. The program would focus on immunizing preschoolers — whose low immunization rates were "startling" if not "inflammatory" — and it would aim to immunize 90 percent of them within three years.

The tracking system would ensure that all children were vaccinated not just at present, but into the future as well.[28] The president initially requested funds to launch the program as part of his economic stimulus package, but when the stimulus bill failed to pass the "gridlocked" Senate, he moved ahead with a legislative proposal for the program instead.[29]

On April 1, 1993, Clinton submitted the Comprehensive Child Immunization Act of 1993 to Congress.[30] The bill addressed the objectives outlined by Shalala in February, and it proposed to meet two of them— lowering the cost of vaccines and increasing opportunities for people to get vaccinated—by making permanent the National Vaccine Injury Compensation Program created in 1986, and by creating a federal universal purchase program. Under such a program, the secretary of Health and Human Services would negotiate bulk discounts with pharmaceutical companies and purchase enough vaccines to immunize every child in the country—regardless of his or her family's ability to pay. The federally purchased vaccines would be available to public and private providers; the most a provider could charge would be a federally determined administrative fee. Administration officials anticipated that universal purchase would be a tough sell, but because the Clintons planned to make immunization a mandated benefit under health reform, it served as an opportunity to set a critical precedent. The proposal was a key litmus test of support for a bigger government role in health care purchasing; one Clinton adviser called it the "lynchpin" of the administration's campaign for health reform overall.[31]

If the initiative was a litmus test, however, the results did not bode well. Even before Clinton submitted the bill, manufacturers voiced strong opposition to the universal purchase component. The president of vaccine maker Connaught Laboratories argued that the initiative would "jeopardize" vaccine supply and "mortgage" future vaccine development.[32] Universal purchase "would just kill innovation," grumbled a representative of the Pharmaceutical Research and Manufacturers Association.[33] While some drug makers protested that universal purchase would cut into profits and hamper research and development, others insisted that the approach was completely misguided. Vaccine price increases might have seemed steep, noted Lederle Laboratories, but in fact they had lagged behind the rate of inflation.[34] Vaccine makers, they insisted, were not to blame.

Clinton had anticipated—and even seemed to relish—the protests of the pharmaceutical industry. The president had campaigned on a promise to take on special interests, and drug companies fit the bill. Administration officials were quick to counter that the prices of some vaccines had climbed nearly 3,000 percent in the last decade; the cost of the DPT vaccine, for one, had climbed from 30 cents in 1981 to $10 in 1991.[35] Such steep increases put vaccines out of reach for many families—and made their target diseases a burden for the country. In one speech, Clinton told the story of Rodney Miller, whose hospital bills for meningitis—a disease that could have been prevented with a new $20 vaccine—totaled more than $46,000.[36] "American taxpayers are getting hit with ten dollars in avoidable health care costs for every one dollar we could spend on immunization," he said.[37] But the drug companies were making it ever harder to spend that one dollar. Vaccine prices, said Clinton, were rising at six times the rate of inflation. Each year, pharmaceutical companies spent a billion more on advertising and lobbying than they spent on new drug development. And in recent years, the industry had been less and less willing to negotiate bulk discounts for the public sector. "Our message to the drug companies today is: Change your priorities. You're not going to profit at the expense of our children," the president told a crowd of reporters. "These practices must stop."[38]

The message had traction. "Clinton Knocks Drug Prices," "President Blasts Cost of Vaccines," "Profits at the Expense of Children: Clinton Calls Pharmacy Prices Shocking," ran front-page headlines across the country the next day.[39] Skyrocketing health care costs had already incurred national outrage, with poll after poll revealing that Americans believed the costs of health care were way too high.[40] Compared to other health care commodities, vaccines were remarkably cheap, but they provided the administration with a clear example of a set of health care prices that could be controlled with government intervention. The full set of childhood vaccines cost well over $200 at the pediatrician's office, but just half that at public clinics. That differential was bad enough; even worse, it was driving a growing number of patients out of private doctors' offices and into public clinics. "We need to stop pediatricians from having to send their patients into already overcrowded public clinics," said CDC director David Satcher.[41] For the administration, the unnecessarily high price of vaccines and the spillover of patients from private into public facilities were two compelling pieces of evidence that the nation's private health care infrastructure was, in the words of one aide, "unraveling."[42]

Moreover, noted Clinton, echoing a point made by Kennedy, there was a bitter irony in the fact that so many children couldn't afford vaccines in the very country that produced and developed the vast majority of the world's vaccines. In the entire Western Hemisphere, only Haiti and Bolivia had preschool immunization rates lower than that of the United States, the president and cabinet members told the press.[43] "In no other Western industrialized country but South Africa is access to so basic a child health service as vaccine directly tied to family wealth," Shalala announced before Congress.[44] Immunization, like clean water, should be treated like a basic public health protection, the secretary argued. Just as government took on the responsibility for providing clean water, clean air, and public education for all, regardless of wealth, it should take on responsibility for providing vaccines. Vaccines should be the right of all Americans—not only because they were developed with American tax dollars and ingenuity, but because they were, like a clean environment, the very foundation of public health.

A former chair of the Children's Defense Fund (CDF), Shalala was one of the administration's most vocal proponents of the idea that government had a responsibility to ensure that every child was vaccinated. CDF, a group that lobbied for children's interests in the capital, had launched an effort to promote childhood vaccination in the eighties. The group's 1987 report, *Who Is Watching Our Children's Health?*, blamed federal funding cuts for poor immunization rates and increases in vaccine-preventable diseases.[45] During the measles epidemic of 1989–91, a survey of community health centers conducted by the organization revealed a shortage of government-funded vaccines; in the years that followed, the group lobbied Congress to increase funding for federal vaccine purchasing.[46] The group's views had direct bearing on the shape of the Clinton immunization initiative, from the content of the proposed legislation to the tactics and discourse used to promote it. "Keep pressure on the drug companies," CDF's founder and president, Marian Wright Edelman, advised the Clintons.[47] Without affordable vaccines, said Shalala, the United States was failing to uphold a basic "tenet of all civilized nations."[48]

The comparison of the country's immunization policies to that of other nations—particularly less "civilized" and non-Western nations—was informed by larger trends in the management and spread of disease. During the measles epidemic, health officials had repeatedly pointed out that the United States' immunization rate was worse than that of nearly every Latin American nation.[49] "Third World Rate Seen in the New York Area," proclaimed the *New York Times*, which noted that preschoolers in Grenada, Uganda, Mexico, Algeria, and El Salvador were immunized at significantly higher rates than those in New York City.[50] Such comparisons were facilitated by the work of the World Health Organization's Expanded Program on Immunization, a global immunization effort that began in the seventies; by the early nineties, the campaign had successfully immunized 80 percent of the world's infants. The remaining 20 percent were concentrated in the world's poorest countries, largely in sub-Saharan Africa.[51] In the realm of international health, a flurry of other activities—the ten-year anniversary of smallpox eradication, the first World Summit for Children, and the founding of the Children's Vaccine Initiative, a cooperative effort of the Rockefeller Foundation, World Health Organization, World Bank, and United Nations—brought even more attention to the stark contrast between immunization rates at home and those abroad. U.S. scientists and funding had eliminated smallpox from the world's most devastatingly impoverished countries and were now making great strides against infectious disease overseas—but in the United States, measles continued to plague American children.

At the same time, the emergence and reemergence of a host of infectious diseases—paired with the discoveries that infectious agents were at the root of some cancers and ulcers, conditions previously thought to be non-communicable—led to a national preoccupation with deadly infections, including HIV, ebola, and tuberculosis. The nineties were witnessing a "resurgence of plagues and pestilences of yesteryear," wrote one reporter.[52] And although many such diseases seemed to originate abroad, they were increasingly turning up on American shores—proof that when it came to disease, local and global were no longer distinguishable. Malaria was spreading in San Diego and North Carolina, and a strange new parasite was circulating on Long Island.[53] The CDC founded a new journal devoted to the topic and tallied some two dozen new infections identified in the previous two decades.[54] The authors of best-selling books like *The Hot Zone* and *The Coming Plague* turned the scientific observations into a cultural obsession.[55] The globalization of disease even made its way into the president's inaugural address: "There is no longer a clear division between what is foreign and what is domestic. The world economy, the world environment, the world AIDS crisis, the world arms race: they affect us all," Clinton said.[56]

In the United States, the emergence of AIDS, the reemergence of tuberculosis among poor and homeless populations in New York, and the epidemic of measles, previously seen at higher rates in Latin America than in the United States, all brought the globalization of disease into stark relief. The epidemics themselves highlighted the United States' poor performance in the provision of health care to its citizens. The country's reactions to the epidemics, meanwhile, suggested a state of moral panic concerning the perceived threat posed by the disease-breeding propensities of predominantly poor, non-white nations. When Hillary Rodham Clinton solicited letters from the public on the nation's health care system in the spring of 1993, many reflected the idea that the country's failure to secure the health of its citizens was a sign that it was headed toward "uncivilized" disease and decrepitude. "The business of health care is not compatible with the public need for services that a civilized nation requires," wrote one citizen. Health care reform is crucial "if the streets of America are not to become like Calcutta," wrote another.[57] The defeat of disease had long been a pillar of Western civilizations. If the United States was to remain "civilized" in a global world (as laden with racist and classist undertones as the term was), then it needed to do more to keep infectious killers at bay. In the nineties, in short, it seemed like good policy and good politics to prevent infectious disease.

"THE KIDS ARE NOT AVAILABLE"

Clinton's Comprehensive Child Immunization Act was designed for a nation frustrated with health care costs, skeptical of greedy drug companies, and worried about the resurgence of infectious diseases. As introduced to Congress, the initiative was a $1.1 billion program that would have vaccinated every child, expanded clinic sites and hours, educated the public, invested in research, and established a computerized system to keep track of the unvaccinated and ensure that they got their shots. Half of the requested amount—about $500 million a year—would have gone toward providing free vaccines to every child. Not surprisingly, it was this aspect of the bill that attracted the most attention and controversy. Newspaper reporters had no trouble locating doctors and other health experts who opposed the plan, claiming that cost was not the number-one barrier preventing infants from getting vaccinated. "Vaccines are available. The problem is, the kids are not available," one pediatrician told the *Washington Post*.[58] In a series of editorials that ran in papers across the country, former Surgeon General C. Everett Koop, who served under Reagan and Bush, denounced the plan as an attempt to treat "the symptoms without trying to cure the disease." Koop pointed out that a number of states had universal purchase programs in effect, and that at 63 percent, their preschool immunization rates were only marginally better than the 58 percent coverage in states without such programs.[59]

Initially, many of the bill's foes were predictable along party lines and interests, foreshadowing the fate that awaited the administration's larger plans for health reform. Republican members of Congress argued that parental ignorance was a bigger problem than cost, pointing out that immunization rates were lowest among the poor, who already had access to free vaccines through Medicaid.[60] Drug companies argued that universal purchase would use taxpayer dollars to vaccinate children in wealthy, insured families.[61] But many of Clinton's allies soon found fault with the proposal, too. Friend and former classmate Kit Ashby suggested that Clinton talk to people at UNICEF: "Their experience has been that the impediments to universal immunization have not been economic but cultural and informational."[62] The president's health care team warned that more money for vaccines "will not suffice"—the measles epidemics had shown that what communities desperately needed was outreach and personnel.[63] CDC scientists, too, attributed the measles outbreaks not to a failure to vaccinate, but to a failure to vaccinate at the right age, and to possible problems with the vaccines themselves.[64] The administration's decision to highlight high drug prices alongside the broken health care system was a calculated political move, designed to capitalize on popular resentment of industry and frustration with health care costs. But it was out of sync with what many health experts perceived to be the true causes of under-immunization—and it was out of sync with how far most Washington politicians were willing to expand government to guarantee health care for children.

Though many on his health care team advised him to focus on funds for immunization personnel, Clinton kept universal purchase front and center. His determination on the issue mirrored his determination on overall health care reform. Clinton preferred "activist" policies over "wonky incrementalism." When advisers provided him with politically expedient proposals instead of the innovative and unprecedented proposals he wanted, he took them off the job.[65] But as Clinton attempted to push the immunization bill through Congress, a stark reminder of American discomfort with big, activist government came from the South. That spring Carter was promoting immunization in his own way, with star power and volunteers in his home state of Georgia. The former president's Atlanta Project, a nonprofit effort dedicated to improving quality of life for urban Atlantans, recruited 7,000 volunteers to knock on 200,000 doors to encourage parents to bring their children in for free shots at immunization drives held across the city. The project was funded by corporate donors, and it administered vaccines already on hand at state and local health departments. It also promised volunteers and vaccine recipients an audience with pop star Michael Jackson at the end of the campaign.[66]

Facing resistance on the immunization bill from congressional Republicans, House and Senate leaders began to work out a compromise with the administration. To preserve the other components of the bill, the administration agreed to replace the universal purchase plan with an entitlement program that would provide free vaccines for all Native American and Native Alaskan children, as well as children without insurance, those covered by Medicaid, and those whose insurance didn't cover vaccines. The Comprehensive Child Immunization Act was ultimately passed as part of the 1993 budget bill. Under the title Vaccines for Children, the modified universal purchase component of the program authorized the federal government to buy enough vaccine to provide immunizations at no cost to the qualifying children. The secretary of Health and Human Services would, as envisioned, negotiate directly with manufacturers, and states that received vaccines through the program would have to make them available to all providers, public and private. Providers were not required to make sure children were eligible for the vaccines, and while they could charge a small fee for administering the vaccines, they couldn't turn away children whose parents were unable to pay.[67] The victory was all in the compromise. In May, as the Clinton administration announced that it would relinquish the universal purchase provision, the Atlanta Project announced that it had immunized 17,000 children, surpassing its goal of 10,000. "We have to say, too, that we hope President Bill Clinton was watching what happened here," wrote the editorial board of the *Atlanta Journal-Constitution*. "If he was, and if he understands it, he may not feel so bad about having to drop his proposal to spend more than a billion dollars in non-existent federal money for mass free immunization."[68]

As Clinton backed down on universal purchase, even the media seemed to change its mind about how much—and what type—of government intervention was need to boost immunization rates. When the Clinton proposal was first announced, the editorial board of the *Washington Post* had supported the idea of universal purchase.[69] In May the board changed its position, positing that federal funds would be better spent on education and outreach.[70] The *New York Times* appeared to agree, running a feature in mid-May that highlighted the stories of poor families that failed to vaccinate their children not because of a lack of free vaccine, but because of the burdens of living in poverty.[71] Clinton's own message would gradually shift, too, in particular as his administration crafted its approach to its next domestic agenda item, welfare reform. Clinton's welfare reform plan—which Congress did enact in

1996—espoused a strong emphasis on personal responsibility. "Governments don't raise children, parents do," was the message the president put forth, in the aftermath of the administration's failure to pass health care reform.[72] Despite the administration's rallying cry against drug companies and the broken health care system, the notion that parents were to blame for their children's well-being—including their immunization status—had displayed remarkable endurance since the Carter era.

During the 1989–91 measles epidemic, Democrats and health officials had loudly voiced frustration over the Bush administration's unwillingness to enforce federal leadership to stem the spread of infection. At the height of the epidemic, President Bush (who did increase federal vaccine funds in response to the outbreaks) told parents that vaccines were there for the taking, and that it was ultimately up to them to make sure their children were properly immunized.[73] Democrats, public health officials, and some lay observers responded by demanding government intervention to control vaccine prices and improve access to vaccines. D. A. Henderson attributed the epidemic to "dangerous complacency" on the part of the federal government.[74] Senator Bumpers, still in Congress, called for more federal dollars to expand clinic capacity and reach out to parents.[75] Parents themselves called the price of measles shots an "outrage" and a "disgrace."[76] Once in office, Clinton responded to these calls by pointing a finger at drug companies and offering a big-government, top-down solution to the problem of low immunization rates.

But as Clinton pushed his immunization initiative through Congress, opinions on the root of the problem, and in fact the very definition of the problem, diverged widely. In the end, as a result, the cornerstone of his proposal, universal purchase, was whittled away. Even so, no sooner was the Vaccines for Children Program passed than it ran headlong into problems and partisan politics. Longtime immunization supporter Senator Bumpers fervidly criticized the program's vaccine storage and distribution plans. A federal report requested by congressional opponents of the administration concluded that the entire program needed to change its focus. Conservative think tanks charged Clinton with "manufacturing" a vaccination crisis and creating a program that was destined for permanent disarray.[77] Republicans in the Senate threatened to repeal the program, and the president's advisers worried that manufacturers would refuse to bid on the federal vaccine contract, forcing the secretary of Health and Human Services to either leave states in the lurch or buy vaccines from other countries.[78]

Despite the threats and challenges, Vaccines for Children remained intact—and even came to be held up as a shining example of the administration's accomplishments. During the president's 1996 reelection campaign, Clinton staffers compiled a long list of health care measures passed during Clinton's first term, and the Vaccines for Children Program was high on the list. By then, 75 percent of two-year-olds were fully immunized, and more than 90 percent of the toddlers had at least one dose of all of the recommended vaccines—a "historic high" that surpassed the president's 1993 goals and that was touted, along with equivalent vaccination rates across children of all racial and ethnic groups, as one of the administration's major accomplishments. The program was an example of how the administration had guaranteed "equal citizenship and opportunity for all" without burdening states and localities.[79] It was also proof that, as the First Lady wrote, "it takes a village" to keep every last child healthy and safe.[80]

Ten years after the program was launched, federal immunization officials themselves would credit Vaccines for Children with helping the United States achieve its highest rates of immunization ever and with reducing disparities in vaccination rates. By then, the program was the largest source of federal funding for vaccines and purchased as much of the nation's childhood vaccine supply as the private sector did. It had also changed the "pattern" of childhood immunization, moving vaccination out of public clinics and into the private sector.[81] And despite the creation of a vaccine entitlement, vaccine research and development had not seemed to suffer. The years after the program went into effect saw the introduction of a series of new vaccines, including immunizations against Hib, hepatitis A, rotavirus, and chicken pox. As each new vaccine came on the market, the federal Advisory Committee on Immunization Practices voted on whether or not it should be included in Vaccines for Children. As the next chapter shows, this aspect of the program would have implications for how some vaccines would come to be used in the future, including the vaccine against hepatitis B.

8

SEX, DRUGS, AND HEPATITIS B

In 1999 young children's vaccination rates reached "record high levels," according to the CDC. Ninety-six percent were protected against diphtheria, pertussis, and tetanus; 93 percent were vaccinated against Hib; 91 percent had shots against measles, mumps, and rubella; 90 percent against polio; and 88 percent against chicken pox and hepatitis B.[1] The figures represented a substantial increase in immunization rates since 1992, the year before Clinton took office—and the year that saw the lowest vaccination rates since the Carter administration. That year, close to 20 percent of children hadn't been fully protected against diphtheria, pertussis, tetanus, or measles, and close to 30 percent hadn't been vaccinated against polio.[2] Figures didn't yet exist for three immunizations still to be added to the childhood schedule: the vaccines against Hib, chicken pox, and hepatitis B.

In fact, in 1999 the vaccine against hepatitis B was a recent addition to the schedule of child immunizations, even though it had been around for close to two decades. When the first hepatitis B vaccine was approved, back in 1981, it was recommended for a narrow segment of the population: drug users, gay men, some health care workers, and select immigrants. By the year 2000, however, forty-seven states had mandated hepatitis B vaccine for all schoolchildren, and federal guidelines recommended immunizing children not before school or preschool, but at birth. In the two intervening decades, hepatitis B infection, like mumps before it, underwent a gradual but dramatic transformation. As in the case of mumps, it was the vaccines against hepatitis B—there were two—that prodded this transformation along.

Hepatitis B, a blood-borne viral infection that attacks the liver, was a foreign obscurity of little direct relevance to most Americans when its first vaccine was unveiled in 1981. In the era of hepatitis B vaccination, however, the infection morphed into an AIDS-like scourge, and then again into a ubiquitous cancer-causing germ. Over the same period of time, the hepatitis B vaccine reflected a rotating set of national hopes and anxieties. It represented the promise of a new era of genetically engineered pharmaceuticals, a bulwark against the lurking threat of infections imported by immigrants, and an impediment to the carelessly assumed hazards of youth sex and fashion trends. Throughout the 1980s and 1990s, these shifting cultural perceptions of the virus and its vaccine continually influenced the evolution of policies regarding who should be vaccinated against hepatitis B, and at what age.

Hepatitis B vaccination policies evolved in the context of reemerging epidemics of vaccine-preventable infections, a national debate over health care, and, of course, Clinton's push to expand federal support for childhood vaccination. They also evolved against a backdrop of accumulating vaccine scares—which prompted *Time* magazine to diagnose the country with a national case of vaccine "jitters" at the end of the twentieth century.[3] For some critics, the hepatitis B vaccine came to signify all that was wrong with vaccines and related policies—precisely because it was originally recommended for select adults and ultimately administered to every newborn. The story of hepatitis B vaccine in the 1980s and 1990s illustrates just how and why the nation increasingly relied on the immunization of its youngest citizens to produce a healthy adult populace, even as vocal resistance to child vaccination continued to mount.

A DISEASE OF "HOMOSEXUALS AND DRUG ADDICTS"

While hepatitis had long been a concern of health officials, the disease wasn't a terribly familiar one to most Americans in the decades prior to the first hepatitis B vaccine's 1981 approval by the Food and Drug Administration. Two distinct forms of hepatitis, A and B—formerly "infectious" and "serum" hepatitis, respectively—had been known since the 1950s. But countless cases of serum hepatitis had gone undetected, in large part because of the vague symptoms it caused. Infection with hepatitis B virus may or may not cause debilitating fatigue, nausea, and loss of appetite. These acute symptoms may or may not be fatal, and those who do recover may or may not become chronic carriers. Carriers, in turn, may or may not become victims, decades later, of hepatitis-induced cirrhosis or liver cancer.[4] Hepatitis B's fairly mundane acute symptoms—its attendant fatigue, nausea, and loss of appetite—meant that for decades it was conflated with other conditions. In the 1970s, however, armed with a new set of diagnostic tools, researchers enumerated for the first time more than 200,000 new cases in the United States each year and more than 200 million carriers worldwide. Concurrent epidemiological studies identified those at highest risk of the disease. In addition to health care workers, the list included hemophiliacs, prisoners, gay men, injection drug users, sex workers, Native Alaskans, and immigrants from sub-Saharan Africa and Southeast Asia.[5]

Those new diagnostic tools emerged from the lab of research physician Baruch Blumberg, who in the 1960s identified a protein in the blood of Australian aborigines that he dubbed the Australia antigen. Australia antigen floated freely in the blood of people infected with serum hepatitis, and in 1967 the researchers determined that Australia antigen was in fact a surface protein on the virus that caused the disease. In related work, Blumberg and his colleagues found that monkeys injected with highly purified Australia antigen did not come down with serum hepatitis; the discovery suggested that non-infectious but still protective material could be separated from the virus itself. The basis for a novel vaccine had been found.[6]

As development of a vaccine against the infection accelerated in the late 1970s, Blumberg and other infectious disease experts predicted that an effective immunization would save hundreds of thousands, if not millions, of lives.[7] But news of the potential vaccine, and of hepatitis B itself, rarely reached lay audiences over the course of the 1970s. The disease had isolated moments in the spotlight: In 1974 the host of the television show *Today's Health* came down with hepatitis B when a surgical patient's blood splashed in his eye; he chronicled in detail the disease's "mean, sneaky malevolence" on TV and in print.[8] Two years later, hepatitis B made headlines again when Blumberg shared the Nobel Prize in Medicine for his work on the disease.[9] The press also reported on the 1979 outbreak of hepatitis B among youths who shared needles to take the recreational drug methylene deoxyamphetamine.[10] Such stories confirmed for the public the picture then emerging from epidemiological studies, which was that the disease posed a risk only to specific subsets of the population, namely, surgeons and drug users.

On the eve of the hepatitis B vaccine's 1981 introduction, most Americans thus had little reason to worry about the virus. Early press reports on the shot affirmed this notion. In 1980 the *CBS Evening News* reported that hepatitis B struck developing countries in Asia and Africa in "epidemic proportions," whereas in the United States it affected mainly "patients on dialysis, medical personnel, and people living in institutions."[11] When news anchor Dan Rather reported on the new vaccine, he announced that it had been approved for "a disease affecting health workers, male homosexuals, and drug addicts."[12] On NBC the evening news anchor told Americans that "hospital workers get it, so do drug addicts, mental patients, homosexuals, and millions of people in Africa and Asia." The network's subsequent segment on the vaccine focused largely on the "medical adventure story" of the virus's discovery, which a reporter recounted over grainy, choppy footage of Aborigines in native attire.[13]

Historian William Muraskin has argued that popular representations of hepatitis B in this period were deliberately constructed by a medical profession acting in self-interest. Having been identified as a high-risk group themselves, health care workers endeavored to define hepatitis B infection as an issue "private" to their profession and outside the public's purview. The media, reliant on the medical community for information about the virus, reported what they were told: that gays, injection drug users, and certain immigrants and refugees were at high risk, and that the spread of hepatitis in hospital settings was controlled through the use of disposable gowns, masks, gloves, and other hygienic practices.[14] The health care profession's internal policy—of voluntary testing for carrier status—became the implicit policy toward the population at large, too. The upshot of this policy was twofold: it prevented "hysteria and discrimination of carriers," but it also hampered public awareness of the extent of the epidemic and the true risk of infection.[15]

Muraskin's assessment downplayed the role of bench scientists, epidemiologists, reporters, and other Americans in constructing hepatitis B's popular image (however faint) in this period. Representations of the virus as one that posed little threat to "average Americans" were common in the late 1970s and the very beginning of the 1980s, but not merely because doctors willed it this way. This portrayal of hepatitis B made sense given the nation's health priorities at the time. Cancer and heart disease were by far the country's top killers; heart attacks alone caused 300,000 deaths a year. Hepatitis, meanwhile, appeared way down the list; in 1981 many more people died of homicide and ulcers than died of hepatitis of any type.[16] A disease doesn't have to cause high mortality, of course, to capture national attention, but hepatitis also failed to align with the nation's other health preoccupations: soaring hospital costs, environmental scares like the meltdown at Three Mile Island, and a relentless "flurry of strange new ailments," including Legionnaires' disease, Lyme disease, Reye's syndrome, and toxic shock syndrome.[17] When a new ailment with characteristics similar to hepatitis appeared in 1981, however, hepatitis's public image underwent a radical reconstruction. In light of AIDS, hepatitis control would take on a new sense of urgency.

TWO NOVEL VACCINES

The hepatitis B vaccine that was approved by the FDA in 1981 was an unusual product in the history of viral vaccine development. The vaccine didn't contain live, weakened virus (like the Sabin polio vaccine) or killed, denatured virus (like the Salk polio vaccine). Instead, Heptavax B, developed by Merck, contained painstakingly purified versions of the antibodies that Blumberg had first discovered, harvested from the blood of people infected with hepatitis B. This novel procedure earned Heptavax B the title of the world's first "subunit" vaccine against a virus— meaning that it stimulated an immune response by using just a part, or subunit, of the virus, and not the virus in its entirety.[18]

In a display of awe and enthusiasm for scientific discovery, the same news reporters who had downplayed the disease's risk for average Americans played up, in the next breath, the new vaccine's development and its novel form. The vaccine, after all, was fairly big news: as news anchor Dan Rather pointed out, it was the "first completely new viral vaccine in ten years"; it was also the "first vaccine ever licensed in the United States that is made directly from human blood."[19] *Newsweek* called its blood-derived antibodies "ingenious," and magazines from *Time* to *Glamour* touted the vaccine as a "medical breakthrough."[20] Fervent reports in popular and scientific journals proclaimed that hepatitis B would soon join such well-known pathogenic villains as smallpox and polio as a problem of the past.[21]

This enthusiastic rhetoric was soon dampened by yet another medical discovery. The clinical trials that had tested the hepatitis B vaccine's efficacy in the late 1970s had included only gay men, who had been identified as being at high risk of the infection.[22] When, in 1982, the federal Advisory Committee on Immunization Practices (ACIP) issued its customary recommendations on who should receive the new vaccine, the list included those considered to be at highest risk for the disease, including health care workers with frequent blood contact; prisoners; patients and staff of institutions for the "mentally retarded"; hemodialysis patients; injection drug users; immigrants from eastern Asian and sub-Saharan Africa; and sexually active gay men.[23] In the flurry of commentaries that followed in the medical literature, several reports highlighted the unique susceptibility of gay men to the infection. An editorial in the *Journal of the American Medical Association* identified the same list of groups to target with the vaccine, but added that "the highest HBsAG [hepatitis B antigen] prevalence in the United States is among male homosexuals. . . . Frequency of intercourse, the number of sexual partners, and the prevalence of anal intercourse all contribute to this."[24]

In fact, gay media outlets had initially reported with pride that hepatitis B–infected gay men were frequent donors of blood for the vaccine. But the plasma-derived vaccine's approval in late 1981 was quickly followed by emergent reports of a mysterious new illness causing "immune system breakdown" in what some experts estimated as tens of thousands of gay men.[25] Within a year, health officials had documented a high rate of hepatitis B infection not just among gay men, but among gay men who were victims of the new illness, which came to be known as AIDS. The announcement spurred fears that the new vaccine was contaminated with the pathogen causing AIDS, increasingly presumed to be a virus.[26] In 1982 and 1983, the press reported that gay men and injection drug users were frequent blood donors for the vaccine, and that many health care workers were refusing the vaccine themselves for this very reason.[27] (Noted one physician, reportedly, to her daughter in confidence: "I know where the vaccine comes from. It comes from the blood of junkies and alcoholics. And who knows what they've got."[28])

The CDC moved quickly to address such fears, announcing in 1983 that of 200,000 individuals vaccinated against hepatitis B since 1982, none had come down with AIDS.[29] There were, however, cases of AIDS in gay men who had participated in the vaccine trials, and the theory that the hepatitis vaccine (among other vaccines) had played a part in AIDS's appearance and spread would gain momentum among certain segments of the population as the 1980s progressed.[30]

In the meantime, however, a new link between hepatitis B and AIDS emerged in mainstream media reports. This new link, a recitation of the similarities between the two viral infections, would persist in popular and scientific discourse for well over a decade. The analogy between AIDS and hepatitis B infection had been drawn early on by epidemiologists working to discover the causative agent of AIDS. Both diseases, scientists noted, appeared to be transmitted sexually and showed a pattern of infection among injection drug users and blood transfusion recipients. When the Centers for Disease Control (CDC) identified in 1983 the groups at "high risk" of AIDS—gay men with multiple sex partners, injection drug users, Haitian immigrants, and hemophiliacs—these closely mirrored the list of those earlier reported to be at high risk of hepatitis B.[31] These parallels were echoed repeatedly by the media. As a 1985 cover story on AIDS in *Time* magazine pointed out, both diseases were scourges of "drug addicts, blood recipients and gay men," and scientists were still uncertain as to whether hepatitis B was a "co-agent of AIDS or merely [a] tagalong infection."[32]

As AIDS gripped the nation's attention, interest in hepatitis also picked up. Media coverage pointed out not only how hepatitis B virus was similar to the virus that caused AIDS, but also how it was worse: fifteen times more prevalent in the population, two hundred times more infectious, far more stable in the environment, responsible for far more deaths, and, unlike the AIDS virus, spread by casual contact.[33] But the public could take solace, said a top infectious disease expert, in the fact that a vaccine existed to keep this "cousin of AIDS" at bay.[34] This message, that the hepatitis B vaccine was a beacon of hope in a time of fear, was oft repeated in the press. There's no cure for AIDS, *Gay Community News* told readers, but there is one for hepatitis B, which kills five times as many people each year.[35] *Mademoiselle* issued the same message: "There is no AIDS vaccine yet, but there are two new ones against hepatitis B."[36]

That second hepatitis B vaccine, widely available by the late 1980s, was a vastly different product from the first, blood-derived vaccine. Recombivax HB, Merck's genetically engineered hepatitis B vaccine approved in 1986, contained viral proteins not harvested from infected patients in the clinical setting, but manufactured by genetically engineered yeast in the lab. The vaccine was hotly anticipated by medical and public health professionals for its potential to address the high cost and "theoretical disadvantages" of plasma-based vaccines.[37] And they weren't the only ones excited about a vaccine made with recombinant DNA.

As news of Recombivax HB's impending approval began to leak, press reports hailed its potential to prove that genetic engineering would revolutionize the pharmaceutical industry.[38] Scientists and the reporters who quoted them called biotech vaccines generally "exciting and imaginative," and referred to the hepatitis B vaccine specifically as a "pioneering product."[39] Researchers told the *New York Times* that biotech shots were "cutting edge weapons" that would globally eliminate not only hepatitis B, but also AIDS and malaria.[40] The business press breathlessly reported on the race among "tiny" California biotech firms to produce the world's first genetically engineered vaccine, and when biotech firm Chiron Corporation's Recombivax was approved, *Venture* magazine crowned it one of the best entrepreneurial ideas of 1986.[41] The approval of Recombivax HB—the first genetically engineered vaccine and the third genetically engineered pharmaceutical to make it to market—was heralded on the front pages of the *New York Times*, the *Los Angeles Times*, the *Wall Street Journal*, and elsewhere for ushering in what FDA commissioner Frank E. Young called a "new era in vaccine production."[42]

By the mid-1980s, that era had been long awaited. Scientists and drug company representatives alike emphasized that the new vaccines would be cheaper and would finally allow for the marketing of hepatitis B vaccine in developing countries, where it was much more sorely needed than in the United States.[43] They were enthusiastic for another reason, as well. As described in earlier chapters, Sabin's oral polio vaccine caused cases of vaccine-associated polio; swine flu vaccine caused neurological disease; pertussis vaccine caused brain damage and sometimes death. Such vaccine-safety scares had been piling up and getting lots of attention. And as a result of lawsuits against oral polio vaccine makers, companies were dropping out of the market, leaving the country reliant on just a few manufacturers for its entire vaccine supply. The genetically engineered hepatitis B vaccine was heralded not only because it held the promise of a new generation of vaccines, but also for its potential to put safety concerns to rest. The *Wall Street Journal* announced that Chiron had brought an end to the days when vaccine development was "an inexact scientific art."[44] Because the new hepatitis vaccine did not contain a whole virus, it "just can't do any damage, period," said a microbiologist at the FDA.[45] Researchers promised that genetically engineered vaccines eliminated the "risk of actually getting herpes, hepatitis B or influenza from the injection, since the viruses themselves are not present in the formula."[46] FDA commissioner Young echoed these sentiments in a press statement he made upon Recombivax's approval: "These techniques should be . . . extended to any virus or parasite." He went on to state that while the plasma-derived vaccine had never posed a risk of AIDS, the new "lab-made vaccine" should further reassure people. He also strongly urged those at high risk of hepatitis B to take advantage of this "new life-saving protection."[47]

A PUSH FOR WIDESPREAD VACCINATION

Young's plea came as public health officials were bemoaning stubbornly low hepatitis B vaccination rates. In the five years since the ACIP had recommended that gay men, injection drug users, health care workers, and select immigrants be vaccinated against the infection, hepatitis B prevalence had not only not decreased—it had increased, with rates particularly high among young adults.[48] Incrementally, federal recommendations evolved in response. Instead of targeting all risk groups, however, the ACIP's new guidelines targeted only those guaranteed to have an encounter with the health care system—namely, pregnant women and their newborns.[49] In 1984 the ACIP had recommended that all "high-risk" pregnant women be screened during prenatal care visits for hepatitis B and, if found positive, that their infants be immunized at birth to prevent them from harboring the virus.[50] But the plan had little impact on total hepatitis B prevalence, because high-risk women were difficult to identify and insurers weren't always willing to cover the cost of screening them. As a result, the ACIP later noted, the United States continued to add another 3,500 chronic hepatitis B carriers—the unimmunized infants of infected mothers—to the population each year.[51]

To get around the difficulty of identifying high-risk women, in 1988 the ACIP recommended that *all* pregnant women be tested for hepatitis B, and, if positive, their infants vaccinated within twelve hours of birth to prevent transmission to the next generation.[52] But disease incidence continued to persist at high rates. And at the same time, the demographics of the infected seemed to be changing: the proportion of homosexuals had decreased significantly since 1982, while the proportion of drug users and heterosexuals with no discernible risk factor had increased. When it released these figures, scientists in the CDC's hepatitis division suggested that the only way to combat the disease would be to immunize all infants, all adolescents, or both.[53] The ACIP agreed, and in 1991 it altered its guidelines yet again, this time recommending that all infants be vaccinated against the disease at birth.[54]

Health officials acknowledged that the new strategy was necessary "because vaccinating persons engaged in high-risk behaviors, life-styles or occupations . . . has not been feasible," but also because many infected people had "no identifiable source for their infections."[55] Most people became infected as either adolescents or adults, the CDC reported; but as with mumps two decades before, vaccinating the youngest citizens offered the most expedient means of ensuring healthy adult citizens.

"We do not feel that targeting adults for vaccination has worked," a CDC official told the *Boston Globe*. "This will be the first time," she went on, "that a vaccine is recommended for children to prevent a disease that primarily occurs in adults."[56] The vaccination of well children to maintain a population of well adults was, of course, by this time a tried-and-tested approach to public health. In this case, however, children were being vaccinated long before they could engage in the types of social activities, like school or play, generally implicated in the spread of contagious disease.

By the early 1990s, the message that just about every American was at risk of hepatitis B came to dominate media reports on the disease. Outlets from the *Philadelphia Tribune* to *Good Housekeeping* reported that a third of people who came down with the disease were not in any of the known risk groups.[57] *Redbook* warned readers that hepatitis was "spreading fast," and the *Boston Globe* noted that the infection was communicated by sharing gum, food, toothbrushes, and razors, and by body piercing.[58] *New York* magazine—in a feature titled "The Other Plague"—recounted the stories of a young woman who contracted a fatal case by getting her ears pierced, a young man who was infected when mugged at knifepoint, and a woman infected at a nail salon.[59] Frequent mentions of the prevalence of asymptomatic carriers heightened the sense of an immediate health threat: in the words of one reporter, anyone could be one of the United States' 1.5 million "Typhoid Marys," unwittingly transmitting hepatitis B to people unaware of the risk.[60]

Health officials at the CDC were meanwhile considering not just revised recommendations to increase hepatitis B vaccination, but a broader program to encourage higher vaccination rates overall. Measles, pertussis, and rubella were all on the rise at the end of the 1980s, and preschoolers' immunization rates, as noted in the previous chapter, were shamefully low; health officials ascribed both situations to problems with the nation's health infrastructure. But while health officials blamed the nation's health care system, the Republican-controlled White House disagreed. "The facilities are there . . . the vaccines are there," said President Bush during the height of the 1989–91 measles epidemic; "make sure your child is immunized," he instructed the nation's parents.[61] As an incentive for the parents deemed most responsible for the rash of epidemics, his administration proposed tying welfare payments to children's immunization status.[62]

A partisan dispute erupted in response to the welfare proposal. Administration officials maintained that individual citizens needed to assume more responsibility for their personal health, while left-leaning members of the public health profession accused the White House of "punishing the poor" and spending more on six hours of the Gulf War than it would take to curb measles.[63] In the end, the administration's welfare proposal was rejected. But while the political battle over measles control raged, some vaccine critics began questioning the measles vaccine itself, as well as the evolving guidelines on who should get it and when. The ACIP had not only added a second vaccine dose to the one previously advised for children; it had also added preschoolers, college students, health care personnel, and international travelers to the list of those who should get the shot. The changes were necessary, the committee wrote, to address the *two* causes of nationwide measles outbreaks: unvaccinated preschoolers and vaccine failure. Among the roughly 17,000 measles cases that had occurred between 1985 and 1988, 42 percent were in vaccinated people; in some school districts, measles outbreaks occurred even though 98 percent of the children were immunized. As noted in chapter 7, scientists had a few explanations for why this might be: vaccine-induced immunity might be fading with time, and some children might be getting vaccinated at too early an age, when maternally inherited measles antibodies could interfere with vaccine response.[64]

But the committee's solution—more shots—struck some as confusing if not downright illogical. "Does it make sense to offer booster shots of any sort if a single shot of the vaccine has not been shown to do the job?" asked one mother, to whom the measles vaccine suddenly seemed "too experimental, too ineffective, and too risky."[65] The risks of the measles vaccine, usually given as the combined MMR vaccine, seemed to be proven by the outbreaks themselves. In its revised guidelines, the ACIP described measles as a "severe" disease that caused encephalitis in 1 of every 1,000 cases and death in 1 of every 1,000 cases.[66] This rate of complications was dramatically higher than that measured when measles vaccination began in the 1960s, noted some attentive critics, who wondered whether the fallible vaccine was actually responsible for having created this more severe disease. Joanne Hatem, a physician who had herself suffered an adverse reaction to rubella vaccination, noted that while measles had become a more serious disease, the vaccine had its own flaws, including a far-from-perfect rate of protection and its own risk of encephalitis and death. She advised a tempered approach to immunization: the MMR shot at fifteen months, and then only boosters against individual infections as needed.[67]

As some parents and physicians received the "more shots" mantra with caution, proponents of a more robust immunization infrastructure found an effective ally in newly elected President Clinton.[68] In addition to launching Vaccines for Children that spring, Clinton signed a proclamation supporting National Preschool Immunization Week, an annual week of coordinated efforts to fully vaccinate preschoolers with all federally recommended vaccines, including the vaccine against hepatitis B.[69] In the context of a national dialogue about the broken health care system, which Clinton kept front and center during his first season in office, the cost-efficiency of vaccination generally, and hepatitis B vaccination in particular, took on new salience. Vaccinating young children against hepatitis B saved more money than efforts to vaccinate any other group could ever save, health economists calculated, simply because it prevented the greatest number of chronic infections.[70]

The enactment of Vaccines for Children coincided with yet another broadened set of hepatitis B vaccine recommendations by the ACIP. The committee now advised that all unvaccinated eleven- and twelve-year-olds be protected against the virus, as well as all children under age eleven who were either Pacific Islanders or who lived in households with immigrants from countries with high rates of hepatitis B. Health officials were blunt in justifying the widespread vaccination of adolescents. While universal infant vaccination would ultimately obviate the vaccination of adolescents and adults, in the meantime, vaccinating preteens would drive down disease incidence more quickly. Targeting immigrant children was necessary, they argued, because such children continued to experience "high rates" of hepatitis B infection: 2 percent became infected each year, and 2 to 5 percent became chronic carriers of the virus.[71] The numbers may not have seemed objectively high, but they did seem excessive to those like Barbara Hahn, a deaf interpreter who, in her inability to trace her own hepatitis B infection, pinned it on immigrant children. "Recently, the immigration policies have brought an increasing number of foreign students into our school systems, and the incidents [sic] of hepatitis are much higher in other countries. Is that how I got this disease?" she wondered. "Did I get it from a child who ran into me on the playground or from the little girl who was upset and bit me while I was working at the Cincinnati public schools?"[72]

This refocused attention on the infectious status of immigrants came at a time when concerns about immigrants, the resurgence of infectious diseases, and the costs of health care were both prominent and intertwined. During his first year in office, as he attempted to overhaul health care generally and access to vaccines in particular, President Clinton entered into a battle with Congress over his campaign-trail promise to overturn a 1987 ban on the immigration of people infected with HIV.[73] Heated opposition to Clinton's plan reflected fears about an impending wave of immigrants from Haiti, which had a large number of people infected with HIV, as well as the nation's resurgence of tuberculosis, which was frequently attributed to "immigrants and travelers."[74] Arguments against the importation of additional infections frequently cited the burdensome costs of providing health care to the chronically ill. Growing resistance to the prospect of adding immigrants to these ranks was embodied by California's passage of Proposition 187, which proposed severely limiting illegal immigrants' access to public services.[75]

Immigration anxieties framed the context in which federal hepatitis B vaccination recommendations took shape, even if they weren't directly cited by the state-level hepatitis B vaccination laws that soon followed. As state health boards and legislatures began taking steps to mandate the hepatitis B vaccine for infants, kindergartners, and seventh-graders, many instead attributed these steps directly to the Vaccines for Children Program. Minnesota's vaccine task force credited the Clinton program for the extra funds and discount pricing that made it feasible to require the hepatitis B vaccine for seventh-graders.[76] State health officials in Colorado, Louisiana, Pennsylvania, and elsewhere also credited the administration for making it possible to require the new vaccine for students, hold school-based drives to encourage vaccination, and enforce new mandates—since students were now guaranteed the vaccine, regardless of their ability to pay.[77]

In addition to new federal funds and guidelines, state-level legislators and health officials had other reasons for requiring hepatitis B vaccination of youths beginning in the early nineties. When states such as Colorado, Idaho, California, and Pennsylvania mandated the vaccine for preteens, health officials and lawmakers cited as justification the growing popularity of tattoos and body piercing.[78] Adult attitudes toward teenage body piercing, a trend that exploded in the nineties, often reflected what historian Paula Fass has referred to as the socially constructed perception of "the rocking, highly sexualized teenager."[79] In countless articles and talk shows devoted to the topic of body art in the late nineties, parents and doctors expressed bewilderment and concern over the trend and its hazards: according to the reports, children as young as eleven and twelve, often influenced by sex-symbol celebrities, were getting pierced and tattooed in record numbers. They were also facing skin rashes, swelling, scar tissue, tetanus, HIV, and hepatitis B as a result. "This fad communicates status, fashion-hipness—and unfortunately, disease," noted *Prevention* magazine.[80] It was also explicitly linked to sex: pop stars like Janet Jackson popularized the idea that because piercing created a "great sensation," it could be "very sexual."[81] Body art—which was also linked in many reports to an increased risk of smoking, alcohol, and drug use—came to epitomize oversexualized youths at risk of a complex set of dangers and diseases. Fortunately, some commentators noted, at least one of these risks could be prevented with a vaccine.

By this time, the rhetoric that had tightly linked hepatitis B to AIDS less than a decade earlier was beginning to diminish, although an association between the two infections persisted in educational materials urging teens to get vaccinated. A 1994 educational campaign by the National Foundation for Infectious Diseases featured "sexpert" Dr. Ruth, who continued to inform audiences that hepatitis B was "100 times more infectious than HIV."[82] But as the 1990s progressed, characterizations of hepatitis B as a sexually transmitted disease increasingly gave way to characterizations of the virus as a preventable infection linked to cancer. "This is a very safe and effective way to avoid what is a terrible disease that causes cancer and other chronic problems," said the head of Colorado's health board regarding shots against hepatitis B.[83] Indeed, this particular portrayal of the infection became increasingly prominent as parental resistance to the vaccine began to emerge—as it did when Colorado attempted to mandate the vaccine for school.

REJECTING HEPATITIS B VACCINE

The majority of state laws and regulations mandating hepatitis B vaccination for children went into effect between 1993 and 1998. Their adoption was largely streamlined by federal enthusiasm for universal vaccination and funding support for recommended vaccines, in addition to cultural preoccupations with the lifestyles of body-pierced youths and disease-harboring foreigners. While many of the laws and regulations were uneventfully adopted, a few minor debates did erupt. When Colorado's health board proposed requiring the shot for kindergartners and seventh-graders in 1996, doctors and health officials were split on the issue. As with mumps nearly three decades before, health and medical experts in the state were neither united nor entirely clear on the urgency of vaccinating children against hepatitis B. Some noted that the American Academy of Pediatrics had advised the immunization of all older children only "where resources permit." Some pointed out that the disease was unlikely to spread among elementary schoolchildren, "unless you have an infected child who's a biter or who has a blood spill."[84] Others trotted out arguments about the risk-taking behaviors of teenagers: "We have a lot of children in this community who feel they're invincible. . . . [T]hey experiment with sex and drugs, then die young because they get chronic hepatitis. . . . Since this is something we can prevent, we should prevent it," said a county health director.[85] But that argument didn't hold water for everyone. "Just because we CAN vaccinate, does that mean we always should?" asked a Colorado state health department hepatitis B expert. "It's a worthy public debate."[86]

Those in favor of the health board's proposal prevailed. By the fall of 1997, parents of all of the state's incoming kindergartners and seventh-graders had to either provide proof of their children's hepatitis B immunization or sign a form claiming a medical, religious, or personal exemption to the requirement.[87] But as the requirement went into effect, popular opposition began to mount. In 1999 Patti Johnson, a member of the Colorado State Board of Education, began a campaign to encourage parents to question the vaccine, citing the small number of hepatitis B cases in young children (279 in 1996) and the large number of purported hepatitis B vaccine-related injuries reported to the federal government (24,776 between 1990 and 1999). Drug companies Merck and SmithKline Beecham hadn't adequately tested the vaccine for long-term safety in children, she charged, and too few had questioned why this inadequately tested vaccine was being given to children to stop a disease that affects "IV drug users, prostitutes, sexually promiscuous persons, health care workers exposed to blood, and babies born to infected mothers."[88] At the *Denver Post*, columnist Al Knight repeatedly chimed in on the vaccine's hazards, informing readers that New Jersey's governor, Christine Todd Whitman, had vetoed a bill to require hepatitis B vaccines for schoolchildren.[89] Whitman had cited the vaccine's unknown duration of protection in her decision to *postpone* signing the bill; in Colorado her "refusal" to sign was described as a bold act that questioned the wisdom of vaccinating the young to prevent a disease brought on by adult behavior.[90]

Vaccine doubts weren't new to Colorado; the state was, after all, the birthplace of *Mothering* magazine. By the late 1990s, several of the state's communities had become renowned for their large numbers of children claiming personal exemptions to vaccine requirements and for the outbreaks of vaccine-preventable diseases, like pertussis, which sometimes ensued.[91] Nor were doubts specific to the hepatitis B vaccine new, either in Colorado or elsewhere. Years before state laws requiring the vaccine for children went into effect, the National Vaccine Information Center worried about the lack of studies examining the shot's long-term effects on children. They also questioned why the vaccine should be given to all infants, most of whom didn't belong to any of the identified risk groups.[92]

But what was new in the last few years of the twentieth century was that suddenly concerns about the hepatitis B vaccine acquired new currency and a large audience, not just in Colorado, but across the country. In 1997 the ACIP again endorsed a "birth dose" of hepatitis B vaccine

for all infants. The recommendation was part of a new strategy to eliminate transmission of the virus entirely in the United States; it also fell
in line with a World Health Organization goal to make infant hepatitis B immunization routine globally before the century's end.[93] But in
1998 the national media reported on France's decision to halt hepatitis B
vaccination because of fears that the shot caused neurological damage,
particularly multiple sclerosis.[94] And early in 1999, the television news
program 20/20 asked whether hepatitis B was "smart preventive medicine or an unnecessary risk." The report featured several adults, including several health care workers, whose neurological and autoimmune
symptoms — resembling multiple sclerosis, arthritis, lupus, and Guillain-
Barré syndrome — set in after getting the vaccine. The broadcast also
focused in-depth on the story of Lyla Rose Belkin, a healthy infant who
went to sleep the night she received the vaccine and never woke up.[95] In
the spring of 1999, stirred by these reports, a House subcommittee held
hearings on hepatitis B vaccine-safety concerns. The hearings, chaired
by Florida Republican John Mica, were called to address charges that
a vaccine now required for nearly all children had been implicated in
a range of diseases and disorders. The testimonies demonstrated how
dramatically the hepatitis B virus and its vaccine had been reframed by
the nation's shifting social and cultural concerns over the course of the
1990s.

HELPING OR HURTING?

On the morning of the hearings in May 1999, Representative Mica informed those in the chamber that they were assembled to answer four
questions: Did the benefits of hepatitis B vaccine outweigh its risks?
Were its hazards adequately disclosed to parents? Were the adverse reactions it caused being adequately studied? And what conflicts of interest existed when the CDC considered how and whether to recommend
a vaccine?

In the testimony that followed, proponents of the vaccine emphasized
the seriousness of hepatitis B infection, its tendency to cause untraceable
infections, and the everyday challenges faced by those living with the

virus in their bloodstream. People who spoke out against the vaccine — primarily people who had been injured themselves, or whose children had been injured following vaccination — emphasized the low infection risk of most infants, the greed of drug companies, and the dismissal they faced from doctors and other health professionals. When witnesses for each side made reference to hepatitis B itself, they seemed to be discussing two different diseases. To officials from the CDC and members of the American Liver Foundation and Hepatitis Foundation, hepatitis B was a lethal disease that infected 1 in 20 Americans and caused 5,000 deaths each year, many of these from liver cancer. To members of Massachusetts Citizens for Vaccination Choice and Parents Requesting Open Vaccination Education — and to the doctors and parents who had witnessed blindness, deafness, seizures, and other effects following vaccination — hepatitis B was instead a rare sexually transmitted infection that threatened drug addicts and foreigners, and posed no risk to American infants from healthy families.[96] Two decades of varied representations of the disease accumulated in the House chamber, a potent illustration of how value-driven perceptions of the disease and its vaccine were destined to make objective answers to Mica's questions a near impossibility.

By the date of the hearings, forty-two states had adopted laws or regulations requiring the vaccine for school or day care, and thousands of side effects following vaccination had been reported to the Vaccine Adverse Event Reporting System established by the National Childhood Vaccine Injury Act. Mica noted that he had called the hearings in part because of a New Hampshire report indicating that the state had had 3 cases of hepatitis B and 48 adverse reactions to the vaccine in children under ten. At his request, FDA statistician Dr. Susan Ellenberg testified that in the entire country in 1997, 95 children under two years of age contracted hepatitis B, and 43 had died following hepatitis B vaccination. But "the problems are all in the interpretation" of those numbers, said Ellenberg, because the reporting system cast a very wide net. Since anyone could report a reaction or death as being probably caused by a recently received vaccine, none of the reactions or deaths were definitively attributable to vaccines until investigated — and in the case of hepatitis B vaccine, that hadn't happened yet. CDC scientist Harold Margolis assured Mica that the agency was conducting several ongoing studies.

But to Mica, all of the present evidence added up to the fact that when parents were asked to vaccinate their babies against hepatitis B, they did so with insufficient—indeed, nonexistent—knowledge of the true risks of the vaccine.

That the perceived danger of the vaccine had begun to overshadow the perceived danger of the disease is a testament to changing attitudes toward HIV and hepatitis B's relationship to that infection. In the 1980s and into the early 1990s, AIDS was a horrific and unmanageable specter, and hopes for an AIDS vaccine were projected onto the hepatitis B vaccine, which came to stand for the promise of triumph over insidious blood-borne infections. But by the late 1990s, the spread of AIDS had begun to come under control in the United States, thanks to campaigns that urged the use of condoms and the effectiveness and availability of antiretroviral drugs. The manageability of AIDS tipped the balance—slightly, but perceptibly—between fears of the disease itself and fears of the vaccine or vaccines that may have caused both AIDS and the rest of the nation's autoimmune diseases. With AIDS under relative control, and with the push for broader hepatitis B immunization requirements, comparisons between the two diseases were no longer convenient for health officials, who increasingly emphasized the fact that anyone—not just drug users and promiscuous individuals—was at risk of hepatitis B. Indeed, in the course of the hearings, only one fleeting mention of the disease's comparability to HIV was made. At the same time, parents of vaccine-injured children continued to quote from CDC publications stating that the disease was sexually transmitted, asserting that it was, as Lyla Rose Belkin's father put it, an infection of "junkies, gays, and promiscuous homosexuals."[97]

The emphasis on defining the precise nature and probability of the hazards posed by the hepatitis B vaccine was also a product of the skyrocketing emotional value of children in the very last decades of the twentieth century. Historians Paula Fass and Mary Ann Mason have argued that this emotional value began to soar in direct response to the breakdown of marriage in the same period: as divorce and nontraditional living arrangements became increasingly common, the bond between parent and child came to exceed the bond between spouses in emotional importance.[98] Spouses, that is, came and went, but children provided a source of emotional gratification that was supposed to last a lifetime. This attitude was evident in the testimony of Marilyn Kirschner, the single mother whose teenage daughter had became incapacitated by seizures, migraines, nausea, and fatigue that grew worse after each of her three hepatitis shots: "This vaccine has ripped out a part of our lives that can't be replaced," she said. The parents whose children suffered from hepatitis B itself felt similarly, as Thelma Thiel, chair of the Hepatitis Foundation International, revealed when she spoke of the loss of her "precious" four-year-old son to cirrhosis. This commonality between parents on opposing sides of the issue was well articulated by Barbara Loe Fisher, who testified in favor of more robust vaccine-safety testing. "Whether death or disability is caused by a disease or a vaccine, the pain is the same," she said. "We are all here because we love our children and we want to protect them from harm."[99]

To parents on both sides of the hepatitis B debate, statistical figures concerning the risk of disease or vaccine injury were meaningless when faced with the lived reality of caring for an irreversibly damaged child. For all the commonalities faced by parents living with sick or disabled children, the origins of their plights led to slightly divergent but ideologically similar attitudes toward state involvement in family health matters. Parents of hepatitis B–positive children spoke of the stigma of the disease, the constant fear that their child would pass the infection to others, and their deep desire that parents in their communities would comply with state rules and have their own children immunized. Parents whose children became ill after vaccination, however, saw in those very same rules a state acting in the interest of itself and its corporate allies with little regard for the welfare of individual children and their families. But that didn't always mean that they wanted less state involvement—like the parents of children infected with hepatitis B, they often wanted more: more oversight, more care, more attention paid to their concerns. Said one mother of a vaccine-injured daughter: "Lindsay, nor anyone [sic], should have to suffer like this because scientific studies weren't done to determine if the vaccine was safe to give to every child. My daughter shouldn't have to suffer like this because government officials and drug company executives didn't do their jobs."[100]

In the hearings, scientists and citizens were given equal time and attention by the assembled lawmakers. That lay and scientific testimonies were equivalently valued on Capitol Hill on that day is just one illustration of the degree to which scientific authority had been eroded over the previous quarter century. This erosion had been accomplished in large part through the social movements that had first gotten under way in the 1960s and 1970s, and whose influence on the opinions of vaccine critics was discernible yet in 1999. The emphasis that Mica and vaccine-injured witnesses placed on the uncertainty of the vaccine's long-term safety and protection were made possible by the now-entrenched risk-oriented rhetoric of the environmental movement. The predominantly female patients who recounted their struggles to get male doctors to believe that their symptoms were real and vaccine-related recalled the anti-hegemonic discourse of the women's movement. The influence of the consumer movement, too, was evident in the pervasive distrust of both government and industry scientists on display. One reportedly vaccine-injured woman, a public health nurse from Indiana who asserted that she was not anti-vaccine, testified that she was troubled to learn that Merck scientists had attended CDC meetings held to assess vaccine safety. "[How] can an employee of a pharmaceutical company that manufactures the vaccine be objective in designing experiments to show fault in a product that generates close to $1 billion in sales for his company?" she asked.[101] The accusation was voiced again and again in the debate over hepatitis B vaccine. For by the late 1990s, the newly vaccine worried and longtime vaccine skeptics alike perceived an abuse of power and violation of trust on the part of industry and government officials engaged in the pursuit of public health.

In the immediate aftermath of the hearings, health officials voiced concern that "antivaccine groups" were gaining ground, getting the media interested, spreading word of vaccine risks on the Internet, and "gaining the ears of state and federal legislatures."[102] One specific fear they voiced was that such groups would successfully repeal hepatitis B requirements in the states where they existed—a logical response to the crisis of faith that seemed to be growing ever more deeply entrenched. But nothing of the sort happened. When New Jersey attempted to mandate the hepatitis B vaccine for its schoolchildren later in 1999, legislators, in response to parental concerns, wrote in a clause permitting parents to exempt their children from the requirement for "personal" reasons. Nervous state health officials, fearing an unenforceable mandate from the legislature, decided to write and adopt their own more restrictive rule while state lawmakers were in recess. The move infuriated vaccine-worried parents.[103] "Even if there is only a slim chance that my perfectly healthy infant might die from a Hepatitis B injection, the fact that we are talking about chances at all is appalling. Since when did the New Jersey State Health Department legitimize gambling with lives?" asked New Jersey mother Laura Maschal.[104] Despite parental worries and the legislative "skirmish," however, the hepatitis B requirement went quietly into effect. By 2001 the vaccine was required of all elementary- and middle-school children in the state.

Maschal's opinion and the venue in which it appeared—the *New York Times*—were testament to the fact that the hepatitis B vaccine was playing a part in keeping debate about the risks of vaccines and government's power and ability to manage them in the public arena. But even as this debate continued and grew—as it would in the 2000s—most Americans seemed willing to accept the state in the role of "superparent," trusting it to determine the best policies for the welfare of their well children.[105] By 2002 all but three states—Alabama, Montana, and South Dakota— had adopted laws or regulations requiring the vaccine for children in day care, grade school, or both. That year 88 percent of the nation's children were vaccinated against hepatitis B. The figure climbed to 92 percent the following year, where it held steady through the early 2010s.[106]

The federal policy recommending the universal vaccination of children against hepatitis B, and the historical moment of which it was born, represented the apex of the new era of vaccination heralded in the late 1960s. The state-level policies requiring the vaccine for all children were made possible by the consolidation of federal authority made manifest in the Vaccines for Children Program. Federal and state policies, meanwhile, embraced the vaccination of infants, placing significant health-citizenship responsibilities on the shoulders of the nation's youngest members. Hepatitis B wasn't ever considered a "childhood" disease, but the presence of an effective vaccine made it possible for health officials and health care providers to treat it like one. This approach represented a convenient route to a healthy populace; as one pediatrician put it, "at least with infants you can capture them because you know you see them at birth."[107] The vaccination of all children, as opposed to those at highest risk of infection, was also more cost-effective than screening for those at high risk of hepatitis B. Moreover, it conformed to the principles of early universal vaccination against childhood diseases that had been worked out decades before, with the administration of vaccines against rubella and mumps to all children at an early age to ensure a healthy adult population in the future.

Policies dictating the hepatitis B vaccine's recommended use evolved in a landscape of shifting federal resources and changing scientific and cultural ideas about the disease itself and the risks it posed. But even as these policies reached the apex described herein, doubts about the hepatitis B vaccine specifically and vaccines generally continued to brew. In the decade that followed, vaccine-skeptical discourse would reach a crescendo pitch in popular media. As a result, when state lawmakers would attempt to pass universal school mandates for another sexually transmitted, oncogenic infection—human papillomavirus (HPV)—in 2006 and 2007, their efforts would meet with public outrage and legislative failure. Looking back, laws mandating the vaccination of all children against hepatitis B appear to have found a temporarily open window. Even as they went into effect, however, the window was already falling shut.[108]

9

VACCINE RISKS AND
THE NEW MEDIA

The headline was damning:

"THE LETHAL DANGERS OF THE BILLION-DOLLAR VACCINE BUSINESS:
With Government Approval, Drug Companies Sell Vaccines That Can Leave
Your Child Brain Damaged, Can Spread Polio From Your Baby To You—And
Can Even Kill."[1]

Ten years after Congress passed the National Childhood Vaccine Injury Act, *Money* magazine—no enemy of the vaccine industry—reported
that while vaccines saved lives, Americans still had plenty of reasons to
be wary about them. DPT shots still caused brain damage and deaths,
the magazine reported, and vaccine makers were holding out on a safer
version. The polio vaccine was now the only cause of the disease in the
United States, and to make things worse, the shot might be causing cancer and other chronic diseases, too. According to the reporters at *Money*,
health officials were systematically downplaying vaccine risks, medical
experts with industry ties were setting policies that padded drug companies' bottom lines, and parents were in the dark about all of it.[2]

The charges were, in effect, a compilation of the vaccine complaints
of the previous two decades, from mothers' concerns that doctors were
withholding vaccine-safety information to Clinton officials' charges that
drug companies were putting profits ahead of people. More importantly,
however, the article signaled the beginning of a trend that would take

off at the turn of the millennium, when reports about vaccine dangers became a regular feature in the news. At first, this media coverage was event-driven and diverse. Stories covered the reported side effects of the hepatitis B vaccine; the dangers of vaccinating against smallpox in post-9/11 America; the risks and benefits of flu vaccination; and societal concerns about the use of a new vaccine against human papillomavirus, or HPV. Over the course of the decade, however, one vaccine-safety story gradually supplanted all others: the story of the relationship between autism and vaccines.

Vaccine-safety activists had long argued that immunization had the potential to affect the immune and nervous systems in ways that might, in some cases, cause autism.[3] In the 2000s, the theory that vaccines caused or contributed to autism gained unprecedented visibility and momentum. Parents, doctors, health experts, lawyers, and journalists examined the theory and picked it apart. The timing was no accident: parents and scientists alike were trying to understand why childhood autism was on a rapid and startling rise. Health and medical experts quickly concluded that there was no link between autism and vaccines. But in the years after they issued their verdicts in the early 2000s, the link only gained more and more attention in the press.

Why did the vaccine-autism story stick around in the news, long after vaccines had been scientifically exonerated as a cause of the condition? As this chapter shows, both journalists and public health experts had a number of reasons for keeping the story afloat. As the new millennium began, vaccines were controversial for a variety of reasons. By consistently pointing out that the question of vaccines' connection to autism had been settled by scientists, however, health officials kept the controversy "within the confines of their scientific and technical expertise."[4] In effect, as historian Mark Largent has shown, keeping the spotlight trained on disproven autism fears allowed health experts to ignore parental vaccine worries that had nothing to do with science. But keeping the vaccine-autism link visible served other public health purposes as well. The majority of media reports referenced the debated link only to refute it.[5] The typical news story reported that parents afraid of autism (generally mothers) weren't vaccinating their kids, and were therefore responsible for the intractability of, and even increases in, vaccine-preventable diseases. This narrative highlighted the importance of listening to science, reason, and experts. It also highlighted a moral duty to vaccinate for community, not just individual, benefit. And it obviated explanations of the complicated and poorly understood scientific reasons for the persistence of vaccine-preventable diseases, despite record-high vaccination rates.

The vaccine-autism link also served a number of journalistic purposes. For journalists under new pressures in a changing media environment, "debate" over the link was a ready source of conflict and tension. It was an easy explanation for the reportedly inadequate vaccination rates behind ongoing outbreaks of vaccine-preventable diseases. It perfectly suited the mass media's role as watchdog for a risk-obsessed society. And it was an appealing attention-getter because it fed upon an abundance of preexisting cultural tensions: over competing claims to expertise between mothers and scientists, about epidemics of disability in children, about chemical contamination of humans and the environment, and about industry and government's perceived failure to protect the public from harm. The debate was a product of the more diffuse set of worries that had begun boiling over in the late 1990s, those itemized by *Money* magazine and a series of exposés, large and small, that got airtime at the end of the millennium. But as the 2000s progressed, the media's tight focus on vaccine-autism fears swept these more diffuse worries under the rug, and ignored the value disputes and cultural battles that lay at their foundation.

MOUNTING SKEPTICISM

At the end of the 1990s, a series of events made vaccine-safety stories a fairly regular feature in the papers and on the evening news. The stories pointed to haste among policy makers, suggested inadequate testing by drug companies, and indicated that even vaccines long assumed to be safe might cause harm in insidious ways.

A new vaccine against rotavirus, a diarrheal disease, garnered such attention. In late August 1998, Wyeth Laboratories' new RotaShield vaccine was licensed for use in infants. Six months later, in March 1999, the ACIP recommended that the three-dose vaccine be given to all infants at ages two, four, and six months. But CDC scientists monitoring the Vaccine Adverse Event Reporting System soon noticed an unusually high number of cases of intussusception, a painful and potentially fatal bowel obstruction, in children who had received the vaccine.[6] Wyeth stopped making the vaccine, and the CDC recommended that parents and pediatricians stop giving it to children. In July 1999, the news made for worrying headlines and evening news stories.[7] When the ACIP met and reviewed the data in October of that year, they issued what one veteran medical reporter called "a rare and embarrassing reversal," recanting their earlier recommendation that all infants be immunized against the disease.[8]

Further chagrin for vaccine scientists came later that same year, with the publication of *The River* by British journalist Edward Hooper.[9] Hooper's massive work painstakingly buttressed a theory, circulating since the eighties, that polio vaccine trials conducted in the fifties had caused the AIDS epidemic. His analysis compellingly suggested that an oral polio vaccine developed using chimpanzee tissue transmitted HIV's simian precursor virus to humans when it was tested on populations in Burundi, Rwanda, and Congo, home to the earliest documented HIV infection, in 1959. The hypothesis triggered a firestorm of debate among scientists. But even experts skeptical of the hypothesis praised the book, in such scientifically revered publications as *Science* and *Nature*.[10] In the popular press, meanwhile, one reviewer called *The River* an "embarrassment" for scientists, since it revealed "that leading researchers kept sloppy records and that prestigious peer-reviewed medical journals published reports that omitted crucial details."[11] Said another, "It could be the biggest 'Oops!' in history."[12] The book's reception gave credence to the notion that vaccines were not just fallible in minor, insignificant ways, but could very well be implicated in the nation's most devastating epidemic in decades.

Hooper's hypothesis was just one more tremor shaking the foundation on which public confidence in vaccines stood in the late nineties. It followed on the FDA's announcement that thimerosal would be removed from vaccines, and it coincided with congressional hearings on hepatitis B vaccine side effects. It also coincided with a revision to the childhood immunization schedule based on a reassessment of the risks and benefits of polio vaccine—yet another news item that drew attention to vaccine hazards. The ACIP and the American Academy of Pediatrics announced that children should now get four polio vaccine injections, instead of the two injections and two oral immunizations they previously received.[13] Although oral polio vaccine (Sabin's vaccine) was more protective against disease, it had been the sole cause of poliomyelitis cases—144 of them—in the United States since 1979. Given the progress of global campaigns to eradicate polio beyond the United States' borders, the ACIP concluded that the more effective but riskier vaccine was no longer justified.[14] Vaccination recommendations always rested on a comparison of risks and benefits, but this time the adjusted calculus made headlines and overlapped with the negative attention that Hooper's book had brought to polio vaccines.[15]

Longtime vaccine critics kept tabs on each new piece of bad news. Late in 1999, the editors of *Mothering* noted that the year had been marked by a "flurry of activity regarding the safety, ethics and politics of vaccines": the withdrawal of RotaShield, the FDA's thimerosal announcement, the replacement of oral polio vaccine with injected polio vaccine, and the CDC's "suspension" of hepatitis B injections for low-risk infants.[16] This time, however, *Mothering* wasn't the only media outlet tallying the nation's vaccine troubles. *ABC Evening News* took a "closer look" at the "hidden dangers in vaccinations," while the news team at CBS reported on the growing number of vaccines recommended for children and the troubling side effects that sometimes followed.[17] The front page of the *San Francisco Chronicle* described an "outbreak of mistrust toward vaccination," and the front page of the *Philadelphia Inquirer* declared that vaccines were "under siege" for countless reasons. Recommendations for new vaccines, like those against hepatitis B and rotavirus, kept being readjusted because of safety concerns. Soldiers argued that vaccines were in part responsible for Gulf War syndrome. Parents were worried that vaccines were causing conditions such as autism, multiple sclerosis, and SIDS. And new evidence kept turning up to suggest that Americans' vaccine worries were justified. Noted the *Inquirer*, "Public-health officials are afraid of what could happen next."[18]

The sum total of this "flurry" of events and the media attention they attracted was not lost on immunization officials, who noted that at the turn of the millennium, the media, Congress, and consumers—a fast-growing number of them on the Internet—were all chattering about vaccine safety. In response, the CDC and the National Institutes of Health asked the Institute of Medicine (IOM) to convene a committee of impartial experts—lacking any financial ties to industry—to carry out an Immunization Safety Review project. At a series of meetings held between 2001 and 2004, the committee analyzed the nation's most prominent vaccine-safety concerns: the purported link between vaccines and sudden infant death; flu vaccine and neurological complications; polio vaccines and cancer; hepatitis B vaccine and neurological disorders; vaccines and immune dysfunction; thimerosal and neurodevelopmental disorders; and MMR vaccination and autism.[19] The IOM's inquiry was broad, reflecting the diverse set of worries articulated by Americans at the end of the nineties. But as the committee's findings were announced, journalists honed in on just one.

AUTISM FEARS

In the spring of 2000, the relationship between vaccines and autism was the subject of emotional and combative congressional hearings called by Republican Representative Dan Burton of Indiana, whose grandson developed autism after he received "nine shots in 1 day." Parents of autistic children testified about their struggles to find treatment, the high costs of care and services, and their observations of the food allergies and "metal toxicities" that seemed to accompany their children's behavioral symptoms. Epidemiologists and statisticians testified; so did federal health officials, autism advocates, vaccination advocates, autism researchers, pediatricians, physicians, and scientists engaged in the latest relevant research. Opinions on the plausibility of a link were sharply divided. To Burton, the growing number of vaccines and their ingredients were both very worrisome, and "research on the potential connection between vaccines and autism" deserved "top priority." To Democrat Henry Waxman of California, who played a supporting role in the hearings, the hearings had been stacked in favor of a link between vaccines and autism. Waxman urged those assembled not to "get ahead of the science or raise false alarms." As yet there was no scientific evidence supporting a causal relationship between vaccines and autism, he stressed, so "why should we then scare people about immunization until we know the facts?"[20]

At that point, the "facts" supporting a link came largely from a paper published by British gastroenterologist Andrew Wakefield, then of the Royal Free Hospital and University College Medical School in London, who also testified at the hearings. A study of twelve autistic children conducted by Wakefield and colleagues purported to have found a possible connection between MMR vaccination, measles virus in the gut, and the development of autism. MMR vaccine, the researchers proposed, caused bowel problems that hampered a child's ability to absorb nutrients, thereby leading to developmental disorders.[21] The study was published in one of the most prestigious medical journals in the world—the *Lancet*—but the scientific community was largely unimpressed by its findings. The American Medical Association, among others, pointed out that the study was based on a sample way too small for drawing grand conclusions. It also suffered from referral bias—that is, its subjects weren't randomly selected. And its conclusions weren't supported by its clinical observations: some of the children had behavioral problems before they had bowel problems. Moreover, a follow-up study by Wakefield's colleagues failed to turn up any evidence of measles virus in the intestines of people with bowel disease.[22]

In the UK and Europe, Wakefield and colleagues' 1998 study was nonetheless the subject of countless sensational headlines, and news of it helped push already declining measles vaccination rates further downhill.[23] "Once again," lamented a British epidemiologist, "the media have succeeded in denting parents' confidence in childhood immunization." In the 1970s, pertussis immunization in England had tumbled from about 80 to 30 percent following media reports of severe side effects; given the headlines about MMR, health experts braced for a repeat performance.[24] In the United States, Wakefield may have been called before Congress in 2000, but his research had actually received only scant attention from the media. His 1998 study made for few headlines and evening news stories in 1998 and 1999; those media outlets that did report on it described it as "controversial" and "methodologically flawed."[25] In fact, the U.S. media gave more time and space to medical experts and research findings that refuted Wakefield's findings, including a Finnish study that tracked 3 million doses of MMR vaccine given to children and did not find a single case of autism among them.[26]

Wakefield's specific findings may have received limited and ambivalent attention from the media when they first came out in the late nineties. But generalized fears of a vaccine-autism link nonetheless permeated the growing number of media reports about autism, on the one hand, and vaccine risks and side effects, on the other. A profile of renowned autism expert Bernard Rimland in the *San Diego Union Tribune*, for instance, mentioned that Rimland believed epidemic autism might be caused by pollution—a term he defined widely enough to include industrial pollution, food additives, and overused pharmaceuticals, including vaccines.[27] Stories about children living with autism—which became increasingly common as more and more children were diagnosed with the condition—often quoted parents who speculated that their children's autism was brought on by vaccines.[28]

While vaccine worries cropped up in articles about autism at the century's end, autism worries appeared in reports on vaccines and vaccine policy. Stories on parents who eschewed chicken pox vaccine, or those who wanted their states to permit philosophical exemptions to school vaccine requirements, quoted parents saying they worried that vaccines caused SIDS, hyperactivity, ADD, and autism.[29] When media reports insisted there was no evidence to support "a well-publicized contention" that the MMR vaccine caused autism, readers wrote in with their objections. "Insulting the immune system" with vaccines could cause autism, Guillain-Barré syndrome, and other autoimmune diseases, wrote a registered nurse to the Albany *Times Union*. "Research has implicated the DPT, MMR, and hepatitis B vaccines in the rising number of autism cases," wrote a New Jersey man to the Newark *Star-Ledger*. "A child became autistic directly after an MMR injection," warned an Erie, Pennsylvania, woman in a letter to the local *Times-News*.[30]

Despite today's claims that Wakefield's 1998 study was single-handedly responsible for fostering parental fears of a vaccine-autism link, such fears didn't originate with Wakefield.[31] If anything, his now-infamous study was a product of parents' worries, not the other way around.[32] Speculations that vaccines might cause autism weren't new at the end of the 1990s; they had been around since at least the early 1980s, as noted in chapter 6. On the other hand, widespread fears of autism generally *were* new at the turn of the millennium, because the condition seemed to be spreading quite suddenly out of control. In the late 1980s, autism affected 1 in 10,000 children. In 2001, studies estimated that it affected as many as 1 in 500, with prevalence projected to climb even further.[33] Scientists called the rise "baffling" and "mysterious." In California, where the increase was particularly acute, researchers studying the trend confessed that they had no idea what was behind it. Genes were known to play a part in the development of autism, but genetics alone couldn't explain such a rapid rise. Nor could publicity, despite suggestions that the 1988 film *Rain Man*, which featured an autistic character played by actor Dustin Hoffman, had fueled autism awareness. And studies showed that neither diagnostic nor demographic changes could fully account for the increase, either. "Experts believe that environmental factors can trigger autism in people with susceptible genes, with suspicions falling at various times on vaccines, infections, heavy metals and other environmental insults," reported the *New York Times* in 2002. "It could take years of study to unravel the widening mystery of autism."[34]

AUTISM REPORTING AND MOTHER WARRIORS

The IOM committee took on the feared link between MMR vaccine and autism as its first order of business, convening a meeting in March 2001 and releasing its conclusions a month later. Looking closely at the existing research, the scientists on the panel didn't find much to implicate the MMR vaccine, but they did find much to exonerate it. To begin with, the MMR vaccine was licensed long before prevalence of autism spectrum disorders began to climb. Eight different epidemiological studies showed no association between MMR vaccination and autism. These studies didn't definitively *disprove* a causal relationship, but at the same time, the single study suggesting a link between vaccines and autism— Wakefield's study—failed to *prove* a causal relationship. Epidemiological evidence aside, there was also no good biological model to explain how MMR vaccines could contribute to autism, in either lab animals or humans. The likelihood of a causal relationship, the committee concluded, seemed remote.[35]

In 2004 the IOM committee revisited the connection between vaccines and autism, in order to take into account the most recent research. This time, they reported that they had found no support for a causal relationship between the two.[36] Media reports had adopted a reassuring tone when the IOM released its 2001 report: "Parents worried about the potential links between one of the most common [vaccines] and autism can rest easier tonight," said network news anchor Tom Brokaw.[37] But the tone of media reports on the occasion of the IOM's 2004 findings reflected a noteworthy shift. Some reports were defensive. On *60 Minutes*, CDC immunization adviser and pediatric infectious disease specialist Paul Offit not only disputed the vaccine-autism link; he emphasized that vaccines were "without question, the safest, best-tested thing we put into our bodies. . . . [T]hey have a better safety record than vitamins, a better safety record than cough-and-cold preparations, a better safety record than antibiotics."[38] Still other reports suggested that scientific assurances were now beside the point. While scientists say it's "clear" that vaccines don't cause autism, said NBC news reporter Robert Bazell, "for some parents, the doubts will always linger."[39] Scientific conclusions, reassuring in 2001, were now no salve for parental fears. The vaccine-autism story, clearly, would not be put to rest. In fact, it only became more prevalent as the decade progressed. U.S. newspapers mentioned the link four hundred times in 2001 and more than three thousand times in 2009. And there were five times the number of evening news stories on the link in 2010 than there had been in 2001.[40]

A number of events helped keep the story in the news. Studies large and small continued to investigate the relationship between vaccines and autism, and a few endorsed the plausibility of a link.[41] Wakefield and his study became news fodder as various conflicts of interest were uncovered and his coauthors withdrew their support for the study; soon after, he was found guilty of misconduct, stripped of his medical license, saw his study retracted from the *Lancet*, and made a new home for himself among supporters in Austin, Texas.[42] The continuing spread of autism did its part, as well. In 2005, autism prevalence rose to 1 in 166 children. A year later, it rose again, to 1 in 110; two more years later, it climbed further still, to an unfathomable rate of 1 in 88.[43]

All the while, parents continued to suspect vaccines. In part, this was likely due to the fact that the media didn't let go of the story; scholars have shown that public concern about a risk increases as news coverage of the risk increases—no matter how small, or unproven, that risk may be.[44] Moreover, as vaccine worries were being amplified in the news, the rise of the Internet created yet another forum for parents' suspicions to circulate and gain momentum.[45] Americans in the early 2000s were flocking to the Internet for all sorts of reasons, including the quest for health and medical information.[46] Physicians and health experts lamented that patients' web research was changing the traditional office visit, and not for the better.[47] But for the parents of autistic children, the online world was a limitless source of information that empowered them to understand and manage their children's needs. A couple in Massachusetts said they spent five hours a day researching autism tips online.[48] A California mom connected with other parents of autistic children online and learned about their successes and failures.[49] Still others went online to diagnose their own children: "[We] put [the kids] to bed and then got on the Web to do the research," said a mother in Illinois. "By the end of the night, we knew [our son] Weston had autism."[50]

The web also gave such parents plenty of reasons to worry about vaccines. In the late 1990s, fledgling autism sites, such as Unlocking Autism and the Autism Autoimmunity Project, noted that vaccines' effects on the immune system could lead to "profound neurological damage," including autism—a connection Andrew Wakefield had "discovered."[51] As time progressed, autism sites not only summed up the evidence supporting a link between vaccines and autism; they also gave parents instructions on how to reduce the risk of vaccine harm. Avoid vaccines with thimerosal, they instructed, and make sure children weren't deficient in vitamin A. Give kids a dose of vitamin C before and after getting vaccinated, they advised, space out vaccines so children didn't receive them all in one day, and opt for separate shots against measles, mumps, and rubella in place of the combined MMR vaccine.[52]

Information about the connections between vaccines and autism multiplied and spread like wildfire on the Internet—just as all information did. At the end of the decade, the CEO of Google estimated that humans were creating as much information in two days as they had since the appearance of *Homo sapiens* through 2003.[53] The abundance of information online—and the countless hours that parents of autistic children spent combing through it all—led autism advocate (and model, television personality, and author) Jenny McCarthy to joke that she "should have a doctorate in Google research."[54] McCarthy began writing popular books about pregnancy and motherhood while pregnant with her son Evan, born in 2002; when Evan was diagnosed with autism two years later, the focus of her work changed. Her 2007 best seller, *Louder than Words*, detailed the trials of seeking treatment for Evan: the frustrations of dealing with the health care system, the doctors who belittled her concerns and dismissed her observations, the friends who couldn't sympathize, even the spouse who pulled away.[55] Only the Internet offered answers and support twenty-four hours a day, seven days a week. Online and through other mothers, McCarthy said she discovered the treatments, therapy, and dietary changes that helped pull her son out of his autistic world.

For parents like McCarthy, the Internet was an invaluable source of community; it was also a powerful tool for creating and disseminating "experiential knowledge," the form of knowledge production cultivated by the women's health movement three decades before.[56] In the late seventies, the feminist authors of *Ourselves and Our Children* had lamented that modern parents were "separated from the accumulated wisdom of other parents," a fact that "deepens our dependence on experts"; this was precisely the problem their book and others like it were designed to address.[57] Three decades later, the Internet provided access to a collective wisdom that was a health feminist ideal, in many ways. Science and medicine had few answers or solutions for the parents of autistic children, but in one another they found abundant expertise, shared across cultures and time zones. Each valuable piece of advice was profoundly treasured, simultaneously deepening the sense that shared lay wisdom was invaluable—and expert wisdom flawed. It was a sentiment McCarthy expressed repeatedly, as when another mother told her about treatment regimens that could free Evan from autism: "Why didn't they tell me all this at the doctors' office?" she bemoaned.[58]

McCarthy and parents like her may have been empowered by the experiential knowledge they shared, but in encounters with health care experts they still found themselves dismissed. "Sometimes mothers instinctively know what works and what doesn't, but the doctor wasn't interested in hearing anything I had to say," noted McCarthy. "It's amazing how easily medical staff ignores crying, yelling mothers."[59] McCarthy wasn't alone in feeling this way. She recounted being thanked by thousands of parents who felt abandoned and belittled in pursuit of treatment for their children; she dedicated her next book, *Mother Warriors*, to them.[60] *Mother Warriors* was dedicated to all parent "warriors" fighting on behalf of their autistic children, but here and elsewhere, mothers were, once again, the primary caretakers of children and the ones uniquely frustrated by health care professionals who disparaged their expertise. "I would just say to the pediatricians, listen to [mothers] sometimes and give us a little bit more respect," said Holly Robinson Peete, actress, autism advocate, and mother of an autistic son. "Our gut is really dead on."[61] Peete made her plea in an appearance on *Oprah*. Across the media—in particular in media targeting women—the fight for answers to autism was portrayed as a mother's fight, and a fight that only mothers could truly understand.[62]

Media representations of the autism epidemic, in short, pitted mothers against experts and institutions that didn't listen to them and showed them little respect. Not surprisingly, a similar representation appeared in media coverage of the autism-vaccine link. McCarthy's own autism advocacy quickly turned into a vaccine-safety crusade, when she concluded that Evan was born with an immune deficiency that was aggravated by vaccines and contributed to his autism. "I am not a doctor, and I am not trying to tell you how to treat your child," she wrote. "But . . . I feel it's good to be aware of the dialogue surrounding a possible link between vaccines and autism."[63] On *Larry King Live*, McCarthy appeared as the sole woman and vaccine critic in a panel discussion on autism. She argued with an all-male panel of health experts over vaccine-safety testing, the number of shots, and pharmaceutical companies' undue influence.[64] Gendered contestations of vaccine risks weren't limited to late-night cable news. On a PBS *Frontline* episode titled "The Vaccine War," groups of mothers described their vaccine hesitations to a male reporter. Vaccine advocates in the film weren't all male, but vaccine fears, it was clear, belonged to a domestic, feminized sphere while rationality and vaccine confidence resided in the masculinized professional domain.[65] Throughout the 2000s, McCarthy kept asking experts to "listen to what the moms are saying"—but the message across the mainstream media was that while mothers were permitted to speak, if they said anything that cast doubt on vaccines, then what they were saying was simply wrong.[66] The vaccine debate had become a gender war—and that was good for ratings.

RATINGS AND RISK

McCarthy aired her views in numerous high-profile media appearances, including interviews with Oprah Winfrey, Diane Sawyer, Ellen De-Generes, and the hosts of *The View*.[67] Three years after the IOM had refuted the claim that vaccines caused autism, McCarthy's celebrity made her take on the matter newsworthy and helped keep the vaccine-autism story alive. Her cause was much broader than the contested link between vaccines and autism, but her media appearances typically served to forge that very link. When, for instance, she appeared on *Larry King Live* in 2008 and again in 2009, she debated pediatricians and health experts on the subject of childhood vaccines, even though the occasion for her appearance was World Autism Day, both times.[68]

The dialogue on *Larry King* partly concerned whether vaccines "contributed" to autism, but it was also about much more than that. The invited health experts denied that vaccines caused autism and stressed the ever-present dangers of the diseases they prevented, including polio, measles, diphtheria, and whooping cough. McCarthy railed against corrupt drug companies and complicit doctors, arguing that too many unsafe vaccines were being forced on children in the name of profit, causing new epidemics in misguided attempts to control overblown ones. The heated and testy conversations on occasion devolved into shouting matches. Debates, of course, are newsworthy; harmonious agreement is not. And newsrooms were, arguably, more in need of debates than ever as the 2000s progressed. The advent of cable, the Internet, and multimedia conglomerates had completely reshaped the news media. Ownership of media outlets had become increasingly concentrated, but at the same time audiences had fragmented, dispersing to hundreds of cable and online sources of news and information. News outlets, as a result, were "losing audience" and under tremendous pressure to keep viewers and readers. It was a "seller's market for information," concluded one report on the changing state of the media.[69]

The debate over the vaccine-autism link was good for ratings and readership precisely because it was so heated, so emotional, and so relevant to contemporary autism concerns. News reports that covered the debate or simply made reference to it found in it a ready source of tension and drama. On the one side stood doctors and public health experts talking about evidence. On the other side stood parents (and sometimes politicians) talking about personal observations, struggles, and beliefs. The debate was also a ready source of an emotional and widely relatable plight: the parent struggling to care for her child as best as she could. Often, as the media reported on scientific findings—another study showing no link between vaccines and autism, another piece of evidence discrediting Wakefield—the news was delivered over images of a parent holding a crying child as a needle slipped into his arm, or over images of autistic children rocking, banging their heads, or flapping their arms as they played alone, their parents powerless to reach them.[70]

On a deeper level, the debate also evoked tension because it wasn't just about science or medicine—it was a contest of values. Its core features included disagreements over the nature of evidence, the battle between reason and emotion, and impossible-to-settle disputes over the kinds of risks that parents should assume and the kinds they should avoid. These themes had deep cultural resonance—as did the trends and stereotypes with which they connected.

The battle between reason and emotion, or irrationality, was one of the most prominent themes in reporting on vaccination and autism in the later 2000s. The theme was well-captured in a 2009 article in *Wired* magazine by science writer Amy Wallace, which drew hundreds of angry comments online. Wallace cast the vaccine-autism controversy as a "war on science," with one side defending data, evidence, and reason while the other side fought for "pseudo-science" and "snake-oil." In defense of science stood Offit, described as a pediatrician living in middle-class modesty, who "from an early age . . . embraced the logic and elegance of the scientific method." Among his enemies were men like autism advocate Curt Linderman, who was reportedly "puffing on a cigarette" as he told Wallace, "We live in a very toxic world." The battle between allegedly rational and irrational actors boiled over into the reader comments. One reader joked that autism was caused by the decline in pirates, since the two were inversely correlated on a graph. Another accused Wallace of being irrational herself: "200 years ago you would've been writing this article on the practice of using leaches to remove 'bad blood,'" he wrote, "eventually the truth does come out . . . hence we know the world is not flat."[71]

That comment hinted at yet another key theme of vaccine-autism reporting: the nature of evidence and expertise. On one level, the debate was about the very claim that science alone formed the pinnacle of knowledge, and that scientists were the only source of such knowledge. "What I'm asking is that people trust their experts," pleaded Offit on *60 Minutes*.[72] McCarthy, meanwhile, asked that doctors and scientists acknowledge the expertise of parents (and mothers in particular). "I believe that parents' anecdotal information *is* scientific information," she argued on *Larry King*.[73] "At home, Evan is my science," she said on *Oprah*. In dispute was the definition of scientific evidence, the validity of that evidence, and such evidence as a basis for expertise. The debate over vaccines and autism took place in the context of a growing popular backlash against the pharmaceutical industry generally, which helped bring scientists' claims to exclusive expertise under scrutiny.[74] "Vaccines make the pharmaceutical industry billions of dollars. They make my business billions of dollars," said UCLA pediatrician and McCarthy supporter Jay Gordon; surely, he added, that would influence how vaccines were used.[75] To McCarthy and other vaccine-skeptical parents, anyone with any connection to Big Pharma's profits could not possibly produce objective evidence. In this context, they argued that parental observations constituted critical evidence not only because parents knew their children best, but also because only their observations were untainted by profit and greed.

But if government, doctors, and the drug industry couldn't be trusted, neither could selfish middle- and upper-middle-class mothers. Representations of these mothers constituted another key theme of vaccine-autism media reports. They were shown caring for autistic children, they were recorded saying that they blamed vaccines for the condition, and, increasingly as the decade progressed, they were held up as the nation's premier vaccine refusers, putting the rest of society at risk of infectious disease epidemics in their narrow-minded quest to protect their own children from vaccine injury. The self-serving, usually white, often liberal mother appeared in media as diverse as *Frontline* and the television drama *Law and Order*—which often drew inspiration from the headlines of the day.[76] In a 2009 episode, the daughter of a working-class mother caught a fatal case of measles from the unvaccinated son of an upper-class mother. When detectives went to the unvaccinated boy's home (a posh New York brownstone), his mother refused to take the blame. "I'm not responsible for other people's kids. It's my family. It's my choice," she said. For choosing not to vaccinate her son, she was called a "lunatic," a "nutcase mom," a "danger to society," and a killer, and ultimately she ended up in court. The jury didn't convict her; her right to refuse vaccination for her son was protected by law. But she was condemned nonetheless for being "selfish" (the title of the episode), and for endangering her community in a misguided and foolish attempt to protect her son from an imagined threat.[77]

This fictional character was a caricature of the educated, well-off, twenty-first-century parent often featured in media representations of the vaccine debate. She was a parent who trusted "alternative medicines, organic food and yoga" and distrusted "Big Pharma and their lackeys in the media."[78] The threat she chose to avoid was a primary concern of many vaccine skeptics in the 2000s (across lines of class, race, or politics): toxic chemicals in vaccines and their artificial stimulation of the immune system. In a 2005 article in *Rolling Stone* magazine, environmental lawyer and activist Robert F. Kennedy Jr. (nephew of President John F. Kennedy) charged that "public health authorities knowingly allowed the pharmaceutical industry to poison an entire generation of American children" with the mercury in vaccines. The controversial article raised the specter of an environmental scare "bigger than asbestos, bigger than tobacco"—invoking the idea that science had been wrong before.[79] McCarthy, too, stressed that she wasn't anti-vaccine but "anti-toxin," and she lent her support to a rising popular movement that marched on Washington in 2008 to demand that government and industry "Green our Vaccines."[80]

"Green Our Vaccines" made the news because it featured celebrities McCarthy and her then-boyfriend and actor Jim Carrey. It also made the news because it spoke to a growing "green" movement in which consumers began seeking environmentally "friendly" cars, food, toys, clothes, and more. The movement was, in many ways, the contemporary embodiment of what anthropologist Mary Douglas and political scientist Aaron Wildavsky called a "widespread, across-the-board concern about environmental pollution and personal contamination." In their study on risk perception, Douglas and Wildavsky argued that in our modern world, "evil" comes in the form of "hidden technological contamination that invades the body of nature and of man."[81] In this world, risks are "hidden, involuntary, and irreversible"—a perfect description of how many vaccine skeptics in the 2000s understood vaccines. Douglas and Wildavsky also argued that each society's view of the environment shapes the risks and dangers it chooses to pay attention to or ignore. In this framework, the search for an objective method to choose between risks is "doomed to failure," because tolerable and intolerable risks are determined not by facts and figures but by a society's commonly held values. In media reports that covered the vaccine-autism link, scientists quoted facts and figures about the very "real" risk of vaccine-preventable diseases, and the minuscule or non-existent risks of vaccines themselves. Parents described the risks they most feared—autism, not polio; toxic chemicals, not diphtheria. Objectivity had little hope of bringing such disparate risk perceptions into alignment. And so the debate continued.

The vaccine-autism debate also persisted because it was, in many ways, the perfect story for what sociologist Ulrich Beck dubbed the "risk society."[82] Concern with risk, Beck argued, is our modern condition. Americans and citizens of other affluent nations are at once acutely conscious of risk and pessimistic about the state's and institutions' abilities to manage risks. They are, as a result, plagued by uncertainty; since risk can't be dependably identified or avoided, one has to assume it is everywhere.[83] This mentality is connected to the increasingly protective form of child rearing prevalent in countries such as the United States, where the economic and emotional value of children continues its upward climb; safety gear and safety precautions for children—from car seats to organic baby food to flame-retardant pajamas—are ubiquitous and ever growing in number.[84] In such a society, the media is a critical venue for identifying, communicating, and evaluating risks.[85] The media certainly embraced this role in the debate over vaccines, covering it attentively, staying focused on the vaccine-autism link long after scientists had dismissed it, and giving voice to parental fears that spoke directly to a lack of confidence in government's—and industry's—ability to protect their children from omnipresent risks.

SERVING PUBLIC HEALTH

It's a journalistic maxim that "good news is no news," and the media's coverage of vaccination in the first decade of the 2000s certainly bore this out. As the debate over vaccines and autism raged, good news about vaccines and childhood vaccination received relatively little attention— though there was plenty of it. Scientists were making strides on entirely new types of vaccines, against chronic diseases such as Alzheimer's, and even addictions including smoking. Diphtheria, rubella, and polio had long since become diseases of the past. Cases of chicken pox were plummeting. Cases of hepatitis B were still steadily declining.[86] Measles was down to record low levels, and, since the late nineties, all instances of the disease appeared to have originated overseas. Measles experts called this a milestone; the disease, they declared, had been "eliminated" from the United States. The achievement was due to high immunization rates: more than 90 percent of all infants and 98 percent of schoolchildren were vaccinated with at least a single dose of measles vaccine.[87] And measles was no mere exception. In 2009 the CDC reported that "immunization rates for vaccines routinely recommended for children remain at or near record highs."[88]

But health officials were still finding cause for concern. Outbreaks of mumps erupted in 2006 and again in 2009.[89] Measles cases were low, but in 2011 the number of cases spiked to more than two hundred, up from an average of sixty cases annually in previous years. Most of the cases, however, were still originating with foreigners or Americans who had traveled abroad: the twenty-four-year-old who brought measles home to Indiana from Indonesia, the Burmese refugee who brought the disease to Los Angeles, the Minnesota toddler who caught the infection on a trip to Kenya.[90] More worrisome was the number of cases of pertussis, which appeared to be making a domestic comeback, particularly on the West Coast. The disease had always peaked in three- to five-year cycles, and cases of the disease had been marching steadily upward since 1980.[91] But in 2004 and 2010, the peaks were much higher than usual. In 2010 pertussis rates in California shot back up to 1940s levels.[92] The same happened in Washington State the following year.[93] Across the country, there were more than 27,000 cases of the disease, a staggering increase over the 13,000 cases reported in 2008.[94]

There were many possible explanations for the outbreaks and increases. To begin with, mumps vaccine was never as protective as the measles and rubella components of the MMR vaccine. And scientists speculated that certain factors—like close contact in close living quarters—could prompt the infection to take hold in even well-vaccinated populations. (The 2006 outbreaks had hit hard in college dorms, and the 2009 outbreak was largely limited to "tradition-observant" Jewish communities.[95]) Moreover, both measles and mumps were more prevalent abroad than in the United States, so imported cases were always to be expected. In the case of pertussis, the situation appeared to be a bit more complicated. Outbreaks might be due to the fact that a new acellular vaccine against the disease, adopted in the nineties, wasn't as effective as the older whole-cell vaccine. They might be due to the fact that the pathogen was evolving, or that new strains of the bacteria were beginning to cause outbreaks. They also might be due to changes in the health of the population; asthma, for example, more prevalent than ever, seemed to increase susceptibility to the disease.[96]

These were the issues debated in scientific meetings and journals. But another explanation—one that covered all of the outbreaks in question—got most of the attention in the popular press: simply put, parents weren't adequately vaccinating their children. Because of unfounded fears of vaccine side effects, parents were delaying, skipping, or altogether avoiding vaccines for their children, and it was these deviations from recommended pediatric care that were behind all rises in vaccine-preventable disease. When three Minnesota children were sickened and one died of meningitis during a Hib vaccine shortage, a CDC spokesperson told the press that the shortage wasn't to blame for the outbreak—the problem was children who weren't vaccinated because of their parents' worries about vaccines.[97] When pertussis cases popped up in Michigan, doctors interviewed by local news media blamed the "growing number" of unvaccinated children and their vaccine-fearing parents.[98] A CNN anchor announced that parental fears about vaccines and autism were behind the country's highest rate of measles in a decade. "An increasing number of parents have been refusing to vaccinate their children against measles because of this fear of a connection," she said.[99]

In media reports on outbreaks of vaccine-preventable diseases, that is, the cause was almost universally distilled to a single point: the misinformed, irrational, vaccine-fearing parent. As described above, this distillation served a multitude of journalistic purposes: it was easy to digest, easy to explain, and at the same time easy to expand into a debate that resonated with deeply held values and concurrent trends. This distillation, however, also served a public health purpose. As researchers struggled to explain the persistence of measles, mumps, and pertussis, 'health officials and now the media emphasized the need for parents to immunize their children. "The rise in whooping cough cases continues because parents are not following through on vaccination," declared a *Seattle Times* editorial, even as other outlets reported that the rise was due to a wholly different reason: fading immunity proffered by the vaccine.[100] The editorial may have been a distortion of the truth, but it served a more general purpose. Pointing a finger at those who refused to vaccinate—and there were particularly robust clusters in Washington State, Colorado, Oregon, and California—helped illustrate the urgency of the need for all members of a community to vaccinate against all preventable infections. "It is of concern when we have these communities in the United States where there's enough people who have made this decision [not to vaccinate] that if the measles virus is imported from overseas, that it could actually spread and cause an outbreak," a CDC scientist told a reporter.[101]

Training the spotlight on vaccine refusers distracted from the short-comings of vaccines while simultaneously building a case for the importance of vaccination. At a time when even record-high vaccination rates failed to fully keep vaccine-preventable diseases at bay, parents who avoided vaccinating their children for personal or philosophical reasons stood as a vivid reminder that so-called herd immunity rested on co-operation of the entire herd. Pointing to, and even demonizing, such parents served to illustrate the importance of vaccination for community and not just individual benefit. Some people were too young or infirm to be safely vaccinated. And among those who could be safely immunized, vaccines were never 100 percent effective, as health officials increasingly pointed out in the press. When everyone who could get vaccinated did get vaccinated, it protected the weaker members of the herd and made up for the individuals whose vaccines didn't "take." And when parents rejected some or all vaccines for personal or philosophical reasons, it put the entire community at risk.

This message was rooted in facts and figures—the number of parents who had vaccine doubts or who had skipped some or all vaccines for their children. According to one study reported in the press, one in five parents feared a link between vaccines and autism, and more than half had some worries about vaccine side effects, even if they didn't believe vaccines caused autism.[102] According to another, 40 percent of parents delayed or refused a vaccine for their kids.[103] To make things worse, the number of vaccine exemptions parents were filing for their children was on the rise (albeit by 0.2 percent).[104] Such reports downplayed (or ignored) the myriad reasons that parents delayed or skipped vaccines—in the context of a growing vaccine schedule, no less. Which meant that the message to vaccinate for the herd was not simply about science but about values. The media's renewed focus on the importance of herd immunity made this clear, as it cast the issue in moralistic terms. Parents who refused vaccines for their children may have had a legal right to do so, but, like the non-vaccinating mother in *Law and Order*, they were portrayed as selfish and immoral. Such parents were "taking advantage of all the parents who get inoculations for their children," declared the *Los Angeles Times*.[105] They were "gambling with the lives of not just their kids, but all the children around them," reported *USA Today*.[106] "By not vaccinating your child you are taking selfish advantage of thousands of others who do vaccinate their children. . . . We feel such an attitude to be self-centered and unacceptable," a Philadelphia physicians group told the press.[107]

The vilification of the philosophical vaccine refuser in the media coincided with a national debate on health care, as President Barack Obama's administration tried to push comprehensive reform, once and for all, through Congress. The political struggle over health reform brought forth familiar and often heated ideological arguments for and against government involvement in health care.[108] The arguments were a reminder that personal responsibility for health is a deeply held American principle. Alone among wealthy democracies, the United States has long opted not to provide insurance for a significant fraction of its population and has opted to keep insurance coverage for most tied to individual or household employment.[109] At the same time, American public health efforts have often emphasized individual behavior and lifestyle choices—regarding diet, exercise, and smoking, for example—as the most important path to good health. At the core of the American health care system, that is, stands an assumption that Americans are fundamentally responsible for their own health and health care. Parents who delayed or avoided vaccines may have been willfully ignoring the exhortation to vaccinate for the good of the community that was echoed across media reports on vaccination, but they were making a choice compatible with deeply ingrained American beliefs about personal responsibility for health.

At the end of the decade, journalist Seth Mnookin, author of a popular book that dissected the debate over the vaccine-autism link, argued that the media was to blame for creating national panic over the issue in the first place.[110] The moment of "original sin," he said, was not Andrew Wakefield's 1998 study, but the media's coverage of the study. He pointed out that the paper had been called the "worst" paper the *Lancet* had ever published, and that there had never been a moment in which the paper seemed like good science. "It's absolutely insane for any journalist to take the results of a twelve-person case series and draw these population-wide conclusions," Mnookin said.[111] Mnookin's mea culpa on behalf of the media reinforced the idea that the media is a key player in scientific debates; Wakefield may have produced bad science, and the *Lancet* may have published it, but the media had a responsibility to see through these systemic failures and get the story—and the risk—right. Mnookin's admission reminded Americans that the media is a key institution in the risk society, one that Americans needed to rely upon as an arbiter of good science and bad.

The media was, in fact, responsible—but not necessarily for peddling a myth about autism and vaccines. The media was responsible for distilling the nation's vaccine worries down to one easily digested, and refuted, reason. It helped create a debate without any shades of gray. At the beginning of the twenty-first century, it often ignored parents who had reservations unconnected to autism. It often ignored other factors that kept children from getting fully vaccinated, such as the challenges of living in poverty, language barriers, and health care system hurdles.[112] It also took the civil liberties conundrum at the foundation of all vaccination efforts and made them an issue of black-and-white morality: people who didn't vaccinate their children were self-serving and dangerous, while those who cared about science, reason, and society's weakest members were the standard by which all Americans should abide.

Moreover, pinning responsibility for a prevalent belief on the shortcomings of the media (or on Wakefield) missed the broader context within which that belief emerged. It overlooked the accumulation of vaccine scares and doubts in the nineties. It paid no heed to the fact that the media always reflects cultural assumptions, widely held values, and ideologies as much as it shapes them. And it sidestepped a set of more fundamental questions about the myriad roles that Americans expect the media to play: government mouthpiece, voice of the citizenry, forum for debate, and a source of entertainment, among others. As the media continued to rapidly evolve throughout this period, its ability to carry out these roles and others continued to shift. And as the debate over vaccines and autism pressed on, another debate, about vaccination against human papillomavirus, or HPV, emerged. This debate belonged wholly to the new media environment, even as it reflected a long history of ideas about vaccines and how Americans should use them.

10

SEX, GIRLS, AND HPV

When Hannah, the twenty-four-year-old main character on the television series *Girls*, went to the doctor in the show's 2012 season, she came away with an unexpected diagnosis: HPV. On learning the news, Hannah cried about having "pre-cancer" and having to have her "cervix scraped out." A friend of hers wept in commiseration, bemoaning the prospect that Hannah might never be able to have children. Another friend, in sharp contrast, took the news in stride. She told Hannah that yet another friend of theirs—the show's most fiercely independent and cosmopolitan character—also had HPV. "Like a couple strains of it," she said, matter-of-fact. "She says that all adventurous women do."[1]

At the end of the episode, Hannah posted the line "All adventurous women do" to her Twitter account. In real life, after the episode aired, Twitter and other social media outlets lit up with the phrase, young women echoing it over and over again as a mantra of life lived boldly and without regret. One fan even tattooed the phrase across her foot.[2] Health experts bemoaned the episode's inaccuracies, seeing in it "a reflection of the real confusion felt by twentysomethings" about HPV.[3] But the show's creator (the twenty-something actress, Lena Dunham, who plays Hannah) seemed to know exactly what she was saying about the sexually transmitted infection. In a single thirty-minute episode, the show rejected the stigma and fear attached to the virus and reconfigured it as a badge of honor.

The episode of *Girls* was just one example of the complicated ways in which human papillomavirus, or HPV, acquired new cultural and scientific meanings in the wake of its vaccine. In a now-familiar pattern, the infection's transformation was sparked by the first vaccine against HPV, Gardasil, which was approved for the U.S. market in 2006. HPV began the new millennium as a widely prevalent but usually harmless sexually transmitted infection most women (and men) had never heard of. A few years later, it was widely known as a cause of cervical cancer. A few years after that, it became a dreaded virus with a range of grave consequences for women and men—even as some Americans (including "Hannah") pushed back at this idea. Though Gardasil's manufacturer, Merck, heavily promoted the vaccine—as well as HPV's connection to cervical cancer—no one party was responsible for these evolutions. As Merck worked to create a market for Gardasil, health officials sought the most cost-effective use of the vaccine, researchers expanded scientific understanding of HPV, and consumers actively took part in a very public conversation about the urgency of HPV prevention. HPV was transformed in response to these overlapping inquiries and activities—all prompted or accelerated by the development of the vaccine.

From the start, public conversations about the new HPV vaccine were highly charged. The year Gardasil came on the market, debate erupted over whether the vaccine should be required for sixth-grade girls. Lawmakers quickly backed down from proposed mandates, but the mandates elicited a series of worries and objections to the vaccine that lingered. At what cost was HPV a health risk worth preventing? Should responsibility for its prevention be distributed across lines of gender—or should it be the responsibility of girls alone? Could industry and government be trusted to develop a safe vaccine and arrive at fair policies to govern its use—or did the public have to be ever on guard? The Gardasil debates took their shape from a conservative trend in sex education, anti-corporate fervor that found specific fault with the pharmaceutical industry, the groundswell of vaccine criticism described in the previous chapter, and the new media.

In fact, the new media environment, and social media in particular, made the Gardasil debates unique in one important way. For the first time, the targets of a childhood vaccine had a public venue in which to speak out. In online forums on HPV and the new vaccine, adolescent girls voiced their opinions on mandates, corporate greed, the pain of vaccination, cervical cancer, gender stereotypes, teenage sexuality, and more. Some embraced the messages they heard in advertising, from doctors and parents, in school, and elsewhere; others rejected them. But all of them fashioned their own meanings of HPV and its vaccine in the process. And whether they cast HPV vaccination as a biomedical miracle, a symbol of lax sexual mores, the epitome of paternalism, or something else altogether, they had one thing in common: they weren't taking authority figures' conceptualizations of the virus at face value. As in the fictional world of *Girls*, in the world of social media, *how* to think about HPV was one thing a girl could still decide for herself.

ENTER GARDASIL

Gardasil became the first vaccine against human papillomavirus licensed in the United States in June 2006. (Gardasil would be joined by a second HPV vaccine, GlaxoSmithKline's Cervarix, in 2009.) At Merck's request, the FDA gave the company's application to sell Gardasil a priority review, an expedited process reserved for new drugs serving an unmet need.[4] Gardasil was the first vaccine to protect against four of more than a hundred strains of HPV: HPV 6, 11, 16, and 18. The first two strains cause about 90 percent of genital warts, and the latter two cause approximately 70 percent of cervical cancers.[5] On the occasion of Gardasil's debut, Merck described the vaccine as the most significant advance in preventing cervical cancer since the Pap test was developed, in the 1940s.[6] The ACIP appeared to agree, moving quickly to issue a set of "provisional recommendations" on how the vaccine should be used: committee members recommended that the vaccine be administered to girls ages eleven to twelve, and that females between the ages of thirteen and twenty-six get a "catch-up immunization."[7] They also voted to make the vaccine available through Vaccines for Children, to ensure that girls from poor, uninsured, or Native American families had access to it.[8]

Three months later, states began considering legislation to make HPV vaccination mandatory for sixth-grade girls. Mandates had key expert support. Reproductive health advocates, such as the Guttmacher Institute and Planned Parenthood, argued that making HPV vaccines mandatory for girls had multiple benefits. Adolescents were notoriously "hard-to-reach" for health care matters; school mandates would ensure that they all got protected and would make sure they got protected before they became sexually active. Vaccination had the added benefit of offering protection where condoms — no guarantee of protection against HPV—fell short.[9] The CDC added that compulsory vaccination would address disparities in cervical cancer rates: in the United States, Hispanic and African American women were two to three times more likely, and Vietnamese women five times more likely, to develop the disease compared with white women.[10] For health experts, there was no doubt that HPV vaccination would be a tremendous boon to women's health.

In September 2006, Michigan became the first state to consider a bill that would require all girls to receive the HPV vaccine before entering sixth grade. A few months later, Texas governor Rick Perry issued an executive order mandating the same. Over the course of 2007, another two dozen states, from Connecticut to California, considered bills to require HPV vaccination for girls enrolling in school.[11] But Michigan's law didn't pass, and Perry's order caused a public outcry and was later overridden by his legislature. Across the country, in fact, lawmakers withdrew or revised their bills as they found themselves accused of promoting liberal sexual mores, encroaching on parental rights, being swayed by corporate dollars, and placing the health of young girls in jeopardy. Ultimately only Virginia and the District of Columbia succeeded in making the HPV vaccine a requirement for girls to attend school.[12]

Lawmakers, medical practitioners, and health officials reacted with fairly uniform surprise at the public opposition to the vaccine. They had good reason: an August 2006 Harris poll had indicated that 70 percent of Americans agreed that "encouraging girls and young women to get the HPV vaccine is a good way to prevent the future spread of cervical cancer."[13] But encouraging, of course, was different from requiring. Proposed mandates drew out beliefs and opinions that were less relevant when the vaccine was simply an option for citizens — and not a proposed requirement of citizenship, let alone one required only of girls.[14] The mandates brought the vaccine itself under broad scrutiny, and the impact of this scrutiny would persist well after talk of mandates diminished.

Opposition to the mandates came from some familiar sources. Vaccine-safety advocate Barbara Loe Fisher argued that mandates were the work of "drug company lobbyists and legislators trying to force young girls to use HPV vaccine without the voluntary, informed consent of parents." To Fisher, such consent was impossible given gaps in knowledge regarding the vaccine's safety and effectiveness.[15] To civil libertarians, the mandates were, like all mandates, an undue encroachment on individual rights. Said the Cato Institute's bioethics director, the decision to get the vaccine was a "health-risk assessment [that] should be left up to individual families . . . because making such determinations rightly rests with families."[16]

Social conservatives took umbrage at proposed mandates, too, but for their own reasons — namely, the acknowledgment of teenage sexual behavior implicit therein. Emboldened by the George W. Bush administration's support for faith-based organizations and abstinence-only sexuality education, self-described "family values" groups, such as Focus on the Family and the Family Research Council, argued that vaccinating girls against a sexually transmitted infection was tantamount to condoning and encouraging premarital sex and promiscuity. They promoted the idea that mandates would undermine abstinence-based sex education — which, in their view, promoted a safer and more effective means of preventing HPV infection than any vaccine.[17] Social conservatives also argued that school mandates were inappropriate for infections that were transmitted sexually — and therefore unlikely to spread in a classroom setting. Conservative radio host Dr. Laura summed up this view: "It makes sense to me to require school children to have immunization to measles, chicken pox, and polio," she said, "because these are highly contagious diseases readily spread in a classroom." But mandating HPV for school girls, she felt, was "patently outrageous."[18] On this point, she (and those who agreed with her) was backed up by vaccination *advocates* who concurred that mandates weren't the right approach to take with Gardasil. Jon S. Abramson, chair of the federal panel that had recommended the vaccine, told a reporter that he opposed the push for mandates. "The vaccines out there now are for very communicable diseases," he said. "A child in school is not at an increased risk for HPV like he is measles."[19]

By targeting children, the floated HPV mandates exposed a lack of consensus on the health-citizenship responsibilities of youths. Were children primarily vaccinated to stop diseases from spreading through schools and communities in the present? Or were they vaccinated to ensure a healthy adult populace in the future? Pediatric care and school enrollment offered established means of implementing and enforcing vaccination requirements. School attendance also offered an obvious means of spreading communicable diseases through a community, and therefore an obvious justification for school-based vaccination laws. But because HPV is exclusively sexually transmitted, proposed school-entry laws lacked a transparent logic; they also evoked tensions historically common to debates over the government's role in the management of sexually transmitted infections.[20] In terms of its likely communicability in the school setting, HPV was not terribly different from hepatitis B—but that virus's vaccine benefited from a tight link to AIDS and attendant cultural anxieties.[21] The HPV vaccine also protected against cancer, but cancer prevention was neither an established component of pediatric care nor an accepted health responsibility of the nation's youngest citizens.[22] Moreover, U.S. children had acquired several new health responsibilities in the form of multiple vaccines added to the childhood schedule since the late 1990s; to many HPV vaccine critics, this was an additional responsibility too many.

Of course, relatively few critics explicitly rejected the HPV vaccine as a civic duty of children. Instead, debate over the proposed mandates focused largely on how safe the vaccine was, how necessary it was, and how to weigh its potential risks against the risks of HPV and cervical cancer. Safety concerns loomed large, which was not surprising given the cultural prominence of drug-safety worries in the mid-2000s. Since the turn of the millennium, Americans had shown a rapidly escalating loss of trust in the pharmaceutical industry. The "Big Pharma" backlash was part of a larger backlash against transnational corporations; the anti-globalization movement of the early 2000s had charged corporations with ignoring environmental, worker, and consumer safety in their quest for ever-larger profits.[23] Merck seemed to exemplify the trend when, in 2004, its widely prescribed arthritis drug, Vioxx, was found to increase the risk of heart attack and stroke only after it had been on the market for five years.[24] The news about Vioxx came on the tail of reports that hormone replacement therapy, used for decades to treat the symptoms of menopause in women, also increased the risk of heart attacks, strokes, and certain types of cancer.

These scares and others inflamed popular cynicism toward the profit-hungry pharmaceutical industry, which in the mid-2000s was suddenly being skewered everywhere from Hollywood to the nation's most venerated medical journals. The villain of one of 2005's top films, *The Constant Gardener*, was a multinational pharmaceutical conglomerate that knowingly tested harmful drugs on poor women and children in Africa—and then murdered without pause to keep its transgressions secret.[25] The film was based on a book by esteemed spy novelist John Le Carré, who said that his research on the drug industry turned up tales so horrifying they made his book seem "as tame as a holiday postcard."[26] The film came out not long after Marcia Angell, former editor in chief of the *New England Journal of Medicine*, authored an exposé that lambasted the industry for its undue influence over government regulators and relentless pursuit of profits.[27] Her book, *The Truth about the Drug Companies*, belonged to a wave of popular books by physicians and investigative journalists that took on Big Pharma in the mid-2000s.[28] One reviewer writing on the trend noted that there was no shortage of material to draw on: in a single week in the summer of 2004, one drug company stood accused of cheating Medicaid, six others were being sued by the city of New York for "inflating costs and defrauding taxpayers," another was paying out billions in settlements to consumers injured by a widely prescribed diet drug, and still another had covered up potentially fatal side effects of a drug for schizophrenia.[29]

Throughout 2004 and 2005, medicine and the media conveyed the message that consumer safety was largely an afterthought for the drug industry. And vaccines, of course, were under their own attack at the time. In fact, as lawmakers debated mandatory HPV vaccines for girls, Jenny McCarthy's *Louder than Words*, with its plea for more cautious government vaccination policies, shot to the top of the nation's best-seller lists.[30] Worries that Gardasil might not be safe bubbled up from the mix of existing misgivings about drug company greed, vaccine side effects, and the speed with which Gardasil was approved for use and being mandated for girls. Asked one critic of HPV vaccine mandates: "What if 90% of all . . . girls are vaccinated within the next five years and then ten or twenty years from now it is discovered that the vaccine made them sterile or . . . caused them to get a different type of cancer . . . ?"[31] Safety worries were all the more prominent because the vaccine had been recommended not for adults but for children. Barbara Loe Fisher, among others, warned that clinical trials had tested the vaccine mostly in older women, who were "biologically different" from young girls. Furthermore, said Fisher, Merck's own product insert disclosed that Gardasil had not been studied for its long-term "ability to affect female fertility, cause other kinds of cancer or be toxic to the genes."[32]

Safety concerns were amplified by the perception that HPV infection wasn't a risk worth taking other risks to avoid. As one mother put it in a letter to the *San Francisco Chronicle*, a disease that killed less than 4,000 U.S. women a year (cervical cancer caused an estimated 3,700 deaths in 2006) didn't seem a big enough problem to warrant vaccinating millions of girls.[33] Her letter was part of a public outcry over an HPV vaccine mandate introduced in California's state legislature. The bill's sponsor, assemblyman Ed Hernandez, attributed the controversy to the fact that so few women saw cervical cancer as a top health priority. "If this was a vaccine to eliminate or reduce breast cancer," he said, "we would not be having this discussion."[34] Decades of consumer-led breast-cancer activism had in fact made breast-cancer prevention a top health priority by the 2000s; broad cultural consensus held that the disease, a leading cause of cancer among women, was worth preventing at just about any cost. Cervical cancer, in stark contrast, invited comparisons to all the causes of death that outweighed it—including breast cancer, which killed more than 40,000 women a year.[35] Moreover, noted critics, Gardasil addressed just one cause of cervical cancer; it did nothing to prevent cervical cancers caused by smoking, DES, and other exposures. And relatively few forms of HPV had a link to cancer; the vast majority of HPV infections are harmless and clear up of their own accord.

The gulf between technocratic and popular risk assessments was widened by the fact that a safe and effective means of preventing cervical cancer—the Pap smear—already existed. To critics of HPV mandates, the Pap smear was proof that the vaccine and its risks were unwarranted. And the argument that mass vaccination would most benefit women who didn't get routine Pap smears didn't always hold water. When D.C. legislators proposed a mandate, *Washington Post* columnist Courtland Milloy questioned the HPV vaccine's safety and white lawmakers' motives in requiring it for a largely black school district. In Milloy's view, the mandate smacked of medical experimentation on black people, bringing to mind the Tuskegee syphilis study and distribution of Norplant to black females convicted of crimes. Affluent and middle-class women could prevent cervical cancer with safe and effective Pap smears, but poor women and women of color were instead being forced to take a pharmaceutical that had been rushed to market before all of its risks were known.[36]

Health experts who spoke out against mandates further fanned the flames of controversy. "I'm not against vaccines, but in Canada and the U.S., women are not dying in the streets of cervical cancer," a health policy professor told the press.[37] "Most deaths from cervical cancer in this country are in women who are not adequately screened (with a simple Pap smear). . . . Unfortunately, there is no lobby for the Pap smear," wrote an obstetrics professor in the *New York Times*.[38] Dartmouth physician Diane Harper, who had tested the HPV vaccine in clinical trials for Merck, told reporters that the vaccine hadn't been tested in girls as young as those targeted by mandates, that there was no proof the vaccine's protection would last into girls' sexually active years, and that it might give young women a false sense of security, prompting them to skip Pap smears even though the tests were still necessary to catch cervical cancers not prevented by the vaccine.[39] "To mandate now is to Merck's benefit, and only to Merck's benefit," she said. "We don't need mandatory vaccinations for little girls."[40]

MALIGNING MERCK

While lawmakers and constituents debated proposed mandates, the media began to scrutinize Merck's activities—and turned up plenty of eyebrow-raising details. Texas governor Rick Perry had ties to a Merck lobbyist and had received hefty campaign contributions from the company. Merck had paid physicians thousands of dollars a pop to give talks endorsing Gardasil and held receptions for politicians to promote the vaccine. The company had also given grants to Women in Government, an organization of female lawmakers that disseminated sample HPV legislation online and whose members introduced HPV-related bills in state legislatures.[41] Mandates, of course, would have ensured the broadest possible market for the vaccine. To make matters worse, at $360 for a series of three shots — or "$400 to nearly $1,000 with markup and office visits" as one reporter pointed out — Gardasil was the most expensive vaccine ever marketed.[42] Even without mandates in place, the vaccine quickly became one of the most profitable, too, earning Merck $1.7 billion in its first full year on the market.[43] (It helped, too, that from 2006 to 2009, Merck sold the only HPV vaccine on the market.)

The press further denigrated the company for its heavy-handed marketing of the new vaccine, not just to physicians, but directly to teenage girls.[44] Taking a cue from the recent popularity of disease-awareness bracelets, Merck sponsored a program called "Make the Commitment," which distributed free "Commitment" bracelets (designed by a celebrity fashion consultant) to girls who took an online pledge to ask their doctor about cervical cancer prevention.[45] Commercials for Gardasil ran before hit movies, during popular TV shows, and on YouTube at any hour of the day. The ads flashed images of strong, healthy, independent, and active young women riding horses, skateboarding, playing soccer, sewing their own clothes, boxing, playing the drums, and jumping double Dutch as they declared that they had chosen to be "one less" victim of cervical cancer, repeating the phrase "one less" like a chant.[46] Consumers didn't need to be told that Merck's marketing was ubiquitous, but the media reinforced the idea anyway. "It's hard to open a magazine targeted to young women and not see an ad for Gardasil," a public health researcher told the *Los Angeles Times*.[47]

But Merck didn't just market Gardasil; it had begun marketing cervical cancer and HPV in 2005 and 2006, before Gardasil was approved by the FDA. Studies in the early 2000s had revealed that most U.S. women had never heard of HPV, and that only half of those who had heard of it knew of its link to cervical cancer.[48] In 2006 Merck ran a national ad campaign featuring the catchphrase "Tell Someone": in television, radio, and print ads, women told each other of the link between HPV and cervical cancer. (Some of the women stood alone; some stood next to their teenage daughters.) The business press called the "disease awareness" campaign "informative and influential," but they also noted that Merck was leaving something out: "What's strangely missing . . . is the key fact that HPV is a sexually transmitted infection."[49] The Gardasil ads were "platonic," but if the CDC recommended the vaccine not just for girls but for boys, too, industry observers predicted, the company was going to have to own up to the fact that the disease was spread through sexual contact.[50]

Merck's marketing—and in particular its distortion of the risk of HPV—prompted a number of health and medical experts to back away from mandates. They accused the company of deceptively "minimizing" the sexual transmission of HPV, "maximizing" the threat of cervical cancer, and ignoring differentials of cervical cancer risk among "subpopulations."[51] They chastised the company for pushing the vaccine on youths in the West even though cervical cancer was a "major killer" in the developing world, especially Africa—where the vaccine was way too expensive to be widely used.[52] But this amplification of risk in the West was precisely what Merck was after. In the early 2000s, Merck's stock had been slumping, and investors "yawned" at the news that the company's lineup of future medicines featured a vaccine for cervical cancer.[53] But the company's first billion-dollar product, a drug to reduce high blood pressure, a risk for cardiovascular disease, suggested to Merck executives a way to approach Gardasil's promotion: market it as a drug with the potential to universally reduce the risk of a disease.[54] The strategy worked and won the company widespread praise: *Pharmaceutical Executive* magazine named Gardasil the 2007 "Brand of the Year" and lauded Merck for making "a market out of thin air."[55]

In this sense, Gardasil fit into a much broader pattern of drug use and marketing. From the 1950s through the 2000s, drug companies marketed—to physicians and consumers—conditions ranging from social phobia to erectile dysfunction, and risk factors including hypertension and high cholesterol, in order to sell their treatments as widely as possible.[56] Historians Robert Aronowitz and Jeremy Greene, among others, have argued that the contemporary moment is in fact defined by the use of drugs to reduce the risk of disease, as opposed to treating disease itself—much like Merck's blockbuster blood pressure drug. Gardasil readily fit into this mold: it doesn't prevent all HPV infections, and it doesn't prevent all cervical cancers, but it effectively reduces an individual's risk of both. Aronowitz notes, however, that this very approach poses a problem: "When vaccines are co-constructed as proprietary drugs against individual risk, they are in danger of losing their appeal as public goods."[57] Vaccine mandates make transparent sense when the objective of vaccination is to eliminate transmission of a disease through a population. But Merck's own marketing approach had highlighted individual risk at the expense of population-level goals. And proposed mandates, by targeting only girls—as opposed to girls and boys—further undermined the idea that the vaccine was meant to benefit all. Which, in turn, further undermined popular acceptance of compulsory HPV vaccination as a health responsibility of all female children.

Many of the acts Merck stood accused of in 2007 and 2008—including dramatizing the complications of an infection and underscoring individual disease risk—had all been committed before, not just by drug companies, but by health officials, medical associations, and other groups looking to increase vaccine uptake in the past. In the 1960s, doctors and health officials had argued that the public needed to be more afraid of measles' dangers, so that they'd be as willing to vaccinate against it as they had been to vaccinate against polio. In the 1970s, health officials had amplified the threat of mumps-induced complications to ensure the vaccination of all children against the disease. In the 1980s and 1990s, the sexually transmitted nature of hepatitis B had been minimized in an effort to justify the widespread vaccination of infants and youths. From the 1960s through the 1990s, in fact, the "marketing" of vaccine-preventable infections' complications and prevalence was often accomplished as much by health officials as by drug companies in an era when pharmaceutical advertising options were more restrained than they became in the 2000s.[58] And the deliberate downplay of differentials of disease risk among subpopulations had been a recurring feature of vaccination programs and policies throughout. In all of these instances, public purchase of new vaccines was dependent on a deliberately constructed awareness of the dangers of a disease the public once widely thought of as harmless or commonplace, if they thought of it at all. The very same was true for HPV.

GIRLS SPEAK OUT

The public may have disparaged Merck in 2007 and 2008, but shareholders and industry observers were enamored with the company. With Gardasil's launch, not only had Merck created a market out of "nothing," and not only had it rebounded from a slump with eye-popping profits— it had also "open[ed] dialogue between mothers and daughters," and "taught girls and young women how to talk about sensitive issues in a forthright, unapologetic way."[59] Such pronouncements may have given Merck's Gardasil campaign a bit too much credit, however—especially as they overlooked the role of new media in creating a space for youths to talk about risqué topics with little oversight or consequence.

Adolescents have rarely been the sole target of vaccine mandates; until very recently, only booster shots were routinely recommended for children in their teens. In 2005, however, the ACIP recommended that all children receive the meningococcal vaccine prior to entering high school. Gardasil thus became the second adolescent vaccine, and the first to draw extensive commentary and resistance from adolescents themselves. By targeting girls, proposed mandates had, not surprisingly, invited charges of discrimination on the basis of gender. "Compulsory vaccination has a legitimate place in our health care system. But why should the government restrict its vaccinations to the victims? Why not include the carriers?" asked a reader in a letter to the New York Times.[60] A physician put it more bluntly when she asked her colleagues, "What about the boys?"[61] The ACIP would answer that question in 2011, when it endorsed the vaccination of eleven- and twelve-year-old boys against HPV.[62] In the meantime, the notion that the HPV mandates reflected and propagated a form of sexual discrimination was expressed in stark terms by teenage girls themselves, who spoke out with candor and passion on the issue online.

Historian Heather Prescott has argued that teens have long desired to be participants in their own medical decision-making and have their opinions on medical matters heard and respected, but that they've rarely been given the opportunity. Historically, when their opinions on medical matters have been solicited, teens have tended to express only what they thought adult authority figures wanted to hear.[63] But social networking websites, which began to take off in the early 2000s, created a histori-cally unique forum for teens to voice their views. The relative anonymity, exposure to diverse opinions, and lack of adult oversight made social media an ideal venue for teenage girls to process the popular debate cen-tered on them and their sexual health. In fact, within a year of Gardasil's approval, the websites MySpace, Facebook, and YouTube boasted thou-sands of groups, videos, blogs, profiles, postings, and forums revealing teens' and young women's reactions to Gardasil, cervical cancer, and all the recent talk of HPV.

For the most part, teens who chimed in on discussion groups and blogs channeled Merck's ads: they urged their peers to spread the word about HPV infection, talk to their doctors, get vaccinated, and order their Commitment bracelets online. On YouTube and elsewhere, teens gushed over how "cool" the skateboarder and drummer in Merck's now-famous "One Less" ads were. Others complained about how incredibly painful (but worth it) the series of Gardasil shots was. Still others—the minority, but no small fraction—engaged in a debate of their own over the merits and drawbacks of proposed HPV vaccination mandates. Teens (male and female alike) called the mandates sexist, offensive, degrading, an imposition, unnecessary, and incompatible with their religious and moral views. Some criticized the high price of the vaccine and Merck's corner on the HPV vaccine market: "F—K One Less!," wrote a com-menter on YouTube. "It's all about 'Lot's more' (of 'Ka Ching' in Mercks' bankaccount [sic] . . .)." Others cited the Vioxx scandal as evidence that Gardasil, too, might someday be proven unsafe; some held up their own side effects as proof that Gardasil was already unsafe. As in the broader debate, safety concerns dominated teens' criticism of the mandates, but here teenage girls expressed a feminist demand for bodily sovereignty that was largely missing from conversations in mass media. "Required vaccination of gardasil shots is a violation of individual rights," wrote one teen; "we are not guinea pigs," wrote another; "its my body Ill make the decision for what goes in it," wrote yet another, typos and all.[64]

In some ways, teens were among the harshest critics of the discriminatory nature of proposed mandates and the gendered implications of Merck's ad campaign. Online, they proved their savvy as media consumers and producers, making spoofs of Merck's ads and exposing the assumptions and norms embedded therein. In one video posted to YouTube, a group of teenage girls parodied the "hip" female youths featured in the "One Less" commercials, replacing them with their own array of "tough" and "independent" women: a sullen housewife, a gang member, a binge drinker, a stripper. The spoof subverted the implications of universal risk inherent not only in the ads, but in federal recommendations and proposed state laws as well. *You think HPV is bad?* it seemed to say. *We can think of worse.*[65] No matter where they stood on the cultural or political spectrum, teens displayed a sophisticated ability to see through the national campaign directed at them. In another video, a teen playing a drug-company spokesperson promised to "drug you up good" and warned against such serious vaccine side effects as "heavy tooting near midnight."[66] Another parody, for the imagined product "Guard Yourself," cautioned that the side effects of abstinence could include the "restoration of family," "true freedom," and the "ability to follow God more closely."[67]

Beyond videos that featured teens themselves, in social media it wasn't always easy to tell who was actually a teen and who was not. Though many online commenters and contributors shared their HPV infection status, diagnosis experiences, and even experiences of cervical cancer, many of these hid behind aliases that made their identities impossible to discern. On the one hand, such sharing foreshadowed Hannah's sharing on *Girls* a few years later—instead of crying about her HPV diagnosis quietly and alone, she shared it with all of her friends, her current boyfriend, and an ex. On the other hand, sharing from behind aliases perhaps suggested that girls weren't as "forthright and unapologetic" about HPV as Merck's admirers said they were. Moreover, sometimes a little digging turned up the fact that the girls chatting about Gardasil and HPV online weren't girls at all. "One less person with a voice. One less person to vote. One less woman to represent us. . . . It's your body! Question everything!!!" wrote a YouTube commenter—who turned out to be not a teenager, but a woman in her forties whose sister had survived stage three cervical cancer but who still opposed the HPV vaccine for girls.[68]

TRANSFORMING HPV

For a variety of reasons, feminist critiques of HPV vaccination for girls were largely muted in the public realm. Before compulsory HPV vaccination became an issue, women's health advocates and researchers had touted HPV vaccines as a miracle drug for women's health. When states began considering mandates, some feminists backed them as an ideal means of making cervical cancer prevention accessible to all women. Relatively few observers considered the approach sexist. In part, this was due to the fact that lawmakers, health experts, and Merck deemphasized HPV's status as a sexually transmitted infection, focusing instead on Gardasil's potential to protect girls and young women from a future health problem that was uniquely theirs: cervical cancer. At the same time, Merck's ads deliberately cultivated HPV vaccination as an act of empowerment and female choice. In online forums, feminist critiques of HPV vaccination as a female responsibility were far more prevalent. But they became more prevalent still, across all types of media outlets, as the policy debate over HPV vaccination evolved.

Ms. magazine's coverage of the HPV vaccine encapsulates this evolution. A 2005 article on the HPV vaccine—then still in development— called the vaccine a "magic bullet" and the "first real advance in women's medical care in 50 years." The article's author predicted that the vaccine would likely be given to girls and boys, both to encourage herd immunity and because HPV caused not just cervical cancer but penile and anal cancers as well.[69] That didn't happen, of course, but *Ms.* contributors didn't seem too troubled by gender-specific mandates—at least not at first. In 2007 the magazine weighed in on the then-raging debate over HPV mandates for girls. The author, who had vaccinated her own fifteen-year-old daughter, wondered how any parent could "say no to the first-ever cancer vaccine." She lamented Gardasil's high price, attributed the defeat of state mandates to conservative resistance, and expressed hope that the vaccine would soon become "as ubiquitous, available and reasonably priced as those for measles and polio."[70] A 2010 article, by contrast, argued that the country needed "frank public talks about the full range of HPV cancers." HPV, noted the author, was an "equal opportunity infector," but men weren't getting vaccinated—in large part because no was one asking them to.[71]

By that time, state lawmakers had effectively abandoned the effort to make HPV vaccination mandatory for girls. But the argument that the nation's approach to HPV prevention was sexist only heated up as the debate over mandatory vaccination for girls burned itself out. One reason *Ms.* writers and others didn't take umbrage at proposed mandates targeting girls was because they assumed that vaccination for boys was just around the bend; as Merck representatives frequently pointed out, the vaccine simply hadn't been adequately studied in them yet. In October 2009, as anticipated, the company received FDA approval to market Gardasil for the prevention of genital warts in boys.[72] But when the ACIP convened later that month, they decided not to include the vaccine in the routine immunization schedule for males. Boys *could* get vaccinated, they concurred, but routine vaccination of boys would be costly, with "minimal benefits."[73] The decision stemmed in part from economic analyses showing that as long as HPV vaccination in girls was sufficiently high, vaccinating boys provided little additional "value."[74]

Online media lit up with accusations of a health care double standard. "This is blatantly sexist," said a *Wall Street Journal* reader.[75] "Thanks for taking all those shots, girls! Now we don't need to do the same to boys," joked a commentator on CBS.[76] "In other words," a writer for *Slate* observed, "boys don't have to get vaccinated for the same reason they don't have to wash dishes, do laundry, buy birth control, or think about other people in general: Girls will do it for them."[77] Feminist blogger Amanda Hess remarked that it was no surprise that this responsibility, like all other sexual health responsibilities, was being forced on female bodies while male bodies were once again "kept at a distance": "I'm forced to receive my annual pap smear and raft of STD tests in order to receive a refill on my birth control prescription, [while] my male sex partner is subject to no such requirement," she wrote. No wonder health experts found it "natural" to add one more health responsibility to women's al-ready long list.[78]

At the same time, the media largely ignored another set of bodies: those of gay men. "Vaccinating women won't help them," noted one of the few journalists to bring the subject of gay men into the debate. "They can't count on somebody else to take care of the problem."[79] The ACIP had in fact worried over what to do for the subcategory of "men who have sex with men," who were, as sociologist Steven Epstein has pointed out, almost wholly invisible in the public debate over HPV vac-cination.[80] "MSM" (the term preferred by public health researchers) suf-fered higher rates of genital warts, anal cancers, and pre-cancers com-pared to heterosexual men, and ongoing trials by Merck scientists were showing that Gardasil was effective in them. A physician representing the Gay and Lesbian Medical Association made an in-person plea to the ACIP to recommend HPV vaccination for all boys, in order to address the barriers that gay youths, especially gay youths of color, already faced in getting health care. At least one member of the committee proposed adding a special clause to the official HPV guidelines to encourage young MSM to get the vaccine. "While it would likely be a departure for ACIP, at least [we] could be adventuresome and consider something different for a change," he said.[81] In the end, however, the committee demurred, citing the need for more data to come in.

The committee didn't have to wait long, because new data on HPV was accumulating at a rapid pace. A 2006 review of the science on HPV and anogenital cancers had noted that there were "few reports" on cancers of the vulva, vagina, penis, and anus, largely because tumors in these sites were so "much rarer" than cancers of the cervix.[82] A few years later, when the same authors revisited the state of the evidence on HPV and cancer, they remarked that in recent years research inquiries on two aspects of HPV—its transmission between men and women, and its prevalence among men—had increased in number.[83] Other recent studies showed that while rates of other cancers were declining or hold-ing steady, two types of HPV-related cancers—oropharyngeal and anal cancers—were on the rise.[84] Even as the ACIP rejected routine HPV vaccination of boys in 2009, a much more robust picture of the burden of HPV in men and women—gay or straight—was already beginning to emerge. Scientists called to present the latest HPV research to the committee reported that men had just as high a rate of HPV infection as women, and possibly higher. Men also suffered genital warts from HPV infection, in addition to a condition called recurrent respiratory papillomatosis, and several forms of cancer, including cancers of the anus, penis, mouth, and throat.[85] In fact, of the more than 17,000 HPV-related cancers affecting men and women in the United States each year, 7,500 affected men. And of those 7,500, about 4,600 were caused by the strains of HPV that Gardasil (and the recently licensed Cervarix) protected against.[86]

Research on HPV and other forms of cancer was directly driven by the vaccine — and its makers. When researchers at the American Cancer Society conducted an annual report on cancer incidence in 2012, they focused on HPV-associated cancers between 1975 and 2009, because they were a "topic of special interest." The researchers found that while cervical cancer rates were decreasing, rates of HPV-associated mouth, throat, anal, and vulvar cancers were increasing. They presented the data alongside statistics on HPV vaccine coverage, arguing that the fact that just 32 percent of girls had received the full series of shots against HPV "underscor[ed] the need for broader interventions to increase vaccine uptake."[87] Whether the same picture of HPV-related cancers would have emerged absent an HPV vaccine is hard to say—but it seems unlikely. Research showing that oropharyngeal cancers caused by HPV were on the rise came from Ohio State University researcher Maura Gillison, who received grant support from Merck and had consulted for Merck and GlaxoSmithKline. A much-reported British study that showed head and neck cancers were on the rise in the developed world, because of the rise in forms of the cancers caused by HPV, came from a research institute that did work for GlaxoSmithKline.[88] Collaborations between industry and academic and government scientists have been the norm for more than a century, but it's quite clear that industry would have had little interest in expanding knowledge on the prevalence of HPV-related cancers if it didn't have a product to promote and a market to expand.[89]

The news that HPV caused more than just cervical cancer made its way into the media, too. When actress Farrah Fawcett died of anal cancer in 2009, some reports noted that her form of cancer was, like cervical cancer, linked to HPV.[90] In fact, noted *Fox News* (among others), anal cancer was on the rise in the United States, "mostly because of the increased spread of HPV."[91] So was oral cancer—for the same reason. "It's something we are very much aware of and look for," a prosthodontist told CNN, in a 2010 story that mentioned that some oral cancer survivors were enrolling in a trial that was investigating whether "a form of the HPV vaccine" could prevent their cancers from returning.[92] A few years later, actor Michael Douglas made headlines when he told a British newspaper that his throat cancer, diagnosed in 2010, was caused by HPV contracted during oral sex (an admission that caused the interviewer to do a double take).[93] Elsewhere, a reporter pointed out that you didn't necessarily need to engage in oral sex to get HPV, which was spread "also by French Kissing or even just sharing a water glass."[94] With media attention like this, HPV began to shed its tight link to cervical cancer and to become a virus that was everywhere, threatening a range of cancers in men and women alike.

In truth, HPV was estimated to be responsible for less than 2 percent of all cancers in North America (compared to more than 14 percent of all cancers in sub-Saharan Africa and more than 15 percent in India).[95] But in light of accumulating evidence on HPV's role in a variety of cancers in men and women, the ACIP returned to the issue of vaccinating boys. In October 2011, the committee recommended that boys be routinely vaccinated with Gardasil at the age of eleven or twelve, and that boys ages thirteen to twenty-one be vaccinated if they hadn't already gotten the vaccine.[96] (Few had; at the time, just 1.4 percent of boys were vaccinated against HPV.[97]) Explaining the reasons behind their decision, the ACIP cited the 7,000 cancers per year in men attributable to HPV 16 and 18, and the higher incidence of anal cancers, pre-cancers, and genital warts among men who have sex with men. There was, as yet, no data on the vaccine's ability to prevent oropharyngeal or penile cancers, but the fact that male vaccination would reduce overall HPV transmission and cancer in women made the policy worthwhile. On the topic of cost-effectiveness, the committee noted that male HPV vaccination was worth the expense when less than half of females were fully vaccinated by age twelve; in 2010 less than half of females were vaccinated by age seventeen.[98] The committee's decision made the front page of the *New York Times* — for those who still read front pages. Beyond that, it garnered relatively little attention. Said a Brown University cancer researcher, "There's been a surprisingly muted reaction. . . . But, you know, sexual activity is something that's almost expected of boys."[99]

Despite low vaccination rates, the prevalence of vaccine-preventable HPV strains among teenage girls fell by more than half between 2007 and 2010, from 11.5 percent to 5.1 percent.[100] Conventional wisdom, based on experience with other vaccine-preventable infections, held that vaccination rates had to reach 90 percent or higher to slow an infection's spread through a population. But HPV levels were declining with just a third of girls, and a tiny fraction of boys, fully protected against infection.

The head of the CDC, Thomas R. Frieden, announced the unexpected and "striking" impact of the first three years of HPV vaccination in a widely covered press briefing in 2013. He noted that HPV was the nation's most common sexually transmitted infection, responsible for a range of cancers — including head and neck cancers — in men and women alike. He called the nation's low HPV vaccination rates "unacceptable" and pointed out that Rwanda had vaccinated far more females (a full 80 percent) than the United States had. He listed a familiar set of reasons why vaccine uptake wasn't higher: doctors weren't doing their part to encourage patients to get vaccinated, some parents didn't think HPV vaccination was necessary, other parents worried the vaccine was unsafe, and for some families the full series of shots against HPV was just too expensive.

Frieden issued the customary entreaty to doctors and reassurances to parents, and noted that under recent health reforms spearheaded by President Barack Obama, the vaccine — like all vaccines — would soon be free for all who wanted it. He added that the HPV vaccine was "an anti-cancer vaccine," and the latest statistics "a wake-up call," to vaccinate youths more widely. "I think we owe it to the next generation, our sisters, our daughters, our nieces, and to those who are our patients to protect them against cervical cancer," he said.[101] His comments were a subtle reminder that vaccination held a position at the core of health reform and was often a unique responsibility of women and children. They were also a reminder that vaccination's success relied on the cooperation of physicians, public health experts, parents, and industry; hinged on popular acceptance of technocratic risk assessments; and was critically dependent on the framing of vaccine-preventable diseases.

As in the past, the first new vaccine against HPV brought unprecedented popular attention to the infection. But the varied objectives of HPV vaccination competed for attention. Was the HPV vaccine primarily a vaccine against HPV, or cervical cancer, or an ever-growing list of cancers in women and men? Frieden's own comments revealed the challenges of framing the target of the HPV vaccine consistently. But one thing was clear: Most average Americans hadn't heard of HPV, or worried about it, before Gardasil entered the scene. Five years later, HPV was widely known, and it was rapidly shedding its old reputations as a ubiquitous but usually harmless infection, a frequent cause of genital warts, and a virus linked solely to one form of cancer.

Gardasil's story contained many of the same elements of its late twentieth-century predecessor vaccines, but its plot unfolded at an accelerated pace. The speed with which Gardasil was approved for market by the FDA, recommended for use by the Advisory Committee on Immunization Practices, and required for school enrollment by proposed state bills reflected an assumption on the part of multiple parties, including Merck, government officials, and state lawmakers: that the new era of vaccination, marked by public acceptance of the mandatory universal vaccination of children, no matter the vaccine or the disease against which it protected, was firmly established. But proposed laws to require HPV vaccination for sixth-grade girls drew loud fire from a broad and diverse group of Americans. And with popular rejection of proposed mandates, HPV vaccination fell out of the mold.

Like other vaccines before it, Gardasil was framed by the cultural and political preoccupations of its time. Proposals to make the vaccine mandatory spotlighted unresolved questions about how vaccines against "milder" diseases of limited communicability and debatable epidemic stature should be deployed. Neither the public, nor scientists, nor policy makers, nor pharmaceutical companies had arrived at a singular agreed-upon answer to the questions of what and who this vaccine was for, and how, ultimately, the health-citizenship responsibilities of the nation's youths should be circumscribed. In the resulting void, the HPV vaccine became the latest object of a long-standing and increasingly urgent debate. After much heated back-and-forth, the nation settled on a set of compromises to answer these questions with respect to Gardasil. The vaccine, like the many that came before it, offered a cutting-edge preventive to a long-standing health threat for those who wished to avail themselves of it. But children, and especially girls, would not be required to take it—at least not at this point in time. Gardasil's fate in the first decade of the twenty-first century was yet one more illustration of the historical contingency of vaccine reception.

CONCLUSION
| | | | | | | | | | | |

In the current era of vaccination, unofficially heralded by the adoption of the Vaccination Assistance Act of 1962, the protection of the adult populace and future generations of Americans against the acute and far-ranging effects of severe and "mild" infections has rested in large part upon the shoulders—or literally, in the arms—of children. This era of vaccination, now roughly five decades old, was marked by several defining characteristics. Federal authority in the area of vaccine recommendations became strong and widely recognized. Public acceptance of the mandatory universal vaccination of children became firmly established. And this acceptance was held in place by public awareness and acknowledgment of a set of health threats worth avoiding, in addition to a patchwork of local laws and regulations making vaccines compulsory for children attending day care, grade school, and sometimes college.[1]

Often the designation of a health "threat" occurred upon the appearance of a vaccine itself. Neither mumps nor chicken pox nor cervical cancer were defined as top U.S. public health priorities in and of themselves before their vaccines were licensed for use. But they were treated as such afterward. Vaccines brought attention to these diseases and others, and in the process refashioned the tools at hand for understanding the infections and their risks. Sometimes, as in the case of measles and hepatitis B, the disease began to appear worse or more widespread than it had before, or it was framed so in order to encourage vaccine uptake. In other instances, vaccines made diseases seem like an unnecessary inconvenience to a productive and comfortable middle-class life-style. This was the view captured in John F. Kennedy's announcement of the Vaccination Assistance Act in 1962, and it was the view that Merck, for one, tried to capitalize on when it advertised its mumps vaccine over a decade later.

Kennedy's promise on announcement of the Vaccination Assistance Act was that Americans no *longer* had to suffer preventable infections; it was also that *no* American should have to suffer. In the period introduced with Kennedy's legislation, vaccines promised to be great equalizers in a nation whose health care system has been fraught with dramatic inequalities in access to care and services. The federal vaccine programs of subsequent decades strived to make vaccines available to all children regardless of family means. Because they are demonstrably cost-effective (for the most part) and easy to administer, and because their use is widely enforceable, vaccines and the policies that govern them came to comprise a not-insignificant portion of this country's universalized health care.

The landmark health reform legislation signed into law by President Barack Obama in 2010 took things even further. The Affordable Care Act, which the president called "the most important reform of our health care system since Medicare passed in the 1960s," placed vaccines, once again, at the heart of reform.[2] The new law extends access to health coverage and reins in still-escalating health costs in a variety of ways, including the promotion of efforts to prevent disease. Vaccination didn't get much attention during Congress's intense imbroglio over proposed health reforms, but the final version of the law requires insurance plans to provide enrollees with all ACIP-recommended vaccines for free. Once the law goes fully into effect, 94 percent of the population—the fraction of the population estimated to be covered by health insurance—will have access to federally recommended vaccines at no cost. The law also increases funding for programs that provide immunization services and provides new funds for states to buy vaccines and encourage children, adolescents, and adults to use them.[3] Federal involvement in vaccination continues to expand, and incentives to vaccinate widely continue to accumulate.

Even before Affordable Care Act changes went into effect, vaccination rates for all of the recommended childhood vaccines were at or near peak levels.[4] But while these overall coverage rates should have been adequate in most cases to establish herd immunity and keep the population free and clear of preventable infections, significant outbreaks of vaccine-preventable diseases continued to occur well into the 2000s. Some were attributed to communities with high concentrations of vaccine refusers.[5] Others, such as the outbreaks of mumps and pertussis in 2010, occurred despite high vaccination rates, prompting health officials to wonder whether vaccine-induced immunity was waning, whether pathogens were evolving, or whether some genetic attribute left certain people still susceptible to disease despite vaccination.[6] Such outbreaks were a reminder that the scientific and political promise made in the 1960s, of an era of freedom from infectious disease through vaccination, was not just far from realization, but an impossibility. Despite this, disease eradication and elimination remain the contemporary expectations of the mass vaccination of children.

The reduction of any disease in a population rests heavily on a broad social consensus that the disease is worth preventing. Since the 1960s, we have achieved this consensus many times over by consistently turning the targets of vaccines into deadly killers, hazards to children, and top public health priorities. But we have vaccinated as much for nonmedical reasons as for medical ones. Over the past half century, we've embraced mass compulsory vaccination in the interest of national security, to slow the ascent of health care costs, to meet the wellness expectations of a wealthy society, to avoid individual discomfort and economic inconveniences, out of faith in biomedicine, out of deference to authority and trust in experts, to placate a host of cultural anxieties, and, as a CDC epidemiologist put it more than five decades ago, because we can. Vaccine rejection, naturally, follows a similar pattern of drawing justifications from across our social, cultural, and political expectations and norms.

This points to the two core lessons that emerge from the social history of vaccination in the United States over the past half-century. First, we have never embraced the mass vaccination of children for strictly medical reasons. Second, new vaccines create imperatives to vaccinate, largely by triggering the transformation of their target diseases into diseases in urgent need of prevention. Measles, mumps, hepatitis B, chicken pox, and HPV—to list only the examples explored in detail here—became vaccine-preventable diseases in the wake of their vaccines' approval for widespread use. This label now holds a unique cultural and scientific meaning. It signifies a disease or infection whose risks are socially prominent, whose complications are manifold, and the urgency of whose prevention has become impossible to dismiss. As new vaccines are developed—such as the vaccines against HIV, Lyme disease, the common cold, smoking, diabetes, and others now in clinical trials—our thinking on these conditions and behaviors, too, will dramatically shift. Whether these and other vaccines are successfully adopted as the targets of mandates supporting the mass vaccination of children, however, will depend, as it always has, on their particular moment in time.

APPENDIX: THE SCIENCE
AND REGULATION OF VACCINES
| | | | | | | | | |

The social history traced in this book took place within a changing scientific and regulatory context. Vaccines protect against disease by prompting the body's immune system to generate antibodies and other immune cells that recognize and attack bacteria and viruses. Vaccines protect individuals, but they also protect communities—when a sufficient fraction of the population is vaccinated against a virus or bacterium, the entire population is protected, because the decreased number of susceptible people means the pathogen no longer has enough "hosts" to sustain itself in that population. The phenomenon of "herd immunity" means that not every individual in a community needs to be vaccinated to control or even eradicate a disease. As long as enough members of the "herd" are protected, the whole herd is protected—including those who didn't get vaccinated or couldn't, for reasons such as, for example, age or conditions affecting the immune system.

Vaccines themselves contain all or part of the target pathogen, sometimes in its live form, and sometimes killed or in a weakened state. The Salk polio vaccine (the polio vaccine currently in use in the United States), for example, consists of inactivated or killed polio virus, which triggers the immune system to manufacture antibodies that are effective against live polio virus. Pertussis vaccine in its original form contained killed whole pertussis bacteria; the "acellular" pertussis vaccine that came into use in the 1990s contains only key proteins from the bacteria. Vaccines against smallpox, measles, mumps, rubella, and rotavirus are live-virus vaccines, meaning they contain a live version of the target virus. In each case, the virus in the vaccine has been weakened or altered, either through passage in animal cells or (more recently) through genetic engineering.[1] The altered virus doesn't cause disease, but it is still familiar enough to the immune system to trigger the formation of antibodies. So-called "recombinant protein" vaccines, such as the vaccines against hepatitis B virus and HPV, contain antigens that were produced by genetically engineered or recombinant microorganisms; the hepatitis B vaccine, for example, contains antigens manufactured by recombinant yeast.[2]

Vaccines are highly effective, but because of their very nature as biological products, they have been, on rare occasions, implicated in the spread of disease. Nineteenth-century smallpox vaccination in the United States and Europe, which in one form involved transferring pus from the arm of one vaccinated person to the next, sometimes transmitted other infections as well, notably syphilis.[3] In the United States, the Biologics Control Act was signed into law in 1902 in response to an episode in which twenty children were sickened and fourteen died after receiving diphtheria immunizations contaminated with tetanus.[4] The act was the first federal effort to oversee the safety of vaccines, which were regulated by the Public Health Service and Marine Hospital Service's Hygienic Laboratory's Biological Control Service.

Yet another contamination episode—in which polio vaccine manufactured by California-based Cutter Labs was contaminated with live polio virus, causing the disease in more than two hundred people—prompted further federal involvement in vaccine regulation in the 1950s. The incident led to the establishment of the Division of Biologics Standards (DBS) within the National Institutes of Health (which had been created two decades earlier from the PHS/MHS Hygienic Laboratory) to more carefully monitor vaccine safety. But following reports in the late 1960s and early 1970s that DBS had failed to properly screen vaccines for safety and effectiveness, oversight of vaccines was strengthened and transferred once again, this time to the Food and Drug Administration's new Bureau of Biologics. The bureau, now called the Center for Biologics Evaluation and Research (CBER), currently regulates and licenses both vaccines and vaccine manufacturers, in addition to other biological products and their makers.[5]

Before licensing a vaccine for use, the FDA's CBER evaluates the vaccine's safety profile, its ability to provoke an immune response, its ability to protect against disease, and its interaction with other drugs and vaccines. Safety and efficacy standards for vaccines are flexible. While vaccines are extremely safe, many do pose small inherent risks of adverse events or side effects, and these risks are weighed against the risks of the target infection before a vaccine is approved.[6] Though federal health officials discussed in the 1960s and 1970s the need for a means of compensating citizens who were harmed by approved and recommended vaccines, no such program was implemented until the National Childhood Vaccine Injury Act was signed into law in 1986. The act was prompted by reports that pertussis vaccine posed non-disclosed risks of encephalitis, brain damage, and death, as described in chapter 5. In addition to establishing a National Vaccine Injury Compensation Program, the 1986 law also created the Vaccine Adverse Event Reporting System. To this day, any individual may submit information on a vaccine reaction to VAERS, which is administered jointly by the FDA and CDC.

ACKNOWLEDGMENTS

| | | | | | | | | | |

This project has benefited from the advice, expertise, and generosity of numerous readers and sources over the past several years. Some read the whole manuscript, some read parts, others listened, still others provided access to the past. Many thanks to Patrick Allitt, Roger Bernier, Liz Bounds, Amy Benson Brown, Julia Bullock, Bob Chen, Claire Clark, James Colgrove, Georganne Conis, Paula Conis, Cynthia Cournoyer, Karen Darling, Barbara Loe Fisher, Kathleen Fullerton, Elizabeth Gallu, Deb Gust, Mary Hilpertshauser, Alan Hinman, Kiersten Israel-Ballard, Robert Johnston, Howard Kushner, Aimee Medeiros, Albert Nason, Peggy O'Mara, Abena Dove Osseo-Asare, Sarah Patamia, Dorothy Porter, Justin Remais, Rob Richards, Tom Rogers, Elizabeth Watkins, members of the Spring 2013 Johns Hopkins University History of Science, Medicine, and Technology Colloquium, the students in my Spring 2013 Drugs and American Culture course at Emory University, the students in Sander Gilman's Spring 2010 Representing Health and Illness course at Emory, the 2007 members of the International Network for Health Policy and Reform, and the anonymous reviewers who read this manuscript.

Special thanks to the institutions that supported my research, including the Department of History, Emory University; the Program in the History of Health Sciences at the University of California, San Francisco; Rollins School of Public Health, Emory University; the Louise M. Darling Biomedical Library, University of California, Los Angeles; the Chemical Heritage Foundation; and the American Association for the History of Medicine.

NOTES

| | | | | | | | | | |

INTRODUCTION

1. Popular debate over vaccination is, as other historians have argued, a proxy for other cultural tensions or societal disagreements. Most recently, Mark Largent argued that the debate over vaccines and autism that took place in the 2000s was a proxy debate about "a complex set of concerns about the modern vaccine schedule." Mark Largent, *Vaccine: The Debate in Modern America* (Baltimore: Johns Hopkins University Press, 2012), 1.

2. Centers for Disease Control and Prevention (CDC), "National, State, and Local Area Vaccination Coverage among Children Aged 19–35 Months—United States, 2009," *Morbidity and Mortality Weekly Report* 59, no. 36 (2010): 1171–77.

3. Scholars agree that a new era of *compulsory* immunization began in the 1960s. See, for example, James Colgrove, *State of Immunity: The Politics of Vaccination in Twentieth-Century America* (Berkeley: University of California Press, 2006); Judith Sealander, *The Failed Century of the Child: Governing America's Young in the Twentieth Century* (New York: Cambridge University Press, 2003); and Michael Willrich, *Pox: An American History* (New York: Penguin, 2011), 339.

4. Colgrove, *State of Immunity*. For more details on how vaccines provide community benefits, see the appendix.

5. Ibid., 38–44, 96–97; Ramunas Kondratas, "Biologics Control Act of 1902," in *The Early Years of Federal Food and Drug Control*, ed. James Harvey Young (Madison, WI: American Institute of the History of Pharmacy and the American Pharmaceutical Association, 1982), 8–27; Willrich, *Pox*, 166–210; Lawrence Gostin, *Public Health Law: Power, Duty, Restraint* (Berkeley: University of California Press, 2000), 66–69.

6. See, for example, Samuel Woodward, "An Argument in Favor of Vaccination, with Statistics of the Incidence of Smallpox in the United States, Its Dependencies and Canada," *New England Journal of Medicine* 202, no. 3 (1930): 122–24.

7. Smallpox, which sickened 21,000 Americans and killed nearly 900 in 1900, claimed its last U.S. victim in 1949. Diphtheria, a top killer in 1900, caused just 628 cases of disease in 1960. The United States often saw upwards of 100,000 cases of

pertussis early in the century; decades later, the number of annual cases had fallen to just over 1,000. Vaccines were not alone responsible for these shifts, but they did play a critical role, along with improvements in nutrition, sanitation, and overall standard of living. Stanley A. Plotkin and Walter A. Orenstein, *Vaccines*, 5th ed. (Philadelphia: Saunders, 2008), 467; CDC, "Impact of Vaccines Universally Recommended for Children—United States, 1990–1998," *Morbidity and Mortality Weekly Report* 48, no. 12 (1999): 243–48.

8. Woodward, "An Argument in Favor of Vaccination."

9. Samuel Woodward, "Arguments in Favor of Compulsory Vaccination for Private School Children," *New England Journal of Medicine* 206, no. 11 (1932): 570–72. See also Robert Johnston, *The Radical Middle Class: Populist Democracy and the Question of Capitalism in Progressive Era Portland, Oregon* (Princeton, NJ: Princeton University Press, 2003), chaps. 12, 13. For other examples, see Colgrove, *State of Immunity*, chap. 2; William J. Reese, *Power and the Promise of School Reform: Grassroots Movements during the Progressive Era* (Boston: Routledge & Kegan Paul, 1986), 232–35; and Michael Willrich, "'The Least Vaccinated of Any Civilized Country': Personal Liberty and Public Health in the Progressive Era," *Journal of Policy History* 20, no. 1 (2008): 76–93.

10. Attempts to make diphtheria immunization compulsory in the 1930s and early 1940s got caught up in this dispute, for instance, and, as a result, few such laws were passed. Paul Starr, *The Social Transformation of American Medicine* (New York: Basic Books, 1982), 196; Evelynn Maxine Hammonds, *Childhood's Deadly Scourge: The Campaign to Control Diphtheria in New York City, 1880–1930* (Baltimore: Johns Hopkins University Press, 2002). See also, for example, G. W. Anderson and G. H. Bigelow, "Diphtheria Immunization in Private Practice," *American Journal of Public Health and the Nations Health* 23, no. 7 (1933): 655–62.

11. Gerald Gross, "U.S. Army Better Equipped Today to Fight Disease than in 1917," *Washington Post*, October 8, 1940, 10; "Influenza Immunization Test Started by Navy," *Los Angeles Times*, July 29, 1941, A5.

12. United Press International, "France Fights Epidemics," *New York Times*, August 5, 1940, 4; Alvin Steinkopf, "Reich at War with Typhoid in Polish State," *Washington Post*, October 14, 1940, 6.

13. Associated Press, "Tell Discovery of a Vaccine to Avoid Measles," *Chicago Daily Tribune*, September 18, 1940, 1. The marginally effective vaccine would be replaced by a more reliable one in the 1960s; see chapter 2.

14. "Adults Are Urged to Be Vaccinated," *New York Times*, May 24, 1942, 28; Adele Bernstein, "U.S. Postwar Epidemics Foreseen," *Washington Post*, October 3, 1943, M12.

15. Leona Baumgartner, "Attitude of the Nation toward Immunization Procedures," *American Journal of Public Health* 33 (1943): 256–60.

16. "Immunization against Smallpox, Diphtheria, Tetanus, and Poliomyelitis," speech given by James L. Goddard before the Clinical Meeting of the American Medical Association, December 3, 1963, Folder: Info 3 Tr.—1963, Box 334065, No. 5, Record Group 442, CDC, Office of the Director Files, NARA Southeast Region. See also Judith Leavitt, "'Be Safe, Be Sure': New York City's Experience with Epidemic Smallpox," in *Sickness and Health in America*, ed. Judith Leavitt and Ronald Numbers (Madison: University of Wisconsin Press, 1997), 407–17.

17. For accounts of the polio vaccine trials and the related activities of the March of Dimes, see, for example, Jane S. Smith, *Patenting the Sun: Polio and the Salk Vaccine*

(New York: William Morrow, 1990); and David M. Oshinsky, *Polio: An American Story* (New York: Oxford University Press, 2005).

18. Otis Anderson, "The Polio Vaccination Assistance Act of 1955," *American Journal of Public Health* 45, no. 10 (1955): 1349–50.

19. Patrick Vivier, "National Policies for Childhood Immunization in the United States: An Historical Perspective" (PhD diss., Johns Hopkins University, 1996).

20. Elizabeth W. Etheridge, *Sentinel for Health* (Berkeley: University of California Press, 1992).

21. *Extension of Poliomyelitis Vaccination Assistance Act, Hearing Before a Subcommittee of the Committee on Interstate and Foreign Commerce, House of Representatives*, 84th Cong. (January 24, 1956), 60.

22. "Surveillance of Poliomyelitis in the United States, 1958–61," *Public Health Reports* 77, no. 12 (1962): 1011–20.

23. *Baby's Milestones: Birth to Seven Years (Completed for "Sarah W." b. June 1964)*, (Norwalk, CT: C. R. Gibson, 1957); Baby Books Collection, Louise M. Darling Biomedical Library, History & Special Collections for the Sciences, UCLA. "Sarah W." is not the baby's real name, though her real name does appear in the baby book cited here.

24. Minn. Stat. 121A.15, https://www.revisor.mn.gov/statutes/?id=121A.15 (accessed June 2012). See *History* 1967 c 858 s 1, 2 at the cited link.

25. These laws are typically referred to as vaccine "mandates," even though exemptions exist for each required vaccine. In all states, children may be exempted from required vaccines for medical reasons; in most states (save Mississippi and West Virginia), they may be exempted for religious reasons. At time of writing, eighteen states also permit "personal" or "philosophical" exemptions. See National Conference of State Legislatures, "States with Religious and Philosophical Exemptions from School Immunization Requirements," February 2012, http://www.ncsl.org/default.aspx?tabid=14376 (accessed August 2012). Though NCSL counts twenty states with philosophical exemptions, Missouri's applies only to preschool children, and New Mexico health officials contend that their religious exemption does not cover philosophical beliefs. Associated Press, "New Mexico Alters Child Vaccination Waiver Form," *SFGate.com*, August 29, 2012; Phaedra Haywood, "State Alters Vaccination Waiver Form, Asking Parents to Cite Religion," *Santa Fe New Mexican*, August 28, 2012.

26. Benjamin Spock, *Baby and Child Care* (New York: Pocket Books, 1964); *Report of the Committee on the Control of Infectious Diseases* (Evanston, IL: American Academy of Pediatrics, 1961).

27. The CDC recommends that all children receive vaccines against hepatitis B, rotavirus, diphtheria, tetanus, pertussis, *Haemophilus influenzae* type b (Hib), pneumococcus, meningococcus, polio, flu, measles, mumps, rubella, and varicella (chicken pox). Some of these are administered as combined vaccines; all of the recommended vaccines (combined or not) are administered in multiple doses, usually totaling between two and four doses. Children with certain risk factors are also advised to be vaccinated against hepatitis A and meningococcus; the latter is recommended for all children beginning at age eleven. CDC, "Recommended Immunization Schedules for Persons Aged 0 through 18 Years—United States, 2012," *Morbidity and Mortality Weekly Report* 61, no. 5 (2012): 1–4.

28. Letter from Lewis Thomas, November 1976, Folder: Immunization, Box 32, Collection JC-DPS: Records of the Domestic Policy Staff, Jimmy Carter Library.

29. My analysis of the framing of vaccines and their target infections is strongly influenced by the work of Charles Rosenberg. See Charles Rosenberg, "Disease in History: Frames and Framers," *Milbank Quarterly* 67, no. S1 (1989): 1–15; Charles Rosenberg, "Framing Disease: Illness, Society, and History," in *Framing Disease: Studies in Cultural History*, ed. Charles Rosenberg and Janet Golden (New Brunswick, NJ: Rutgers University Press, 1997), xxi–xxvi; and Charles Rosenberg, "What Is Disease?" *Bulletin of the History of Medicine* 77 (2003): 491–505.

30. This articulation borrows from Siddhartha Mukherjee's eloquent explanation of the process by which diseases are framed: "Every era casts illness in its own image. Society, like the ultimate psychosomatic patient, matches its medical afflictions to its psychological crises; when a disease touches such a visceral chord, it is often because that chord is already resonating." Siddhartha Mukherjee, *The Emperor of All Maladies: A Biography of Cancer* (New York: Scribner, 2010), 182.

31. For these reasons, I do not use the terms "anti-vaccinationist" or "anti-vaccine" to describe individuals or groups active in the last century unless those groups or individuals defined themselves by their opposition to *all* vaccines.

32. On the subject of health citizenship, see, for example, Dorothy Porter, *Health, Civilization, and the State* (London: Routledge, 1999).

33. Some such critics believe, for instance, that vaccines are behind a subtle, chronic deterioration of the human immune system that is passing from one generation to the next. Such fears are based partly on the fact that vaccines replace natural immunity with vaccine-induced immunity. These fears also fit the pattern of the modern "risk society" articulated by sociologist Ulrich Beck, in *Risk Society: Towards a New Modernity* (Los Angeles: Sage, 1992).

34. Arthur Allen, *Vaccine: The Controversial Story of Medicine's Greatest Lifesaver* (New York: Norton, 2007), 15; Colgrove, *State of Immunity*, 8; Sabrina Tavernise, "Washington State Makes It Harder to Opt Out of Immunizations," *New York Times*, September 20, 2012, A18.

35. See note 25.

CHAPTER ONE

1. Press Conference No. 9 of the President of the United States, April 12, 1961, Digital Identifier JFKPOF-054-011, Press Conference Series, Papers of John F. Kennedy, John F. Kennedy Library.

2. James Colgrove provides an overview of the act in the context of 1960s campaigns to eliminate polio in *State of Immunity: The Politics of Vaccination in Twentieth-Century America* (Berkeley: University of California Press, 2006), 144–47.

3. Testimony by Walter Orenstein on the *Immunization Grant Program of the PHS Act, Before the Senate Committee on Labor and Human Resources, Subcommittee on Public Health and Safety*, May 6, 1997, Assistant Secretary for Legislation, Department of Health and Human Services, http://www.hhs.gov/asl/testify/t970506a.html (accessed May 2013).

4. A. R. Hinman, W. A. Orenstein, and L. Rodewald, "Financing Immunizations in the United States," *Clinical Infectious Diseases* 38, no. 10 (2004): 1440–46. Since the

early nineties, a separate program—passed under Clinton and discussed in chapter 7—has provided the lion's share of public-sector support for child immunization.

5. Historians and medical experts have debated whether Roosevelt actually suffered from polio or another disease with neurological effects, but as historian David Oshinsky points out, what matters is that in his own time, Roosevelt's disease was considered to be polio. David M. Oshinsky, *Polio: An American Story* (New York: Oxford University Press, 2005), 28. On Roosevelt's illness, see, for instance, A. S. Goldman et al., "What Was the Cause of Franklin Delano Roosevelt's Paralytic Illness?," *Journal of Medical Biography* 11, no. 4 (2003): 232–40; and Barron H. Lerner, "Crafting Medical History: Revisiting the 'Definitive' Account of Franklin D. Roosevelt's Terminal Illness," *Bulletin of the History of Medicine* 81, no. 2 (2007): 386–406.

6. Oshinsky, *Polio*.

7. For accounts of the Salk and Sabin vaccine trials and the activities of the National Foundation for Infantile Paralysis, see, for example, Jane S. Smith, *Patenting the Sun: Polio and the Salk Vaccine* (New York: William Morrow, 1990); and Oshinsky, *Polio*.

8. See Associated Press, "48 to Check Polio Vaccine Black Market," *Washington Post*, August 18, 1955, 32; and "Racket Is Feared in Polio Vaccine," *New York Times*, March 29, 1955, 26.

9. David Blumenthal and James A. Morone, *Heart of Power: Health and Politics in the Oval Office* (Berkeley: University of California Press, 2009), 109; Jill S. Quadagno, *One Nation, Uninsured: Why the U.S. Has No National Health Insurance* (New York: Oxford University Press, 2005), 43–44.

10. C. F. Trussell, "House Approves Polio Vaccine Aid," *New York Times*, August 2, 1955, 26; Associated Press, "2 Rows Delay Congress Windup," *Chicago Daily Tribune*, August 1, 1955, 3.

11. Otis Anderson, "The Polio Vaccination Assistance Act of 1955," *American Journal of Public Health* 45, no. 10 (1955): 1349–50.

12. "U.S. Gives States Full Control over Polio Vaccine Distribution," *New York Times*, August 1, 1955, 1; Trussell, "House Approves Polio Vaccine Aid."

13. Colgrove, *State of Immunity*, 113–48.

14. Oshinsky, *Polio*, 268.

15. Philip J. Smith, David Wood, and Paul M. Darden, "Highlights of Historical Events Leading to National Surveillance of Vaccination Coverage in the United States," *Public Health Reports* 126, no. S2 (2011): 3–12.

16. Colgrove, *State of Immunity*, 138.

17. Sydney A. Halpern, *American Pediatrics: The Social Dynamics of Professionalism, 1880–1980* (Berkeley: University of California Press, 1988), 14.

18. Joseph M. Hawes and N. Ray Hiner, *American Childhood: A Research Guide and Historical Handbook* (Westport, CT: Greenwood Press, 1985); Joseph M. Hawes and N. Ray Hiner, eds., *Children in Historical and Comparative Perspective* (Westport, CT: Greenwood Press, 1991); Hugh Cunningham, *Children and Childhood in Western Society since 1500* (New York: Longman, 1995).

19. George Rosen, *A History of Public Health*, MD Monographs on Medical History, no. 1 (New York: MD Publications, 1958), 360–61.

20. Ibid., 363–64. On the child health and welfare movement generally, see Richard A. Meckel, *Save the Babies: American Public Health Reform and the Prevention of*

Infant Mortality, 1850–1929 (Ann Arbor: University of Michigan Press, 1998); and Alexandra Stern and Howard Markel, *Formative Years: Children's Health in the United States, 1880–2000* (Ann Arbor: University of Michigan Press, 2002).

21. Jeffrey Brosco, "Weight Charts and Well-Child Care: How the Pediatrician Became the Expert in Child Health," *Archives of Pediatric Adolescent Medicine* 155, no. 12 (2001): 1385–89.

22. Halpern, *American Pediatrics.*

23. "Vaccination Assistance Act of 1962," *Congressional Record—House* (1962): 11739–53. States could also apply for grants to provide immunizations through title V of the Social Security Act and Section 314(c) of the Public Health Service Act. U.S. Congress, House Committee on Interstate and Foreign Commerce, *Intensive Immunization Programs* (Washington, DC: U.S. Government Printing Office, 1962), 73.

24. U.S. Congress, House Committee on Interstate and Foreign Commerce, *Intensive Immunization Programs*, 2.

25. "Vaccination Assistance Act of 1962," 11745.

26. "Memorandum for the President, from the Secretary for Health, Education, and Welfare," not dated, Folder: Health, Education, and Welfare 1/62–6/62, Box 79A, President's Office Files, Departments and Agencies, John F. Kennedy Library.

27. John F. Kennedy, "Annual Message to the Congress on the State of the Union," January 11, 1962, American Presidency Project, UCSB.

28. News Roundup, "Kennedy Calls for 'Mass Immunization' against Diseases; No Details Supplied," *Wall Street Journal*, January 12, 1962, 2.

29. John F. Kennedy, "Special Message to the Congress on National Health Needs," February 27, 1962, American Presidency Project, UCSB.

30. John F. Kennedy, "Special Message to the Congress on the Nation's Youth," February 14, 1963, American Presidency Project, UCSB.

31. Kennedy, "Special Message to the Congress on National Health Needs."

32. Kennedy, "Special Message to the Congress on the Nation's Youth."

33. Allan M. Brandt, *No Magic Bullet: A Social History of Venereal Disease in the United States since 1880* (New York: Oxford University Press, 1985).

34. The 1955 figure is taken from Oshinsky, *Polio*, 255. The 1961 figure is taken from the "Fact Book Relating to the Vaccination Assistance Act of 1962," Folder: Information 3—Immunization, 1963, Box 334062, No. 2, Record Group 442, CDC, NARA Southeast Region.

35. For figures on vaccination coverage rates in 1962 and 1963, see "Memo from Albert Sabin to William Seidman, August 16, 1976," Folder: CDC Liability Proposal, Box 8, Swine Flu Immunization Program Files, Record Group 442, CDC, NARA Southeast Region. "Sabin Sundays" were chronicled in many media outlets; see, for example, "Medicine: Wiping Out Polio," *Time*, July 6, 1962, http://content.time.com/time/maga zine/article/0,9171,940023,00.html (accessed May 2013). Sabin's race to develop an alternative to the Salk vaccine is described in Oshinsky, *Polio*; and Smith, *Patenting the Sun.*

36. Lawrence O'Kane, "Broad Search on in Smallpox Case," *New York Times*, August 21, 1962, 21; Associated Press, "3,000 Vaccinated as Result of Smallpox Scare in the East," *Los Angeles Times*, August 21, 1962, 3; Nate Haseltine, "Is America Safe from Smallpox?," *Washington Post*, September 2, 1962, E7.

37. News release from Louisiana State Board of Health, September 19, 1963, Folder:

Information 3—1963, Box 334062, Office of the Director Files, No. 2, Record Group 442, CDC, NARA Southeast Region.

38. "Vaccination Assistance Act of 1962."

39. Benjamin Spock, *The Common Sense Book of Baby and Child Care* (New York: Duell, Sloan and Pearce, 1960), 221.

40. News Roundup, "Kennedy Calls for 'Mass Immunization' against Diseases."

41. Disease incidence figures are taken from "Fact Book Relating to the Vaccination Assistance Act of 1962." A search of seventeen different newspapers—including the *Atlanta Constitution, Chicago Tribune,* and *Pittsburgh Courier*—turned up just ten articles that mentioned either tetanus, diphtheria, or whooping cough in 1962; for 1963, the same search turned up just four mentions of these three diseases.

42. "Diphtheria Immunization: A Survey of Existing Legislation," *International Digest of Health Legislation* 8 (1957): 171–98.

43. Ibid.

44. Gaylord West Anderson and Margaret G. Arnstein, *Communicable Disease Control: A Volume for the Health Officer and Public Health Nurse* (New York: Macmillan, 1953), 92.

45. "Vaccination Assistance Act of 1962," 11740.

46. U.S. Congress, House Committee on Interstate and Foreign Commerce, *Intensive Immunization Programs,* 74.

47. *Vaccination Assistance Act of 1962, Senate Report Submitted by Mr. Hill, to Accompany H.R. 10541, August 22, 1962, 87th Congress, 2nd Session* (Washington, DC: U.S. Government Printing Office), 2.

48. Other scholars have made this same point. See Patrick Vivier, "National Policies for Childhood Immunization in the United States: An Historical Perspective" (PhD diss., Johns Hopkins University, 1996); and Colgrove, *State of Immunity,* 144–47.

49. *Vaccination Assistance Act of 1962.*

50. U.S. Congress, House Committee on Interstate and Foreign Commerce, *Intensive Immunization Programs,* 46–47.

51. "Health and Social Security for the American People, a Report to President-Elect John F. Kennedy, January 10, 1961," Folder: Health and Social Security Task Force Report, Box 1071, Pre-Presidential Papers of John Fitzgerald Kennedy, John F. Kennedy Library. The report was also excerpted in "Task Force on Health and Social Security Reports," *Journal of the American Osteopathic Association* 60 (1961): 502–6.

52. W. H. Foege, "Centers for Disease Control," *Journal of Public Health Policy* 2, no. 1 (1981): 8–18.

53. F. Robert Freckleton, "Federal Government Programs in Immunization," *Archives of Environmental Health* 15 (1967): 514.

54. Meeting Minutes, Meeting No. 1 of the Advisory Committee on Immunization Practices, May 25–26, 1964, Folder: Info 3 ACIP Immunization 1964–5, Box 334062, No. 2, Record Group 442, CDC, NARA Southeast Region. The committee's formation was prompted by disputes over the relative risks and benefits of the Salk and Sabin polio vaccines. See Elizabeth W. Etheridge, *Sentinel for Health* (Berkeley: University of California Press, 1992), 147.

55. Edward Shorter, *The Kennedy Family and the Story of Mental Retardation* (Philadelphia: Temple University Press, 2000), 36–37, 41; Robert Dallek, *An Unfinished Life: John F. Kennedy, 1917–1963* (Boston: Little, Brown, 2003), 71–73.

56. Eunice Kennedy Shriver, recorded interview by John Stewart, May 7, 1968, John F. Kennedy Library Oral History Program.

57. Ibid.

58. Ibid. On the work and objectives of the task force, see also Wilbur J. Cohen, recorded interview by Charles T. Morrissey, November 11, 1964, John F. Kennedy Library Oral History Program; and Robert E. Cooke, recorded interview by John F. Stewart, March 29, 1968, John F. Kennedy Library Oral History Program. Cooke's interview also suggests that the task force championed age-related approaches to health research and health care in part because they found the approach novel and innovative. He, like Shriver, felt that child health in particular had been overlooked in federally supported health research. Elevating the stature of children's health as an area of scientific research was, for them, one means of increasing support for mental retardation, which Eunice called "the single most frequent and serious disabling problem of childhood."

59. "Remarks at the National Association for Retarded Children Convention," October 24, 1963, Digital Identifier JFKPOF-047-040, Speech Files Series, Papers of John F. Kennedy, John F. Kennedy Library.

60. Patrick's diagnosis at the time was hyaline membrane disease. Eunice Kennedy Shriver, recorded interview by John Stewart.

61. Dallek, *An Unfinished Life*, 27.

62. Ibid., 489–90.

63. List of major proposals for 1962, Folder: Legislative Files 5/1–18/62, Box 50, President's Office Files, Legislative Files, John F. Kennedy Library.

64. Memo, not dated, Folder: Legislative Files 5/1–18/62, Box 50, President's Office Files, Legislative Files, John F. Kennedy Presidential Library.

65. Kennedy, "Special Message to the Congress on the Nation's Youth."

66. "Fact Book Relating to the Vaccination Assistance Act of 1962."

67. Speech given by Assistant Secretary of HEW Boisfeuillet Jones, University of Pennsylvania, April 29, 1963, Folder: FA5 12-16-62–4-30-63 General, Box 99, White House Central Files, John F. Kennedy Presidential Library.

68. Editorial, "Two Kinds of Immunity," *Christian Science Monitor*, May 12, 1962, 16; Editorial, "Health Insurance, Plus Mass Inoculations," *Christian Science Monitor*, February 28, 1962, 14; Editorial, "A Subsidy for Medical Compulsions," *Christian Science Monitor*, March 10, 1962, 16.

69. Hugh White, "Vaccination Act," *Christian Science Monitor*, August 2, 1962, 16.

70. Letter to the President, Leslie Gampp, Laura Oard et al., June 16, 1962, Folder: LE/HE–LE/HE 6, Box 481, White House Central Files, John F. Kennedy Presidential Library.

71. Letter from Mr. and Mrs. James De Haan, July 12, 1962 Folder: LE/FA 5, Box 473, White House Central Files, John F. Kennedy Library.

72. "Nationwide Mass Vaccine Program," *Congressional Record—Senate* (1962): 8379–80.

73. Otherwise, the AMA said it endorsed "the principle" of the bill. U.S. Congress, House Committee on Interstate and Foreign Commerce, *Intensive Immunization Programs*, 129–30.

74. "Immunization against Smallpox Diphtheria Tetanus and Poliomyelitis," speech given by James L. Goddard before the Clinical Meeting of the American Medical Asso-

ciation, December 3, 1963, Folder: Info 3 Tr.—1963, Box 334065, No. 5, Record Group 442, CDC, Office of the Director Files, NARA Southeast Region; emphasis in original.

75. U.S. Congress, House Committee on Interstate and Foreign Commerce, Subcommittee on Health and Safety, *Polio Vaccines* (Washington, DC: U.S. Government Printing Office, 1961); "Vaccination Assistance Act of 1962," 11744–45.

76. "Vaccination Assistance Act of 1962," 11745–46.

77. U.S. Congress, House Committee on Interstate and Foreign Commerce, *Intensive Immunization Programs*, 50–51.

78. Freckleton, "Federal Government Programs in Immunization."

CHAPTER TWO

1. Details of the outbreak and the eleven-year-old girl's symptoms appear in T. E. Corothers and Gabriel S. Zatlin, "An Outbreak of Diphtheria: A Story of Investigation and Control," *Clinical Pediatrics* 5, no. 1 (1966): 29–33.

2. "End to Measles Possible, Says MD," *AMA News*, October 31, 1966.

3. U.S. Department of Health, Education, and Welfare, "Polio Packet," 1959, Object 2011.12.5, CDC Museum Collection.

4. "Preliminary Report—Rhode Island Poliomyelitis Epidemic 1960," Folder: Communicable Disease Center 34, Joseph Stokes Papers, American Philosophical Society.

5. "Polio Surveillance Report," January 13, 1961, Folder: Communicable Disease Center 33, Joseph Stokes Papers, American Philosophical Society.

6. Ibid.

7. U.S. Department of Health, Education, and Welfare, "Polio Packet."

8. Luther Terry, "The City in National Health," November 16, 1961, Folder: Speeches and Conferences, Box 20, Luther L. Terry Papers, 1957–1995 (MS C 503), NLM.

9. Kenneth T. Jackson, *Crabgrass Frontier: The Suburbanization of the United States* (New York: Oxford University Press, 1985), 8.

10. David R. Farber and Beth L. Bailey, *The Columbia Guide to America in the 1960s* (New York: Columbia University Press, 2001), 263–64.

11. "Preliminary Report—Rhode Island Poliomyelitis Epidemic 1960."

12. U.S. Department of Health, Education, and Welfare, "Polio Packet"; U.S. Congress, House Committee on Interstate and Foreign Commerce, *Intensive Immunization Programs* (Washington, DC: U.S. Government Printing Office, 1962).

13. See, for example, Charles E. Rosenberg, *The Cholera Years: The United States in 1832, 1849, and 1866* (Chicago: University of Chicago Press, 1962); Marilyn Chase, *The Barbary Plague: The Black Death in Victorian San Francisco* (New York: Random House, 2003); G. B. Risse, "'A Long Pull, a Strong Pull, and All Together': San Francisco and Bubonic Plague, 1907–1908," *Bulletin of the History of Medicine* 66, no. 2 (1992): 260–86; James H. Jones, *Bad Blood: The Tuskegee Syphilis Experiment* (New York: Maxwell McMillan, 1993); Susan Reverby, *Examining Tuskegee: The Infamous Syphilis Study and Its Legacy* (Chapel Hill: University of North Carolina Press, 2009); and Naomi Rogers, *Dirt and Disease: Polio before FDR* (New Brunswick, NJ: Rutgers University Press, 1992).

14. U.S. Congress, House Committee on Interstate and Foreign Commerce, *Intensive Immunization Programs*.

15. U.S. Department of Health, Education, and Welfare, "Polio Packet."

16. See, for example, U.S. Congress, House Committee on Interstate and Foreign Commerce, *Intensive Immunization Programs*, 131.

17. "No Pockets of Polio," *Science News-Letter* 85, no. 24 (1964): 374.

18. Lola M. Irelan, *Low-Income Life Styles* (Washington, DC: U.S. Department of Health, Education, and Welfare, Welfare Administration, Division of Research, 1966); Regional Field Letter No. 695, August 1, 1966, Folder: Department of Health, Education, and Welfare Regional and Field Letters Nos. 560–699, December 30, 1963–August 29, 1966, Box 108, General Records of the Department of Health and Human Services, NARA College Park.

19. I. M. Rosenstock, M. Derryberry, and B. K. Carriger, "Why People Fail to Seek Poliomyelitis Vaccination," *Public Health Reports* 74, no. 2 (1959): 98–103.

20. In a striking turn of events decades later, the affluent and their lifestyles would similarly be held to blame for the persistence of vaccine-preventable disease; see chapter 9.

21. U.S. Department of Health, Education, and Welfare, "Polio Packet."

22. This very pattern had likely biased the outcome of the field trials that tested the Salk vaccine in the first place; families who volunteered their children to participate in the trials were wealthier and better educated than those families whose children constituted the control group. David M. Oshinsky, *Polio: An American Story* (New York: Oxford University Press, 2005), 204.

23. U.S. Department of Health, Education, and Welfare, "Polio Packet."

24. See, for example, John Kenneth Galbraith, *The Affluent Society* (Boston: Houghton Mifflin, 1958); Michael Harrington, *The Other America: Poverty in the United States* (New York: Macmillan, 1962); and Dwight MacDonald, "Our Invisible Poor," *New Yorker*, January 19, 1963.

25. David Mark Chalmers, *And the Crooked Places Made Straight: The Struggle for Social Change in the 1960s* (Baltimore: Johns Hopkins University Press, 1991), 65–67.

26. Press release dated July 18, 1961, Folder: June–August 1961, Box 1, US HEW Secretaries' Speeches 1961–1979 (MS C 388), NLM.

27. Address by Anthony Celebrezze, 3rd Anniversary Celebration of the North Carolina Joint Council on Health and Citizenship, Nov. 10, 1963, Folder: November–December 1963, Box 1, US HEW Secretaries' Speeches 1961–1979 (MS C 388), NLM.

28. Lyndon B. Johnson, "Remarks upon Signing the Higher Education Facilities Act," December 16, 1963, American Presidency Project, UCSB.

29. Lyndon B. Johnson, "Remarks at the University of Michigan, May 22, 1964," American Presidency Project, UCSB; Lyndon B. Johnson, "Annual Message to the Congress on the State of the Union," January 4, 1965, American Presidency Project, UCSB.

30. Speech by Luther Terry before Political Science Club, May 27, 1961, Folder: January–May 1961, US HEW Secretaries' Speeches 1961–1979 (MS C 388), NLM.

31. Luther Terry, Conference on the Super-City of Tomorrow, October 30, 1963, Folder: Speeches and Conferences, Box 20, Luther L. Terry Papers, 1957–1995 (MS C 503), NLM.

32. Regional Field Letter No. 699, August 29, 1966, Folder: Department of Health, Education, and Welfare Regional and Field Letters Nos. 560–699, December 30, 1963–August 29, 1966, Box 108, General Records of the Department of Health and Human Services, NARA College Park.

33. Joe William Trotter, Earl Lewis, and Tera W. Hunter, *African American Urban Experience: Perspectives from the Colonial Period to the Present* (New York: Palgrave Macmillan, 2004), 1.

34. Regional Field Letter No. 682, May 2, 1966, Folder: Department of Health, Education, and Welfare Regional and Field Letters Nos. 560–699, December 30, 1963–August 29, 1966, Box 108, General Records of the Department of Health and Human Services, NARA College Park.

35. Community Health Services Extension Amendments of 1965, Public Law 89-109, 89th Cong., 1st sess. (August 5, 1965), 436.

36. U.S. Congress Senate, Committee on Labor and Public Welfare, Subcommittee on Health, *Public Health Grants and Construction of Health Research Facilities Hearing, 89th Congress, 1st Session, on S. 510 and S. 512. January 27, 1965* (Washington, DC: U.S. Government Printing Office, 1965), 20–21.

37. In some states, Wellbee promoted causes other than immunization: dental health, community sanitation, smoking, and even gun safety. ("A hunter climbed over a fence with his gun cocked," said a wry Wellbee in Kentucky. "He is survived by his widow, three children, and a rabbit.") Box of Wellbee materials, Object No. 1995.464, CDC Museum Collection.

38. Luther Terry, Speech before the Political Science Club, May 27, 1961, Folder: January–May 1961, US HEW Secretaries' Speeches 1961–1979 (MS C 388), NLM.

39. *Chicago Board of Health Newsletter*, June 15, 1965, Object 1995.464.37, CDC Museum Collection; emphasis added.

40. "Right from the Start," press release, November 25, 1963, Folder: Information 3—1963, Box 334062, Record Group 442, CDC, NARA Southeast Region.

41. In the 1950s, a live attenuated vaccine was developed by Harvard scientist John Enders; it was often given with serum (also called immune globulin) to reduce its frequent side effects. J. Stokes et al., "Efficacy of Live, Attenuated Measles-Virus Vaccine Given with Human Immune Globulin," *New England Journal of Medicine* 265, no. 11 (1961): 507–13; Correspondence Concerning Measles Vaccine, 1958, Folder: Measles 4, Joseph Stokes Papers, American Philosophical Society.

42. Stanley A. Plotkin and Walter A. Orenstein, *Vaccines*, 5th ed. (Philadelphia: Saunders, 2008), 359–62.

43. Harold Schmeck, "Measles Have Just about Had It," *New York Times*, March 26, 1967, 151.

44. "Current Status of Measles Immunization," *Journal of the American Medical Association* 194, no. 11 (1965): 1237–38; "Measles Immunization," *Journal of the American Medical Association* 198, no. 8 (1966): 837–38.

45. S. Dandoy, "Measles Epidemiology and Vaccine Use in Los Angeles County, 1963 and 1966," *Public Health Reports* 82, no. 8 (1967): 659–66.

46. "Twelve Million Children Immunized against Measles; Cases Drop Sharply," *Journal of the American Medical Association* 196, no. 8 (1966): 29–30, 38–39.

47. Harris D. Riley, "Recent Advances in the Study of Measles and Measles Vaccination," *Journal of the American Medical Association* 174, no. 15 (1960): 1968–69.

48. Dandoy, "Measles Epidemiology and Vaccine Use in Los Angeles County."

49. "A Measles Death," Philadelphia Department of Public Health Reported Communicable Diseases Weekly Report, 1961, Folder: Measles Vaccination 3, Joseph Stokes Papers, American Philosophical Society.

50. Editorial, "Vaccination against Measles," *Journal of the American Medical Association* 194, no. 11 (1965): 185.

51. Luther Terry, Speech Given at International Conference on Measles Immunization, November 7, 1961, Folder: November 1961, Box 1, US HEW Secretaries' Speeches 1961–1979 (MS C 388), NLM.

52. *Merck Review* 24, no. 2 (Fall 1963). For a history of Merck's vaccine development efforts in this period, see Louis Galambos and Jane Eliot Sewell, *Networks of Innovation: Vaccine Development at Merck, Sharp & Dohme and Mulford, 1895–1995* (New York: Cambridge University Press, 1995).

53. National Communicable Disease Center, "Immunization against Disease 1966–67," 1968, Folder: Information 3—Imm 1967, Box 334062, No. 2, Record Group 442, CDC, NARA Southeast Region.

54. *Merck Review* 24, no. 2.

55. Norman Lewak, "Importance of Vaccination against Measles," *Pediatrics* 34, no. 3 (1964): 438–39.

56. N. A. Harvey, "Measles Vaccine," *Pediatrics* 35, no. 4 (1965): 719–20; Seymour Musiker, "The Silence Is Deafening," *Pediatrics* 36, no. 1 (1965): 145.

57. Minutes from Meeting No. 1, Advisory Committee on Immunization Practices, May 25–26, 1964, Folder: Info 3 ACIP Immunization 1964–5, Box 334062, No. 2, Record Group 442, CDC, NARA Southeast Region.

58. Minutes from Meeting No. 4, Advisory Committee on Immunization Practices, Folder: Information 3—Imm 1964–1965, Box 334062, Record Group 442, CDC, NARA Southeast Region.

59. Lyndon B. Johnson, "Remarks at the Signing of the Health Research Facilities Amendments of 1965," August 9, 1965, American Presidency Project, UCSB; Nan Robertson, "President Signs $280 Million Bill for Health Study," *New York Times*, August 10, 1965, 1.

60. Merck Sharp and Dohme Research Laboratories, *By Their Fruits* (Rahway, NJ: Merck & Co., 1963).

61. E. Harold Hinman, "How Much Control of Communicable Diseases?," *American Journal of Tropical Medicine and Hygiene* 15, no. 2 (1966): 125–34.

62. Donald Hopkins, *The Greatest Killer: Smallpox in History* (Chicago: University of Chicago Press, 1983), 304.

63. Horace G. Ogden, *CDC and the Smallpox Crusade* (Washington, DC: U.S. Government Printing Office, 1987). Eradication was made feasible by the existence of the WHO and technological developments including freeze-dried vaccine and a jet-injector, which made it possible to immunize 1,000 people per hour. "Smallpox Eradication and Measles Control in Africa," Brochure Published by the National Communicable Disease Center and the Agency for International Development, 1967, Folder: Information 3 SE-1966, Box 334065, No. 5, Record Group 442, CDC, Office of the Director Files, NARA Southeast Region.

64. Hopkins, *The Greatest Killer*, 302–3; Ian Glynn and Jenifer Glynn, *The Life and Death of Smallpox* (New York: Cambridge University Press, 2004).

65. Hopkins, *The Greatest Killer*, 305. See also Donald Hopkins, *Princes and Peasants: Smallpox in History* (Chicago: University of Chicago Press, 1985). On the eradication campaign, see Donald A. Henderson, *Smallpox: The Death of a Disease* (New York: Prometheus Books, 2009).

66. "Measles Eradication 1967," Supplement to the *Morbidity and Mortality Weekly Report*, April 15, 1967, Folder: Information 3 Imm 1964–1967, Box 343357, Record Group 442, CDC, NARA Southeast Region. See also National Communicable Disease Center, "Immunization against Disease 1966–67."

67. David J. Sencer, H. Bruce Dull, and Alexander Langmuir, "Epidemiologic Basis for Eradication of Measles in 1967," *Public Health Reports* 82, no. 3 (1967): 253–56.

68. Historian Elizabeth Etheridge also notes that when the Vaccination Assistance Act was renewed in 1965, federal funds became available expressly for measles vaccination. "Thus, a disease previously given little thought worked its way to the top of the health agenda." Elizabeth W. Etheridge, *Sentinel for Health* (Berkeley: University of California Press, 1992), 169.

69. A. D. Langmuir, "Medical Importance of Measles," *American Journal of Diseases of Children* 103 (1962): 224–26.

70. "Immunization: Theory and Practice," Report by V. F. Guinea, D. S. Martin, and Other Members of the CDC Immunization Seminar Services Committee, Folder: Info 3 Tr.—1963, Box 334065, No. 5, Record Group 442, CDC, NARA Southeast Region.

71. There were certainly those who dissented from this view, including most notably René Dubos, who warned against the hubris and misguided intents he saw in eradication programs. See René J. Dubos, *Man Adapting* (New Haven, CT: Yale University Press, 1965).

72. Radio-TV Bulletin, the Advertising Council, March–April 1967, Folder: Information 3 Imm 1964–1967, Box 343357, Record Group 442, CDC, NARA Southeast Region.

73. "Ann Landers: Get Kids Vaccinated Against Measles Now!," April 4, 1967, Folder: Information 3 Imm 1964–1967, Box 343357, Record Group 442, CDC, NARA Southeast Region; press release, November 1, 1966, Folder: Information 3 Imm 1964–1967, Box 334062, No. 2, Record Group 442, CDC, NARA Southeast Region. Surgeon General William H. Stewart argued that there was "no excuse for needlessly prolonging the fight against this disease, which for centuries has attacked virtually all children and left many of them mentally retarded." This particular comment of his played to parental fears of disability exacerbated by the recent rubella epidemic of 1964–65, which caused birth defects in some 20,000 children.

74. Quoted in Galambos and Sewell, *Networks of Innovation*, 115.

75. Trudy Stamm, "Kim Aids Measles Drive," *Children Limited* 15, no. 5 (1966): 1. For a description of the inaugural polio poster child, see Oshinsky, *Polio*, 82–86; and Heather Green Wooten, *The Polio Years in Texas: Battling a Terrifying Unknown* (College Station: Texas A&M University Press, 2009), 92–94. Kim was also the 1966 poster girl for the National Association for Retarded Children. Joy Miller, "Kim Is Example: Measles Can Cause Retardation," *Free-Lance Star*, May 9, 1967, 9.

76. Lewak, "Importance of Vaccination against Measles."

77. "Spot Prevention" coloring book, U.S. Department of Health, Education, and Welfare, Public Health Service, not dated, Folder: Information 3—Imm 1967, Box 334062, Record Group 442, CDC, NARA Southeast Region.

78. Brochure: "Emmy Immunity and the Dirty Disease Gang," not dated, Folder: Information 3 Tr.—1963, Box 334065, Record Group 442, CDC, NARA Southeast Region; William Lakeman, "Immunizations Stir Concern," *Free-Lance Star*, October 23, 1974, 26.

79. "Memo to Vaccination Assistance Program Field Personnel from Immuniza-

tion Activities Office," CDC, June 25, 1965, Folder: Information 3 Imm 1964–1967, Box 334062, No. 2, Record Group 442, CDC, NARA Southeast Region.

80. Edward F. Gliwa and Harold Horoho, "The Vaccination Assistance Act," *Delaware Medical Journal* 38, no. 9 (1966): 275–76.

81. National Communicable Disease Center, "Immunization against Disease 1966–67."

82. "Measles Remain Serious Health Menace Despite Proven Success of Rubeola Vaccine," *Journal of the Medical Association of the State of Alabama* 36, no. 6 (1966): 663–65.

83. CDC, "Current Trends—Measles," *Morbidity and Mortality Weekly Report* (1967): 2.

84. Etheridge, *Sentinel for Health*, 174.

85. Letter to Hermann Rinne, February 5, 1970, Folder: General Correspondence—Dr. Wallace, Box 338638, Record Group 442, CDC, NARA Southeast Region; Robert J. Bazell, "Health Programs: Slum Children Suffer Because of Low Funding," *Science* 172, no. 3986 (1971): 921–25.

86. Letter from John Witte to Adolf Karchmer, February 9, 1970, Folder: General Correspondence—Dr. Abrutyn, Box 338638, Record Group 442, CDC, NARA Southeast Region.

87. Letter to Hermann Rinne.

88. "Measles—United States, Epidemiologic Year 1969–70," reported in *MMWR* 19, no. 6 (February 14, 1970), Folder: General Correspondence—Dr. Wallace, Box 338638, Record Group 442, CDC, NARA Southeast Region.

89. Secretary of Health, Education, and Welfare Wilbur J. Cohen (he occupied the post from 1968 to 1969) pointed a finger at "the interrelated problems of poverty, slums, discrimination, ill health, inadequate education, disgraceful housing, despair, neglect and raised but unfulfilled aspirations," for a crisis that had shaken all institutions, including the health care system. How could the nation deliver health resources to children so poor and bereft that when shown a picture of a teddy bear, they identified it as a rat, he asked. Wilbur Cohen, "A New Day in Health Care," December 14, 1968, Folder: December 14–January 16, 1969, Box 3, US HEW Secretaries' Speeches 1961–1979 (MS C 388), NLM.

90. Note from John Witte to Robert Wallace, not dated, Folder: EPI-70-40-2, Measles, Chicago, Illinois, Box 338638, Record Group 442, CDC, NARA Southeast Region; G. E. Hardy Jr. et al., "The Failure of a School Immunization Campaign to Terminate an Urban Epidemic of Measles," *American Journal of Epidemiology* 91, no. 3 (1970): 286–93.

91. Under Nixon, states were typically given block grants to spend as they wish, instead of funds targeted for specific health programs, such as immunization. Bazell, "Health Programs"; K. A. Johnson, A. Sardell, and B. Richards, "Federal Immunization Policy and Funding: A History of Responding to Crises," *American Journal of Preventive Medicine* 19, no. S3 (2000): 99–112.

92. Letter from Alfred Sommer to Donald Putnoi, October 30, 1969, Folder: General Correspondence—Dr. Sommer, Box 338638, Record Group 442, CDC, NARA Southeast Region; Notes, JP Friedman—EIS Conference, 1969, not dated, Folder: "The Simultaneous Administration of Multiple Live Virus Vaccines," Box 343357, Record

Group 442, CDC, NARA Southeast Region; "Multiple Antigen" manuscript, not dated, Folder: Multiple Antigen Manuscript by Dr. Karchmer, Box 343357, Record Group 442, CDC, NARA Southeast Region.

93. "Current Trends: Measles—United States," *Morbidity and Mortality Weekly Report* 26, no. 14 (1977): 109–11; letter to Dr. Frank Perkins, August 10, 1970, Folder: General Correspondence—Dr. Wallace, Box 338638, Record Group 442, CDC, NARA Southeast Region; letter from Robert Wallace to Harold Yates, December 1, 1969, Folder: General Correspondence—Dr. Wallace, Box 338638, Record Group 442, CDC, NARA Southeast Region.

94. Letter from Alfred Sommer to Robert Israel, October 31, 1969, Folder: General Correspondence—Dr. Sommer, Box 338638, Record Group 442, CDC, NARA Southeast Region; note, "Regarding Death of 2½ Year Old Negro Female, October 1968," Folder: General Correspondence—Dr. Abrutyn, Box 338638, Record Group 442, CDC, NARA Southeast Region.

95. James Colgrove argues that this policy shift marks a new era of vaccination in the late 1960s. For more on the adoption of school laws in the 1970s, see James Colgrove, *State of Immunity: The Politics of Vaccination in Twentieth-Century America* (Berkeley: University of California Press, 2006), 174–78.

96. Note from John Witte to Robert Wallace.

97. "End to Measles Possible, Says MD."

CHAPTER THREE

1. The live virus mumps vaccine, first licensed in 1968, is referred to simply as the mumps vaccine or Mumpsvax from here forward. "Additional Standards—Mumps Virus Vaccine, Live," *Federal Register* 33, no. 14 (1968): 744–46.

2. Faye Marley, "Vaccine for Mumps Not Widely Used," *Los Angeles Times*, May 30, 1969, B4.

3. This analysis is influenced by Rosenberg's theoretical work on the framing of disease. See, for example, Charles Rosenberg, "Disease in History: Frames and Framers," *Milbank Quarterly* 67, no. S1 (1989): 1–15; Charles Rosenberg, "Framing Disease: Illness, Society, and History," in *Framing Disease: Studies in Cultural History*, ed. Charles Rosenberg and Janet Golden (New Brunswick, NJ: Rutgers University Press, 1997), xxi–xxvi; and Charles Rosenberg, "What Is Disease?" *Bulletin of the History of Medicine* 77 (2003): 491–505.

4. On the rubella vaccine, which is not addressed at length in this work, see Jacob Heller, *The Vaccine Narrative* (Nashville: Vanderbilt University Press, 2008); and Leslie J. Reagan, *Dangerous Pregnancies: Mothers, Disabilities, and Abortion in America* (Berkeley: University of California Press, 2010).

5. Maurice Hilleman, credited with developing several dozen vaccines (against human and animal infections), is a legendary figure in the fields of vaccine development and public health. His measles vaccine, notably, was licensed the same month his daughter came down with mumps. Hilleman's work at Merck is recounted in Louis Galambos and Jane Eliot Sewell, *Networks of Innovation: Vaccine Development at Merck, Sharp & Dohme and Mulford, 1895–1995* (New York: Cambridge University Press, 1995).

A popular account of his life and scientific work is given in Paul A. Offit, *Vaccinated: One Man's Quest to Defeat the World's Deadliest Diseases* (Washington, DC: Smithsonian Books, 2007). See also Lawrence Altman, "Maurice Hilleman, Master at Creating Vaccines, Dies at 85," *New York Times*, April 12, 2005, A1; and Associated Press, "Vaccine Researcher Saved 'Millions of Lives,'" *Chicago Tribune*, April 12, 2005, 7.

6. The development of the mumps vaccine is described in several popular accounts, including Arthur Allen, *Vaccine: The Controversial Story of Medicine's Greatest Lifesaver* (New York: Norton, 2007); Offit, *Vaccinated*, 20–25; Richard Conniff, "A Forgotten Pioneer of Vaccines," *New York Times*, May 6, 2013, D1. See also Galambos and Sewell, *Networks of Innovation*, 100–101.

7. Clara Councell, "War and Infectious Disease," *Public Health Reports* 56, no. 12 (1941): 547–73.

8. Milton Levine, "A Sponsored Epidemic of Mumps in a Private School," *American Journal of Public Health* 34, no. 12 (1944): 1274–76. See also Councell, "War and Infectious Disease." Flu and measles were of equal importance to mumps as causes of illness among troops during the First World War; measles and flu, but not mumps, were among leading causes of death in that war.

9. A. C. McGuiness and E. A. Gall, "Mumps at Army Camps in 1943," *War Medicine* 5 (1943): 95.

10. C. D. Johnson and E. W. Goodpasture, "Investigation of Etiology of Mumps," *Journal of Experimental Medicine* 59 (1934): 1–19.

11. Karl Habel, "Cultivation of Mumps Virus in the Developing Chick Embryo and Its Application to Studies of Immunity to Mumps in Man," *Public Health Reports* 60, no. 8 (1945): 201–12.

12. J. E. Gordon and L. Kilham, "Ten Years in the Epidemiology of Mumps," *American Journal of the Medical Sciences* 218, no. 3 (1949): 338–59.

13. J. F. Enders et al., "Immunity in Mumps I: Experiments with Monkeys (Macacus Mulatta). The Development of Complement-Fixing Antibody Following Infection and Experiments on Immunization by Means of Inactivated Virus and Convalescent Human Serum," *Journal of Experimental Medicine* 81, no. 1 (1945): 93–117; J. F. Enders, S. Cohen, and L. W. Kane, "Immunity in Mumps II: The Development of Complement-Fixing Antibody and Dermal Hypersensitivity in Human Beings Following Mumps," *Journal of Experimental Medicine* 81, no. 1 (1945): 119–35; L. W. Kane and J. F. Enders, "Immunity in Mumps III: The Complement Fixation Test as an Aid in the Diagnosis of Mumps Meningoencephalitis," *Journal of Experimental Medicine* 81, no. 1 (1945): 137–50.

14. Habel, "Cultivation of Mumps Virus"; John Enders et al., "Attenuation of Virulence with Retention of Antigenicity of Mumps Virus after Passage in the Embryonated Egg," *Journal of Immunology* 54 (1946): 283–91.

15. K. Habel, "Vaccination of Human Beings against Mumps: Vaccine Administered at the Start of an Epidemic. I. Incidence and Severity of Mumps in Vaccinated and Control Groups," *American Journal of Hygiene* 54, no. 3 (1951): 295–311. Although the experiments were performed in 1945 and 1946, Habel did not publish the results until 1951.

16. Enders et al., "Attenuation of Virulence."

17. See, for example, H. R. Morgan, J. F. Enders, and P. F. Wagley, "A Hemolysin Associated with the Mumps Virus," *Journal of Experimental Medicine* 88, no. 5 (1948): 503–14.

18. David M. Oshinsky, *Polio: An American Story* (New York: Oxford University

Press, 2005), 122–30. Polio virus cultivation ultimately led to Salk's polio vaccine and won Enders a Nobel Prize.

19. See, for example, A. A. Smorodintsev and N. S. Kliachko, "[Specific Prevention of Mumps; Preliminary Communication]," *Zh Mikrobiol Epidemiol Immunobiol* 11 (1954): 6–11; and N. S. Kliachko and L. K. Maslennikova, "[Specific Prevention of Mumps. II. Study of Safety and Immunogenicity of Living Attenuated Mumps Vaccine by Intradermal Immunization of Children]," *Vopr Virusol* 2, no. 1 (1957): 13–17.

20. Council on Drugs, "New and Nonofficial Drugs: Mumps Vaccine," *Journal of the American Medical Association* 164, no. 8 (1957): 874–75.

21. Ibid.

22. Edward B. Shaw, "Mumps Immunization," *Journal of the American Medical Association* 167, no. 14 (1958): 1744.

23. T. R. Van Dellen, "How to Keep Well: Salivary Gland Enlargement," *Chicago Daily Tribune*, February 26, 1950, 20.

24. Ibid.

25. "Mumps" (Albany: NY State Department of Health, 1955), *Mothering* collection.

26. ANP, "He Can't Whistle — He's Got the Mumps!," *Atlanta Daily World*, November 21, 1961, 6; United Press International, "Not Immune," *Chicago Defender*, May 18, 1963, 11; "Bridegroom Missing, They Wed by Phone," *Los Angeles Times*, June 30, 1953, A1.

27. Frank Colby, "Take My Word for It," *Los Angeles Times*, June 2, 1950, A5; United Press International, "It Only Hurts When We Swallow," *Daily Defender*, December 31, 1963, 4.

28. Associated Press, "Tony, the Boxer," *Chicago Daily Tribune*, March 25, 1957, C10.

29. See, for example, "World of Sports," *Washington Post and Times Herald*, February 14, 1958, D3; Associated Press, "VPI Tackle Richards Sidelined with Mumps," *Washington Post and Times Herald*, August 30, 1955, 15; Associated Press, "Mumps Bench Ram Star," *Chicago Daily Tribune*, December 21, 1955, B1; and United Press International, "Albert Has Mumps," *Los Angeles Times*, July 13, 1958, C2.

30. "Redskin Takes His Mumps," *Washington Post*, September 28, 1950, 12.

31. United Press International, "Mumps Hits F. Robinson," *Daily Defender*, April 23, 1968, 24.

32. United Press International, "Mumps Vaccine Gains Government Approval," *Los Angeles Times*, January 5, 1968, 12.

33. Peter Yates, dir., *Bullitt* (Warner Brothers, 1968).

34. United Press International, "Mumps Vaccine Gains Government Approval"; Harold Schmeck, "A Mumps Vaccine Is Licensed by U.S.," *New York Times*, January 5, 1968, 72. Among the nation's papers, only the *Washington Post* placed the story on the front page: Associated Press, "Vaccine for Mumps Licensed," *Washington Post*, January 5, 1968, A1.

35. Advisory Committee on Immunization Practices, "Recommendation of the Public Health Service Advisory Committee on Immunization Practices," *Morbidity and Mortality Weekly Report* 16, no. 51 (1967): 430–31.

36. "New Mumps Vaccine Not for Everyone," *Consumer Reports*, July, 1968, 377.

37. Editorial, "Mumps Vaccine: More Information Needed," *New England Journal of Medicine* 278, no. 5 (1968): 275–76.

38. Editorial, "Vaccination against Mumps," *Lancet* 292, no. 7576 (1968): 1022–23.

39. Adolf Karchmer, "Mumps: A Review of Surveillance, Vaccine Development, and

Recommendations for Use," Folder: Paper for Immunization Conference, Box 343357, Record Group 442, CDC, NARA Southeast Region.

40. Thomas Shope, Adolf Karchmer, and F. Robert Freckleton, "Immunizations in the Future," *Journal of the Oklahoma State Medical Association* 62 (1969): 111–15.

41. Harris D. Riley, "Current Concepts in Immunization," not dated, Folder: Info 3 Tr. — 1963, Box 334605 No. 5, Record Group 442, CDC, NARA Southeast Region.

42. Marley, "Vaccine for Mumps Not Widely Used." Some commentators cited a set of five principles outlined by British doctor G. S. Wilson at an international immunization conference held in 1961. Wilson argued that (1) vaccines should be harmless to the healthy; (2) they should cause no more disturbance (fever, discomfort) than the disease itself; (3) they must be easy to administer; (4) they must provide both herd and individual benefit; and (5) the immunity conferred should not require frequent revaccination. Cited in Samuel Katz, "Immunization with Live Attenuated Measles Virus Vaccines: Five Years' Experience," Paper Selected for Distribution at CDC Seminars on Immunization, not dated, Folder: Info 3 Tr. — 1963, Box 334605 No. 5, Record Group 442, CDC, NARA Southeast Region.

43. See the appendix for a brief overview of vaccine licensing and approval.

44. Editorial, "Vaccination against Mumps."

45. Notes, March 12–15, 1968, Folder: Paper for Immunization Conference, Box 343357, Record Group 442, CDC, NARA Southeast Region.

46. Karchmer, "Mumps: A Review."

47. CDC, "Current State of Mumps in the United States," *Journal of Infectious Diseases* 132, no. 1 (1975): 106–9.

48. Memo, January 14, 1968, Folder: Epi Aid 58-51-1, Box 343357 No. 1, Record Group 442, CDC, NARA Southeast Region.

49. Documents pertaining to these outbreaks are located in folders titled Epi Aid 68-51-1; Epi Aid Memo 68-57-1; West Virginia Study; Epi 68-20-1; and Epi Aid 68-20-2, all in Box 343357 No. 1, Record Group 442, CDC, NARA Southeast Region. The term "mentally retarded" is used here as it was used at the time, as an accepted medical diagnosis.

50. Notes, Fort Custer EIS Study, not dated, Folder: Fort Custer — Work Sheets and Potpourri, Box 343357 No. 1, Record Group 442, CDC, NARA Southeast Region.

51. On the subject of popular attitudes toward mentally retarded children in the 1950s and 1960s, see Steven Noll and James W. Trent, *Mental Retardation in America*, The History of Disability (New York: New York University Press, 2004), part 4.

52. Manuscript, November 4, 1967, Folder: MMWR — Mumps at Fort Custer, Box 343357 No. 1, Record Group 442, CDC, NARA Southeast Region.

53. Abstract, not dated, Folder: Fort Custer — Presentation and Abstracts, Box 343357 No. 1, Record Group 442, CDC, NARA Southeast Region.

54. A reviewer at the CDC noted that while many often claimed that it was a challenge to immunize teens and adults, few had adequately produced quantitative support for this claim. Reviewer comments, "Re: Public Acceptance of Mumps Immunization," not dated, Folder: Public Acceptance of Mumps, Box 343357, Record Group 442, CDC, NARA Southeast Region.

55. Phillip Jones, "Public Acceptance of Mumps Immunization," *Journal of the American Medical Association* 209, no. 6 (1969): 901–5.

56. Ibid.

57. Historians have argued that earlier diphtheria immunization campaigns achieved this effect as well. See chapter 3 in James Colgrove, *State of Immunity: The Politics of Vaccination in Twentieth-Century America* (Berkeley: University of California Press, 2006); and Evelynn Maxine Hammonds, *Childhood's Deadly Scourge: The Campaign to Control Diphtheria in New York City, 1880–1930* (Baltimore: Johns Hopkins University Press, 2002).

58. Stuart Auerbach, "D.C. Has Rash of 261 Measles Cases," *Washington Post*, February 10, 1970, C1.

59. T. R. Van Dellen, "How to Keep Well: Mumps Vaccine," *Chicago Tribune*, September 30, 1967, B14.

60. Brochure, not dated, Folder: 1964 Rubella Epidemic Cost Analysis, Box 343357, Record Group 442, CDC, NARA Southeast Region.

61. Meeting Minutes, Meeting No. 1 of the Advisory Committee on Immunization Practices, May 25–26, 1964, Folder: Info 3 ACIP Immunization 1964–5, Box 334062 No. 2, Record Group 442, CDC, NARA Southeast Region.

62. Memo from Adolf Karchmer to Martin D. Skinner, "Draft of Rubella Control Program," December 2, 1968, Folder: Rubella Control Program—Montana, Box 343357, Record Group 442, CDC, NARA Southeast Region.

63. Ibid.

64. "Rubella Vaccination Seen for School Children," *New York Amsterdam News*, September 13, 1969, 6.

65. See, for example, Harry Meyer, Paul Parkman, and Hope Hopps, "The Control of Rubella," *Pediatrics* 44, no. 1 (1969): 5–23.

66. Heller, *The Vaccine Narrative*, 62–63; Reagan, *Dangerous Pregnancies*, 180–81.

67. On perceptions of mental retardation in this period, see Katherine Castles, "Nice Average Americans: Postwar Parents' Groups and the Defense of the Normal Family," in *Mental Retardation in America*, ed. Steven Noll and James W. Trent (New York: New York University Press, 2004), 351–70. See also Edward Shorter, *The Kennedy Family and the Story of Mental Retardation* (Philadelphia: Temple University Press, 2000).

68. Reagan, *Dangerous Pregnancies*, 191.

69. See, for example, R. E. Weibel et al., "Live Attenuated Mumps-Virus Vaccine. 3. Clinical and Serologic Aspects in a Field Evaluation," *New England Journal of Medicine* 276, no. 5 (1967): 245–51; and J. Stokes Jr. et al., "Live Attenuated Mumps Virus Vaccine. II. Early Clinical Studies," *Pediatrics* 39, no. 3 (1967): 363–71.

70. "Mumps War Is Declared at Halsted UPC," *Daily Defender*, June 22, 1967, 5.

71. Several studies from the 1960s and 1970s found meningo-encephalitis to occur in about 15 to 18 percent of cases, without lasting effect other than headache and "nervousness." See, for example, H. G. Murray, C. M. Field, and W. J. McLeod, "Mumps Meningo-Encephalitis," *British Medical Journal* 1, no. 5189 (1960): 1850–53; P. H. Azimi, H. G. Cramblett, and R. E. Haynes, "Mumps Meningoencephalitis in Children," *Journal of the American Medical Association* 207, no. 3 (1969): 509–12; and CDC, "Mumps Vaccine," *Morbidity and Mortality Weekly Report* 26, no. 48 (1977): 393–94.

72. Sherwood Schwartz, "Never Too Young," *The Brady Bunch*, season 5, episode 4 (American Broadcasting Company, 1973).

73. Karchmer, "Mumps: A Review." See also Editorial, "Mumps Vaccine."

74. Ronald Kotulak, "New Lease on Life Given Man: A Tale of Nine Future Vaccines," *Chicago Tribune*, December 24, 1967, 6.

75. Galambos and Sewell, *Networks of Innovation*, 121.

76. Joan Beck, "It'll Be Vintage Year for Babies Born in the U.S.," *Chicago Tribune*, January 3, 1967, B1.

77. Mary Ann Mason, "The State as Superparent," in *Childhood in America*, ed. Paula Fass and Mary Ann Mason (New York: New York University Press, 2000), 549–54.

78. "Immunization: Theory and Practice," report by V. F. Guinea, D. S. Martin, and Other Members of the CDC Immunization Seminar Services Committee, Folder: Info 3 Tr. — 1963, Box 334065 No. 5, Record Group 442, CDC, NARA Southeast Region. An inverted version of this adage would later be embraced by vaccine critics in order to emphasize the significance of the small risks posed by vaccines. See chapters 8, 9, and 10.

79. Paula S. Fass, *Children of a New World: Society, Culture, and Globalization* (New York: New York University Press, 2007), 181–83.

80. Ibid., 182.

81. See, for example, "Mumps May Be on Its Way Out," *Daily Defender*, June 27, 1966, 2; and Van Dellen, "How to Keep Well: Mumps Vaccine."

82. "A Shot in Time," *Mademoiselle*, July, 1977, 126.

83. National Communicable Disease Center, *Immunization against Disease* (Washington, DC: U.S. Department of Health, Education, and Welfare, 1967).

84. "Vaccinations Everyone Ought to Have," *Changing Times*, September, 1974, 11–13. For a similar representation of mumps, see also "Immunization = Self Defense," *Current Health*, October, 1979, 23–25.

85. Nicholas Fiumara, "Use of Mumps Vaccine," *New England Journal of Medicine* 278, no. 12 (1968): 681–82.

86. Merck, "The First Live Mumps Vaccine," *British Medical Journal* 2, no. 5910 (1974), advertisement in front matter.

87. The Merck sales figure comes from Galambos and Sewell, *Networks of Innovation*, 115. Population immunization figures are cited in CDC, *United States Immunization Survey* (Washington, DC: U.S. Department of Health, Education, and Welfare and Bureau of the Census, 1971).

88. CDC, "Current State of Mumps in the United States."

89. E. B. Buynak et al., "Combined Live Measles, Mumps, and Rubella Virus Vaccines," *Journal of the American Medical Association* 207, no. 12 (1969): 2259–62.

90. "Mumps Vaccine," *Morbidity and Mortality Weekly Report* 21, suppl. (1972): 13–14.

91. CDC, "Mumps Vaccine."

92. Colgrove makes a similar point about pertussis and the DPT vaccine. Colgrove, *State of Immunity*, 112.

93. Galambos and Sewell, *Networks of Innovation*, 118.

94. "7 Health Sins Listed; Immunizations Urged for County Children," *Los Angeles Times*, December 6, 1973, OC-A16.

95. See, for example, Dan Kaercher, "Immunization: A Call to Action," *Better Homes and Gardens*, September, 1979, 70.

96. Midge Lasky Schildkraut, "The New Threat to Your Children's Health," *Good Housekeeping*, August 1978, 219–20.

97. Calls to eradicate mumps were in fact infrequent, but see, for example, "Georgians Play Major Role in Developing New Vaccine," *Atlanta Daily World*, January 7, 1968, 1; and Reviewer comments, "Re: Public Acceptance of Mumps Immunization." In 1968 Massachusetts health officials announced that they had unanimously decided

to eradicate mumps; the announcement sparked a debate between health officials and physicians, who argued that the vaccine's use should be left to their discretion. See, for example, Fiumara, "Use of Mumps Vaccine"; and T. C. Peebles et al., "Use of Mumps Vaccine," *New England Journal of Medicine* 281, no. 12 (1969): 679.

98. "Mumps Vaccine Now Ready for Public," *Los Angeles Sentinel*, March 28, 1968, E7; Associated Press, "Vaccine for Mumps Licensed"; Schmeck, "A Mumps Vaccine Is Licensed by U.S."

99. Dodi Schultz, "Why Childhood Diseases Are Coming Back," *New York Times Sunday Magazine*, May 7, 1978, 35.

100. "Measles, Mumps and Rubella Threaten the Unprotected," *Atlanta Daily World*, January 3, 1978, 3.

101. Schildkraut, "The New Threat to Your Children's Health."

102. Interestingly, Merck displayed relatively little interest in portraying mumps as a serious disease at this time. With MMR on the market and endorsed by the ACIP, company materials pointed out the dangers of measles and rubella, but tended to say little about the hazards of mumps. See, for example, *Fifty Years of Research: Merck Sharp & Dohme Research Laboratory* (Rahway, NJ: Merck Sharp & Dohme, 1983).

103. U.S. Department of Health, Education, and Welfare, *Protect Your Child*, DHEW Publication No. OHDS 78-02027 (Washington, DC: Office of Human Development Services, 1978).

104. CDC, "Current Trends Mumps—United States, 1984–1985," *Morbidity and Mortality Weekly Report* 35, no. 13 (1986): 216–19.

105. Ibid.

106. Mumps vaccination in the United States thus exemplified what historian Dorothy Porter has described as the modern democratic, free-enterprise state's configuration of the individual citizen as a "political and economic unit of a collective whole." Dorothy Porter, *Health, Civilization, and the State* (London: Routledge, 1999), 57.

107. Jimmy Dean, *To a Sleeping Beauty* (Columbia Records, 1962).

CHAPTER FOUR

1. Joseph A. Califano, "Immunizing Our Children," *Parents*, November, 1977, 122.

2. U.S. Congress, Senate Committee on Labor and Human Resources, *Immunization and Preventive Medicine* (Washington, DC: U.S. Government Printing Office, 1982), 44.

3. On Carter's campaign and core messaging, see, for example, Patrick Anderson, *Electing Jimmy Carter: The Campaign of 1976* (Baton Rouge: Louisiana State University Press, 1994); Peter G. Bourne, *Jimmy Carter: A Comprehensive Biography from Plains to Post-Presidency* (New York: Scribner, 1997); and Frye Gaillard, *Prophet from Plains: Jimmy Carter and His Legacy* (Athens: University of Georgia Press, 2007).

4. Edward D. Berkowitz, *Something Happened: A Political and Cultural Overview of the Seventies* (New York: Columbia University Press, 2006), 104–10.

5. Jimmy Carter, "Inaugural Address, January 20, 1977," American Presidency Project, UCSB.

6. Jimmy Carter, Address to the Nation on Energy and National Goals: "The Malaise Speech," July 15, 1979, American Presidency Project, UCSB.

7. On the inflation of health care costs, see "The Problem of Rising Health Care

Costs," Executive Office of the President Council on Wage and Price Stability Staff Report, April 1976, Folder: Health Costs, Box 305, Carter Presidential Papers, Jimmy Carter Library.

8. "Healthy People: The Surgeon General's Report on Health Promotion and Disease Prevention, 1979," Folder FG 22-10 1/20/77–1/20/81, Box FG-136, White House Central Files, Jimmy Carter Library.

9. Memo from Stu Eizenstat to Jody Powell, February 28, 1977, Folder: HE 1/20/77–7/22/77, Box HE-1, White House Central File, Jimmy Carter Library.

10. David Blumenthal and James A. Morone, *The Heart of Power: Health and Politics in the Oval Office* (Berkeley: University of California Press, 2009), 261; Bourne, *Jimmy Carter*, 432.

11. Blumenthal and Morone, *The Heart of Power*, 270–71.

12. Ibid., 275.

13. Letter from Leila West Lehde to Mrs. Nell Balkman, February 6, 1973, Folder: Children's Immunization Program, 2/77–12/78 [2], Box 7, First Lady's Office — Projects Office — Cade Subject File, Jimmy Carter Library.

14. Memorandum, Jimmy Carter to Heads of Executive Offices and Agencies, April 7, 1977, Folder: Immunization, Box 32, Collection JC-DPS: Records of the Domestic Policy Staff, Jimmy Carter Library.

15. "World Health Day," *International Review of the Red Cross* 17, no. 193 (1977): 218–19; Memorandum, Jimmy Carter to Heads of Executive Offices and Agencies; "Memo to Health Cluster Members from Donna Brown," National Council of Organizations for Children and Youth, March 2, 1977, Folder: Children's Immunization Program, 2/77–12/78 [1], Box 7, Collection JC-FL: Records of the First Lady's Office, Jimmy Carter Library. The theme for World Health Day reflected the formalization of a 1974 World Health Assembly decision to adopt an Expanded Programme on Immunization, which aimed to provide immunization services against diphtheria, polio, pertussis, tetanus, measles, tuberculosis, and smallpox globally by the year 1995. K. Keja and R. H. Henderson, "Expanded Programme on Immunization: The Continuing Role of the European Region," *WHO Chronicle* 39, no. 3 (1985): 92–94.

16. "President Carter's Proposed Statement for World Health Day," April 7, 1977, Folder: Trips/Events, Health Center for Mothers and Children in DC, 1977, Box 42, Collection JC-FL: Records of the First Lady's Office, Jimmy Carter Library.

17. Letter from George Gallup to Attorney General Griffin Bell, October 11, 1978, Folder: WE 1 Welfare, Children, Box 138, Collection JC-FL: Records of the First Lady's Office, Jimmy Carter Library.

18. Bourne, *Jimmy Carter*, 432.

19. Blumenthal and Morone provide a detailed analysis of Carter's personal views on both medicine and government involvement in health care. Blumenthal and Morone, *The Heart of Power*, 252.

20. U.S. Congress, House Committee on Appropriations, *Supplemental Appropriations for Fiscal Year 1977: Hearings before Subcommittees of the Committee on Appropriations, House of Representatives, 95th Congress, 1st Session, Part 2* (Washington, DC: U.S. Government Printing Office, 1977), 572.

21. James Colgrove, *State of Immunity: The Politics of Vaccination in Twentieth-Century America* (Berkeley: University of California Press, 2006), 172.

22. U.S. Department of Health, Education, and Welfare, "Secretary Califano's Ad-

dress to the Second National Immunization Conference [sic]," press release, April 6, 1977, Folder: Immunization, Box 32, Collection JC-DPS: Records of the Domestic Policy Staff, Jimmy Carter Library.

23. U.S. Congress, House Committee on Appropriations, Subcommittee on Departments of Labor and Health, Education, and Welfare and Related Agencies, *Departments of Labor and Health, Education, and Welfare Appropriations for 1978: Hearings before a Subcommittee of the Committee on Appropriations, House of Representatives, 95th Congress, 1st Session* (Washington, DC: U.S. Government Printing Office, 1977), 443; U.S. Congress House Committee on Appropriations, *Supplemental Appropriations for Fiscal Year 1977*, 571.

24. On Merck's advertising efforts, see Louis Galambos and Jane Eliot Sewell, *Networks of Innovation: Vaccine Development at Merck, Sharp & Dohme and Mulford, 1895–1995* (New York: Cambridge University Press, 1995), 115.

25. Colgrove, *State of Immunity*, 175–76.

26. Many cited an outbreak of measles in the "divided" city of Texarkana when making the case for such laws. Many more cases of measles occurred on Texarkana's Texas side, which had no measles vaccination law, than on the Arkansas side, which did have such a law. See for example, P. J. Landrigan, "Epidemic Measles in a Divided City," *Journal of the American Medical Association* 221, no. 6 (1972): 567–70.

27. Colgrove, *State of Immunity*, 177; "Current Trends: Measles—United States," *Morbidity and Mortality Weekly Report* 26, no. 14 (1977): 109–11.

28. Fact Sheet, "Childhood Immunization Initiative," Folder: Immunization, Box 32, Collection JC-DPS, Jimmy Carter Library; "Current Trends: Measles—United States."

29. Fact Sheet, "Childhood Immunization Initiative."

30. U.S. Department of Health, Education, and Welfare, "Secretary Califano's Address to the Second National Immunization Conference [sic]."

31. U.S. Congress, House Committee on Appropriations, *Supplemental Appropriations for Fiscal Year 1977*, 574.

32. Letter and Fact Sheet to Parents and Volunteers from Betty Bumpers, "Every Child by 74," 1974, Folder: Children's Immunization Program, Box 7, Collection JC-FL: Records of the First Lady's Office, Jimmy Carter Library.

33. "Memo to Health Cluster Members from Donna Brown."

34. "The Immunization of Children," *Congressional Record—Senate* (1977): 2679.

35. Immunization promotional materials, Folder: Immunization, Box 32, Collection JC-DPS: Records of the Domestic Policy Staff, Jimmy Carter Library.

36. U.S. Department of Health, Education, and Welfare, "Secretary Califano's Address to the Second National Immunization Conference [sic]."

37. "The Immunization of Children," 2680.

38. For safety reasons, children under three were excluded. Memo from Ted Cooper to Richard Friedman, May 5, 1976, Folder: CDC Influenza Conference, Box 8, Swine Flu Immunization Program Files, Record Group 442, CDC, NARA Southeast Region; Minutes from May 6–7 meeting on flu, Advisory Committee on Immunization Practices, 1976, Folder: Flu ACIP Meeting 5/6–7/76, Box 8, Swine Flu Immunization Program Files Record Group 442, CDC, NARA Southeast Region.

39. On the 1918 pandemic, see John M. Barry, *The Great Influenza: The Epic Story of the Deadliest Plague in History* (New York: Penguin, 2005), 397.

40. Editorial by Sidney Wolfe, April 11, 1976, Folder: ACIP Meeting of June 21–22,

1976, Box 8, Swine Flu Immunization Program Files, Record Group 442, CDC, NARA Southeast Region; Letter from Ronald Hattis to David Sencer, not dated, Folder: ACIP Meeting of June 21–22, 1976, Box 8, Swine Flu Immunization Program Files, Record Group 442, CDC, NARA Southeast Region.

41. On the swine flu immunization program, see Elizabeth W. Etheridge, *Sentinel for Health* (Berkeley: University of California Press, 1992), chap. 18; Richard E. Neustadt and Harvey V. Fineberg, *The Swine Flu Affair: Decision-Making on a Slippery Disease* (Washington, DC: U.S. Department of Health, Education, and Welfare, 1978); and Richard E. Neustadt and Harvey V. Fineberg, *The Epidemic that Never Was: Policy-Making and the Swine Flu Scare* (New York: Vintage, 1983).

42. G. Timothy Johnson, "Immunizations Important Despite Swine Results," *Chicago Tribune*, March 18, 1977, A9.

43. Letter from Ronald Hattis to David Sencer.

44. Harris Survey, Activity under Secretary of HEW Joseph Califano, March 7, 1977, Folder: Immunization, Box 32, Collection JC-DPS: Records of the Domestic Policy Staff, Jimmy Carter Library.

45. Note from Kathy Cade to Rosalynn Carter, Folder: Children's Immunization Program, 2/77–12/78 [1], Box 7, Collection JC-FL: Records of the First Lady's Office, Jimmy Carter Library.

46. Even before the swine flu immunization campaign was launched, the CDC envisioned using the mobilized volunteers to promote widespread childhood immunization after the swine flu campaign concluded. Immunization Action: Every Child in 76/77—Draft, Centers for Disease Control Immunization Division, Folder: Children—Immunization Program, 2/77–12/78 [1], Box 7, Collection JC-FL: Records of the First Lady's Office, Jimmy Carter Library.

47. "HEW Developing Plan to Immunize 20 Million Youths," *Congressional Record—Senate* (1977): 10508.

48. Associated Press, "HEW Plans Drive to Inoculate Children," *Chicago Tribune*, April 5, 1977, 10.

49. "President Carter's Proposed Statement for World Health Day."

50. "The Immunization of Children," 2679.

51. U.S. Department of Health, Education, and Welfare, "Secretary Califano's Address to the Second National Immunization Conference [sic]."

52. Joseph A. Califano, *Inside: A Public and Private Life* (New York: Public Affairs, 2004), 127–28.

53. Robert G. Kaiser, "HEW's Califano: Flashy First Four Months," *Washington Post*, May 15, 1977, 1.

54. Rudy Abramson, "Joe Califano: 1-Man Band in LBJ Style," *Los Angeles Times*, May 16, 1977, B1.

55. "President Carter's Proposed Statement for World Health Day."

56. Ibid.; U.S. Congress, House Committee on Interstate and Foreign Commerce, Subcommittee on Oversight and Investigations, *National Child Immunization Programs: Hearing before the Subcommittee on Oversight and Investigations of the Committee on Interstate and Foreign Commerce, House of Representatives, 95th Congress, 1st Session, August 31, 1977* (Washington, DC: U.S. Government Printing Office, 1977), 7.

57. *Star Wars* Immunization Poster, 1977, Object 2005.005.03, CDC Museum Collection.

58. U.S. Department of Health, Education, and Welfare, "Secretary Califano's Address to the Second National Immunization Conference [sic]."

59. Ibid. The AMA, in response, answered the campaign's entreaties with a seemingly token gesture: a proposal to alter the children's game of hopscotch, by giving each square in the game the name of a different disease and labeling the home square "Immunization." "The American Medical Association's Immunization Campaign," *Congressional Record—Extension of Remarks* (1977): 38877.

60. "HEW Developing Plan to Immunize 20 Million Youths," 10508.

61. U.S. Department of Health, Education, and Welfare, "Secretary Califano's Address to the Second National Immunization Conference [sic]."

62. Victor Cohn, "HEW Developing Plan to Immunize 20 Million Youths: Vaccination of 20 Million Eyed," *Washington Post*, April 5, 1977, A1.

63. Califano, "Immunizing Our Children."

64. "Feature," not dated, Folder: Immunization, Box 32, Collection JC-DPS: Records of the Domestic Policy Staff, Jimmy Carter Library.

65. Rosalynn Carter, Notes, December 1978, Folder: Children—Immunization, Box 7, Collection JC-FL, Records of the First Lady's Office, Jimmy Carter Library.

66. Helen H. Doane, "Letters to the Times," *Los Angeles Times*, May 11, 1977, E4.

67. *ABC News Special: A Conversation with the Carters*, December 14, 1978, Jimmy Carter Library.

68. Letter from Mrs. James Conrad McLarnan to Rosalynn Carter, 5/22/79, Folder WE 1, Box 138, Collection JC-FL, Records of the First Lady's Office, Jimmy Carter Library.

69. U.S. Department of Health, Education, and Welfare, "Secretary Califano's Address to the Second National Immunization Conference [sic]."

70. Immunization promotional materials.

71. Vernon Thompson, "Red Measles Outbreak Alarms Health Officials," *Washington Post*, February 5, 1977, B7. The outbreaks were a focus of congressional hearings held in Los Angeles in August 1977. U.S. Congress, House Committee on Interstate and Foreign Commerce, Subcommittee on Oversight and Investigations, *National Child Immunization Programs*.

72. Harry Nelson, "Measles Ultimatum: L.A. Students to Be Ousted If They Don't Have Shots," *Los Angeles Times*, March 31, 1977, A1.

73. Jack McCurdy and Harry Nelson, "Schools Bar Thousands Lacking Measles Shots," *Los Angeles Times*, May 3, 1977, C1.

74. "Warn Parents on Shots," *Chicago Tribune*, February 11, 1978, F9; Carla Hall, "Pupils Suspended in Crackdown on Measles Shots," *Washington Post*, December 2, 1977, B8; Associated Press, "5 N.Y. Students Suspended in Immunization Crackdown," *Washington Post*, September 24, 1977, A4.

75. Mumps was a less common addition to state laws than rubella was. CDC, *State Immunization Requirements for School Children* (Atlanta: Department of Health and Human Services, Public Health Service, 1981).

76. "Remarks by Secretary Califano," Conference on Immunization Held December 12, 1978, Folder: Children—Health, Education, and Welfare (HEW) Conference—Childhood Immunization, 12/12/78, Box 7, Collection JC-FL: Records of the First Lady's Office, Jimmy Carter Library.

77. Ibid.

78. "Policy Perspectives," National Immunization Conference, NIH, November 12–14, 1976, Folder: Immunization, Box 32, Records of the Domestic Policy Staff, Jimmy Carter Library. For an extended discussion on informed consent debates at this time, see Colgrove, *State of Immunity*, 187–97. On the patients' rights movement and health care consumer movements, see, for example, David J. Rothman, *Strangers at the Bedside: A History of How Law and Bioethics Transformed Medical Decision Making* (New Brunswick NJ: Aldine Transaction, 2008); and Beatrix Hoffman, *Health Care for Some: Rights and Rationing in the United States since 1930* (Chicago: University of Chicago Press, 2012), 142–63.

79. Cohn, "HEW Developing Plan to Immunize 20 Million Youths."

80. A Department of Health, Education, and Welfare study concluded that one case of paralytic polio occurred in every 11.5 million recipients of the oral polio vaccine, which contained live vaccine. A World Health Organization study found that in some countries, there were two cases for every million people vaccinated. Before Congress, Jonas Salk (not necessarily an objective source) testified that one case of the disease occurred for every 300,000 children fully vaccinated with OPV. Elena O. Nightingale, "Recommendations for a National Policy on Poliomyelitis Vaccination," *New England Journal of Medicine* 297, no. 5 (1977): 249–53; "The Relation between Acute Persisting Spinal Paralysis and Poliomyelitis Vaccine (Oral): Results of a WHO Enquiry," *Bulletin of the World Health Organization* 53, no. 4 (1976): 319–31; "Testimony before Subcommittee on Health, Committee on Labor and Public Welfare, U.S. Senate, by Jonas Salk," *Congressional Record—Senate* (1977).

81. Department of Health, Education, and Welfare, "Parents' Guide to Childhood Immunization," October 1977, Folder: Children's Immunization Program, 2/77–12/78 [2], Box 7, Collection JC-FL: Records of the First Lady's Office—Cade Subject File, Jimmy Carter Library.

82. U.S. Congress, House Committee on Interstate and Foreign Commerce, Subcommittee on Oversight and Investigations, *National Child Immunization Programs*, 9.

83. For a more detailed discussion on the issues of informed consent and liability, see Colgrove, *State of Immunity*, chap. 6.

CHAPTER FIVE

1. "Mrs. Carter's Remarks," HEW Immunization Conference, December 12, 1978, Folder: Children, HEW Conference, Childhood Immunization, Box 7, Collection JC-FL, Jimmy Carter Library.

2. The phrase "vaccine-safety movement" was employed by the activists themselves and accurately describes their worries and predominant demands. They were not anti-vaccinationists in the historical sense, as they did not oppose all vaccines or all vaccination. For a related argument on the politics of contemporary "anti-vaccinationism," drawn from a different set of episodes in modern vaccination history, see Robert Johnston, "Contemporary Anti-Vaccination Movements in Historical Perspective," in *The Politics of Healing: Histories of Alternative Medicine in Twentieth-Century North America*, ed. Robert Johnston (New York: Routledge, 2004), 259–86.

3. A few historians have pointed out that social movements, including women's movements, have historically given momentum to vaccination resistance. Michael

Willrich, for instance, has noted that women's rights advocates were among those who threw their support behind Progressive Era anti-vaccinationism. Colgrove has pointed out the link between the social movements of the sixties and seventies, broader challenges to the paternalistic authority of science and medicine, and growing popular discontent with respect to vaccines. An in-depth exploration of the specific influence of feminism or gender norms on vaccination policies and reception, however, was not the objective of either scholars' work. Michael Willrich, *Pox: An American History* (New York: Penguin, 2011), 252–53; James Colgrove, *State of Immunity: The Politics of Vaccination in Twentieth-Century America* (Berkeley: University of California Press, 2006), 187–91, 236.

4. Edward D. Berkowitz, *Something Happened: A Political and Cultural Overview of the Seventies* (New York: Columbia University Press, 2006).

5. Jane S. Smith, *Patenting the Sun: Polio and the Salk Vaccine* (New York: William Morrow, 1990), 77.

6. Immunization Project Meeting Agenda, Governor's Mansion, Little Rock, AR, March 8, 1973, Folder: Children's Immunization Program, 2/77–12/78 [2], Box 7, Collection JC-FL, Jimmy Carter Library.

7. Walter Alvarez, "Epidemic of Measles Feared," *Los Angeles Times*, January 6, 1972, F12.

8. Rebecca Jo Plant, *Mom: The Transformation of American Motherhood* (Chicago: University of Chicago Press, 2010), 2–17.

9. Ann Landers, "Ann Landers: Consequences," *Washington Post*, April 14, 1975, B5.

10. Stuart Auerbach, "D.C. Has Rash of 261 Measles Cases," *Washington Post*, February 10, 1970, C1.

11. Editorial, "Measles' New Muscle," *Chicago Tribune*, August 3, 1971, 10.

12. Walter Alvarez, "Poverty, Ignorance Halting Vaccination," *Los Angeles Times*, April 19, 1973, G20.

13. Harold Schmeck, "Health Strategy for U.S. Urged to Reduce Unnecessary Illness," *New York Times*, March 12, 1976, 47.

14. Jonathan Spivak, "Measles Resurgence Sparks New Campaign to Immunize Children," *Wall Street Journal*, February 20, 1970, 1.

15. Michael Stern, "Immunizations Lag Called Peril in City," *New York Times*, June 17, 1971, 1.

16. Dodi Schultz, "Why Childhood Diseases Are Coming Back," *New York Times Sunday Magazine*, May 7, 1978, 35.

17. "645 Measles Cases Reported by State," *New York Times*, February 1, 1974, 64.

18. Philip Jones, "Public Acceptance of Mumps Vaccination," *JAMA* 209 (August 11, 1969): 901–5.

19. Alvarez, "Poverty, Ignorance Halting Vaccination." For more details on the promotion of rubella vaccination, see Jacob Heller, *The Vaccine Narrative* (Nashville: Vanderbilt University Press, 2008), 57–83, chap. 3; and Leslie J. Reagan, *Dangerous Pregnancies: Mothers, Disabilities, and Abortion in America* (Berkeley: University of California Press, 2010), esp. 180–220.

20. Reagan, *Dangerous Pregnancies*, 181, 194, 207.

21. Philip Jenkins, *Decade of Nightmares: The End of the Sixties and the Making of Eighties America* (New York: Oxford University Press, 2006). 28; Gail Collins, *When Everything Changed: The Amazing Journey of American Women from 1960 to the Present* (New York: Little, Brown, 2010), 96–100.

22. Auerbach, "D.C. Has Rash of 261 Measles Cases"; Stern, "Immunizations Lag Called Peril in City"; Rudy Johnson, "Paterson Fights Rise in Measles," *New York Times*, December 27, 1973, 78.

23. Meg Rosenfeld, "Many Va. Children Not Getting Shots," *Washington Post*, April 27, 1975, 1.

24. "List of Lead Voluntary Organizations," Folder: Children—Health, Education, and Welfare (HEW) Conference—Childhood Immunization, 12/12/78, Box 7, Collection JC-FL, Jimmy Carter Library.

25. Rhoda Gilinsky, "Volunteerism and Women: A Status Report," *New York Times*, November 12, 1978, WC16.

26. Diane M. Blair and Shawn Parry-Giles, "Rosalynn Carter: Crafting a Presidential Partnership Rhetorically," in *Inventing a Voice: The Rhetoric of American First Ladies of the Twentieth Century*, ed. Molly Meijer Wertheimer (Lanham, MD: Rowman & Littlefield, 2004), 341–64.

27. Notes from an interview conducted by Suzanne Wilding, Folder: Suzanne Wilding, *Town and Country Magazine* Interview with RSC, November 16, 1978, Box 7, Collection JC-FL, Jimmy Carter Library.

28. Letter from Mrs. James Conrad McLarnan to Rosalynn Carter, 5/22/79, Folder: WE 1, Box 138, Collection JC-FL, Jimmy Carter Library.

29. Landers, "Ann Landers: Consequences."

30. Ann Landers, "Ann Landers," *Washington Post*, June 1, 1976, B4.

31. Ann Landers, "Ann Landers," *Washington Post*, February 9, 1979, C5.

32. Ann Landers, "Ann Landers," *Washington Post*, November 8, 1977, B7.

33. Elaine Fein, "Immunization: Is Your Child Protected?" *Harper's Bazaar*, July 1979, 75–76.

34. Morris Wessel, "Immunizations Are Important," *Parents*, December 1979, 28.

35. See, for instance, Wendy Kline, *Bodies of Knowledge: Sexuality, Reproduction, and Women's Health in the Second Wave* (Chicago: University of Chicago Press, 2010); Sandra Morgen, *Into Our Own Hands: The Women's Health Movement in the United States, 1969–1970* (New Brunswick, NJ: Rutgers University Press, 2002); and Sheryl Ruzek, *The Women's Health Movement: Feminist Alternatives to Medical Control* (New York: Praeger, 1978).

36. Elizabeth Siegel Watkins, *On the Pill: A Social History of Oral Contraceptives, 1950–1970* (Baltimore: Johns Hopkins University Press, 1998), 103–31; Elizabeth Siegel Watkins, "Doctor, Are You Trying to Kill Me?: Ambivalence about the Patient Package Insert for Estrogen," *Bulletin of the History of Medicine* 76 (Spring 2002): 84–104.

37. Susan Speaker, "From 'Happiness Pills' to 'National Nightmare': Changing Cultural Assessment of Minor Tranquilizers in America, 1955–1980," *Journal of the History of Medicine* 52 (July 1997): 338–76.

38. Boston Women's Health Collective, *Ourselves and Our Children: A Book by and for Parents* (New York: Random House, 1978), 217.

39. *Mothering* readers in general—and those who debated vaccines—embraced varied feminisms. However, as with all of the vaccine critics in this chapter, whether they self-identified as one type of feminist or another was not always obvious from their thoughts on vaccines. But as historian Wendy Kline has noted, women could identify as feminists or adopt feminist ideas without participating in organized feminist groups. Kline, *Bodies of Knowledge*, 25.

40. See letters to the editor in *Mothering* issues 1–47 (1979–86), held at *Mothering's* offices in Santa Fe, New Mexico. Selected correspondence and features on immunization were compiled in Peggy O'Mara, *Vaccinations*, 3rd ed. (Santa Fe, NM: Mothering Magazine, 1989); and Peggy O'Mara, *Vaccination: The Issue of Our Times* (Santa Fe, NM: Mothering Magazine, 1997).

41. Carol Horowitz, "Immunizations and Informed Consent," *Mothering*, Winter 1983, 37–41.

42. Peggy O'Mara, "Editorial," *Mothering*, Summer 1996, 25.

43. As a result, pertussis vaccination rates declined in those countries. Jeffrey P. Koplan et al., "Pertussis Vaccine: an Analysis of Benefits, Risks and Costs," *New England Journal of Medicine* 301, no. 17 (1979): 906–11.

44. The figure of 1 in 7,000 appeared in government vaccination promotion materials, which noted that 1 in 7,000 children could suffer a "serious" side effect of the DPT vaccine, such as high fever or convulsion. The same sources listed the risk of encephalitis or brain damage as occurring "more rarely," roughly once in every 100,000 doses. See, for example, Department of Health, Education, and Welfare, "Parents' Guide to Childhood Immunization," October 1977, Folder: Children's Immunization Program, 2/77–12/78 [2], Box 7, Collection JC-FL: Records of the First Lady's Office, Jimmy Carter Library. The risk of severe brain damage or death following pertussis vaccination was widely disputed in the seventies and early eighties; later estimates ranged from 1 in 174,000 to 1 in 1 million shots. See Roy Anderson and Robert May, "The Logic of Vaccination," *New Scientist* 96 (November 18, 1982): 410–15.

45. Lea Thompson, producer, *DPT: Vaccine Roulette* (Washington, DC: WRC-TV, April 19, 1982).

46. U.S. Senate, Committee on Labor and Human Resources, *Immunization and Preventive Medicine: Hearing before the Subcommittee on Investigations and General Oversight of the Committee on Labor and Human Resources, 97th Congress, 2nd Session* (Washington, DC: U.S. Government Printing Office, 1982), 7.

47. Donna Hilts, "The Whooping Cough Vaccine: A Protector or a Killer?" *Washington Post*, April 28, 1982, Va2.

48. U.S. Senate, Committee on Labor and Human Resources, *Immunization and Preventive Medicine*.

49. Robert Galano, "Crusading Camera'd Champions of the Consumer," *Washington Post*, March 30, 1980, TV3.

50. Susan Okie, "How Two Angry Mothers Beat Uncle Sam at His Own Game," *Washington Post*, October 11, 1980, A3.

51. Kline, *Bodies of Knowledge*, 1–4.

52. Thompson, *DPT: Vaccine Roulette*.

53. Ibid.

54. Harris L. Coulter and Barbara Loe Fisher, *DPT: A Shot in the Dark* (New York: Harcourt, Brace, Jovanovich, 1985), 61. The book was reissued in 1991, by a different publisher. Harris L. Coulter and Barbara Loe Fisher, *A Shot in the Dark: Why the P in the DPT Vaccination May Be Hazardous to Your Child's Health* (Garden City, NY: Avery Pub. Group, 1991).

55. Ibid., 13. The parents interviewed in Coulter and Fisher's book are identified by first names only. I use Janet's last name here because she spoke out under her full name in various arenas, even though she is identified only as "Janet" in the book.

56. On maternalist activism, see, for example, Amy Hay, "Recipe for Disaster: Motherhood and Citizenship at Love Canal," *Journal of Women's History* 21, no. 1 (2009): 111–34; and Molly Ladd-Taylor, *Mother-Work: Women, Child Welfare, and the State, 1890–1930* (Urbana: University of Illinois Press, 1994).

57. Suzanne Arms, *Immaculate Deception: A New Look at Women and Childbirth in America* (Boston: Houghton Mifflin, 1975); Gail Sforza Brewer and Tom Brewer, *What Every Pregnant Woman Should Know: The Truth about Diet and Drugs in Pregnancy* (New York: Penguin, 1977); Gena Corea, *The Hidden Malpractice: How American Medicine Treats Women as Patients and Professionals* (New York: William Morrow, 1977).

58. Arms, *Immaculate Deception*, 22–23.

59. Sandra Sugawara, "Two Parents Groups Speak out against Multiple DPT Vaccine; Side Effects Blamed for Brain Damage," *Washington Post*, February 8, 1985, D5.

60. Ivan Illich, *Medical Nemesis: The Expropriation of Health* (New York: Pantheon, 1976), 16.

61. Robert S. Mendelsohn, *Male Practice: How Doctors Manipulate Women* (Chicago: Contemporary Books, 1981), 188–91.

62. Robert S. Mendelsohn, *Confessions of a Medical Heretic* (Chicago: Contemporary Books, 1979), 143–45. Vaccine resisters were mostly but not exclusively mothers; among the men who spoke out against vaccines were unorthodox physicians like Mendelsohn, adherents of natural or alternative healing methods, and some fathers.

63. Britain, by contrast, faced a pertussis outbreak in 1977–79 that health officials blamed on a vaccination rate that had fallen 50 percent since 1970. Three deaths and seventeen cases of brain damage resulted. See Anderson and May, "The Logic of Vaccination." In 1998 a group of resarchers contested the argument that Sweden, West Germany, and other countries did not face increased rates of pertussis in the wake of popular vaccine resistance; see Eugene J. Gangarosa et al., "Impact of Anti-Vaccine Movements on Pertussis Control: The Untold Story," *Lancet* 351 (January 31, 1998): 356–61.

64. Coulter and Fisher, *DPT: A Shot in the Dark*, 9.

65. Ibid., 40.

66. By 1980 upwards of 96 percent of all children entering school were vaccinated against measles, rubella, polio, diphtheria, pertussis, and tetanus, achieving some of the highest rates of vaccine coverage the country had ever seen. U.S. Senate, Committee on Labor and Human Resources, *Immunization and Preventive Medicine*, 4.

67. U.S. Senate, Committee on Labor and Human Resources, *Task Force Report on Pertussis: Hearing before the Committee on Labor and Human Resources, United States Senate, 98th Congress, 1st session* (Washington, DC: U.S. Government Printing Office, 1983), 77.

68. U.S. Senate, Committee on Labor and Human Resources, *Immunization and Preventive Medicine*, 41.

69. Thompson, *DPT: Vaccine Roulette*.

70. On the women's movement's commitment to "informed medical consumerism," see Morgen, *Into Our Own Hands*, 95; and Elizabeth Watkins, *The Estrogen Elixir: A History of Hormone Replacement Therapy in America* (Baltimore: Johns Hopkins University Press, 2007), 130.

71. Horowitz, "Immunizations and Informed Consent," 37.

72. Thompson, *DPT: Vaccine Roulette*.

73. Coulter and Fisher, *DPT: A Shot in the Dark*, 10.

74. Ibid., 408.

75. Watkins, *The Estrogen Elixir*.

76. Coulter and Fisher, *DPT: A Shot in the Dark*, 408.

77. Thompson, *DPT: Vaccine Roulette*.

78. U.S. Senate, Committee on Labor and Human Resources, *Task Force Report on Pertussis*, 74–87.

79. Barbara Loe Fisher, "Why Not Use a Safer Vaccine?," *Washington Post*, May 9, 1988, A14.

80. Elsa Walsh, "State Defends Vaccine," *Washington Post*, June 30, 1982, MD6.

81. Coulter and Fisher, *DPT: A Shot in the Dark*, 83.

82. U.S. Senate, Committee on Labor and Human Resources, *Immunization and Preventive Medicine*, 6.

83. Ibid., 55; emphasis in original. The parents' group DPT later adopted this phrase as a motto, placing it on the front page of its newsletters.

84. See chapter 3, note 78.

85. Jenkins, *Decade of Nightmares*, 108–33.

86. Susan Brownmiller, *In Our Time: Memoir of a Revolution* (New York: Dial Press, 1999).

87. Jenkins, *Decade of Nightmares*, 109, 256–60.

88. Thompson, *DPT: Vaccine Roulette*. For more details on the types of exemptions included in school vaccination laws, see note 25 in the introduction.

89. Peter Kelley, "Whooping Cough Vaccine's Tragic Side Effects Unmasked," *Patriot-News*, October 8, 1986, C1.

90. Sonia Nazario, "A Parental-Rights Battle Is Heating up over Fears of Whooping-Cough Vaccine," *Wall Street Journal*, June 20, 1990, 17.

91. Emily Isberg, "No School for Non-Immunized Child," *Montgomery County Sentinel*, February 22, 1979.

92. U.S. Senate, Committee on Labor and Human Resources, *Immunization and Preventive Medicine*, 137.

93. Coulter and Fisher, *DPT: A Shot in the Dark*, 406.

94. Gerri Cohn, "DPT Pushes Vaccine Bill in Maryland," *Dissatisfied Parents Together News* 1, no. 1 (1983): 1.

95. Jeff Schwartz, "Support for Vaccine Damage Compensation Bill Grows," *Dissatisfied Parents Together News* 1, no. 1 (1983): 10–11.

96. The coalition formed in favor of the National Vaccine Injury Compensation Act was fragile by the time the act was signed. Although he openly opposed the act, Reagan signed it upon inclusion of a provision allowing pharmaceutical companies to sell abroad drugs that were unapproved for use in the United States. See Robert Pear, "Reagan Signs Bill on Drug Exports and Payment for Vaccine Injuries," *New York Times*, November 15, 1986, 1; and "President Reagan Signs Vaccine Injury Compensation and Safety Bill into Law," *Dissatisfied Parents Together News* 3, no. 1 (1987): 1.

97. Public Law 99-660, 99th Congress, November 14, 1986.

98. "CDC Unclear about DPT Vaccine Death Definition," *Dissatisfied Parents Together News* 3, no. 1 (1987): 9; "New Jersey Passes Law Limiting Most Vaccine Injury Law-

suits," *Dissatisfied Parents Together News* 3, no. 2 (1987): 1; "House Votes to Appropriate Funds to Compensate Children Injured by Vaccines before October 1988," *Dissatisfied Parents Together News* 4, no. 1 (1988): 1.

99. Barbara Loe Fisher, "Editorial," *Dissatisfied Parents Together News* 3, no. 2 (1987): 11.

100. NVIC's mission included "1) informing the public about childhood diseases and vaccines in order to prevent vaccine injuries and deaths; 2) assisting those who have suffered severe reactions to vaccinations; 3) representing the vaccine consumer by monitoring vaccine research and development, vaccine policymaking, and vaccine related federal and state legislation; 4) working to obtain the right of parents to choose which vaccines their children will receive; and 5) promoting the development of safer and more effective vaccines." "NVIC/DPT," *NVIC News* 2, no. 2 (1992): 3.

101. Lameiras, "Fighting for a Choice," 36.

102. Barbara Loe Fisher, "Editorial," *Dissatisfied Parents Together News* 1, no. 1 (1983): 17.

103. Personal communication, June 15, 2012.

104. Maria M. Lameiras, "Fighting for a Choice: Vaccination—One Mother's Crusade," *Today's Chiropractic Lifestyle*, August 2006, 31–36, 32.

105. Coulter and Fisher, *DPT: A Shot in the Dark*, 40–41.

CHAPTER SIX

1. Robert S. Mendelsohn, *The Risks of Immunization and How to Avoid Them* (Evanston, IL: The People's Doctor, 1988).

2. See, for example, Arthur Allen, *Vaccine: The Controversial Story of Medicine's Greatest Lifesaver* (New York: Norton, 2007), 52, 58; Nadja Durbach, *Bodily Matters: The Anti-Vaccination Movement in England, 1853–1907* (Durham, NC: Duke University Press, 2005), 131–32; Gareth Williams, *Angel of Death: The Story of Smallpox* (Basingstoke, UK: Palgrave Macmillan, 2010), 265; and Michael Willrich, *Pox: An American History* (New York: Penguin, 2011), 93–94, 154.

3. Durbach, *Bodily Matters*, 41.

4. Martin Kaufman, "The American Anti-Vaccinationists and Their Arguments," *Bulletin of the History of Medicine* 41, no. 5 (1967): 463–78. Not all subscribers to these schools of healing thought opposed vaccines. See, for example, Nadav Davidovitch, "Negotiating Dissent: Homeopathy and Anti-Vaccinationism at the Turn of the Twentieth Century," in *The Politics of Healing: Histories of Alternative Medicine in Twentieth-Century North America*, ed. Robert Johnston (New York: Routledge, 2004), 11–28.

5. See James Colgrove, *State of Immunity: The Politics of Vaccination in Twentieth-Century America* (Berkeley: University of California Press, 2006), chap. 2; and Michael Willrich, "'The Least Vaccinated of Any Civilized Country': Personal Liberty and Public Health in the Progressive Era," *Journal of Policy History* 20, no. 1 (2008): 76–93.

6. See Robert Johnston, *The Radical Middle Class: Populist Democracy and the Question of Capitalism in Progressive Era Portland, Oregon* (Princeton, NJ: Princeton University Press, 2003), part 4.

7. Colgrove, *State of Immunity*, 74. For details on the activism of Lora Little, see Johnston, *The Radical Middle Class*, chap. 14. For more on Higgins, see Colgrove, *State*

of Immunity, 52–54; and Kaufman, "The American Anti-Vaccinationists and Their Arguments."

8. Annie Riley Hale, *The Medical Voodoo* (New York: Gotham House, 1935).

9. Eleanor McBean, *The Poisoned Needle: Suppressed Facts about Vaccination* (Mokelumne Hill, CA: Health Research, 1957).

10. Alfred Russel Wallace, *The Wonderful Century: Its Successes and Its Failures* (New York: Dodd, Mead, 1898).

11. McBean, *The Poisoned Needle*, 28.

12. Ibid.

13. Ibid., 27.

14. Ibid., 44; emphasis in original.

15. Eleanor McBean, *The Poisoned Needle*, 3rd ed. (Mokelumne Hill, CA: Health Research, 1974). See message from the publisher, inside front cover. According to Nikki Jones Roberts, current co-owner of Health Research Books, now located in Pomeroy, Washington, the book was probably reissued in 1974 because that is when the first print run ran out, but no specific sales or distribution figures from that time remain. Personal communication, July 2011.

16. See, for example, Harry Nelson, "1.1 Million Immunizations: Clinics Plan Massive Attack on German Measles Threat," *Los Angeles Times*, September 11, 1970, 9A; "Center Will Give Shots to Children," *Los Angeles Times*, July 2, 1972, SF B3; and "1st in Decade: Polio Case Suspected in L.A. County," *Los Angeles Times*, July 19, 1973, OC3.

17. Samuel Hays and Barbara Hays, *Beauty, Health, and Permanence: Environmental Politics in the United States, 1955–1985* (New York: Cambridge University Press, 1987), 175; Thomas R. Dunlap, *DDT: Scientists, Citizens, and Public Policy* (Princeton, NJ: Princeton University Press, 1981).

18. Ida Honorof and Eleanor McBean, *Vaccination: The Silent Killer* (Sherman Oaks, CA: Honor Publications, 1977); Eleanor McBean, *Swine Flu Expose* (Los Angeles: Better Life Research Center, 1977); Eleanor McBean, *Vaccinations Do Not Protect* (Yorktown, TX: Life Science, 1980); John Crawford, "The Poisoned Needle," *Mothering*, Winter 1979, 40.

19. Hays and Hays, *Beauty, Health, and Permanence*.

20. Ibid.; Joseph Melling and Christopher Sellers, "Introduction," in *Dangerous Trade: Histories of Industrial Hazard across a Globalizing World*, ed. Joseph Melling and Christopher Sellers (Philadelphia: Temple University Press, 2011), 1–14.

21. Allan M. Brandt, *The Cigarette Century: The Rise, Fall, and Deadly Persistence of the Product that Defined America* (New York: Basic Books, 2007).

22. On the evolution of popular understanding of environment risks in this period, see Gerald Markowitz and David Rosner, *Deceit and Denial: The Deadly Politics of Industrial Pollution* (Berkeley: University of California Press, 2002); and Melling and Sellers, "Introduction."

23. Elena Conis, "Debating the Health Effects of DDT: Thomas Jukes, Charles Wurster, and the Fate of an Environmental Pollutant," *Public Health Reports* 125, no. 2 (2010): 337–42; Dunlap, *DDT*.

24. Daniel A. Lander, "On Immunization," *Mothering*, Fall 1981, 32–35; Daniel A. Lander, *Immunization: An Informed Choice* (Glen Cove, ME: Dr. Daniel Lander, Family Chiropractor, 1978).

25. Cynthia Cournoyer, *What about Immunizations?* (Canby, OR: Concerned Parents for Information, 1983).

26. Richard Moskowitz, "Immunizations: The Other Side," *Mothering*, Spring 1984, 32–37.

27. Peter A. Coates, *Nature: Western Attitudes since Ancient Times* (Berkeley: University of California Press, 1998).

28. Crawford, "The Poisoned Needle," 40.

29. Barry Commoner, *Science and Survival* (New York: Viking Press, 1966).

30. Richard Knox, "A Shot in Arm, a Shot in Dark," *Boston Globe*, December 26, 1976, A2.

31. Moskowitz, "Immunizations."

32. Robert S. Mendelsohn, *How to Raise a Healthy Child . . . in Spite of Your Doctor* (Chicago: Contemporary Books, 1984), 232.

33. Marian Tompson, "Viewpoint," *The People's Doctor* 6, no. 12 (1982).

34. Leonard Jacobs, "Eating Well—the Best Vaccine," *Mothering*, Fall 1978, 17.

35. Ibid.

36. Patricia Savage, "A Mother's Research on Immunization," *Mothering*, Fall 1979, 76.

37. Victor LaCerva, "Letter to the Editor," *Mothering*, Spring 1979, 6.

38. James C. Whorton, *Nature Cures: The History of Alternative Medicine in America* (New York: Oxford University Press, 2002).

39. Brown's letter was published in a later compilation issued by *Mothering* magazine: Sue Brown, "More on Immunizations," in *Vaccinations*, ed. Peggy O'Mara (Santa Fe, NM: Mothering Magazine, 1988), 21.

40. Lander, "On Immunization." Chiropractors have a long tradition of rejecting vaccines. See, for example, Mark Largent, *Vaccine: The Debate in Modern America* (Baltimore: Johns Hopkins University Press, 2012), 42–51.

41. Jacobs, "Eating Well—the Best Vaccine."

42. Cynthia Cournoyer, *What about Immunizations?: Exposing the Vaccine Philosophy*, 5th ed. (Santa Cruz, CA: Nelson's Books, 1991).

43. Marian Tompson, "Viewpoint (1982)," in *The Risks of Immunization and How to Avoid Them*, ed. Robert S. Mendelsohn (Evanston, IL: The People's Doctor, 1988), 31. It is unclear where Tompson found these figures, or whether they refer to immunization generally or against a specific infection.

44. Diane Rozario, *The Immunization Resource Guide*, 2nd ed. (Burlington, IA: Patter Publications, 1994).

45. Google's Ngram Viewer (https://books.google.com/ngrams) produces a vivid illustration of the use of these two terms over time. Use of the term "poison" declined steadily from 1900 to 2000, whereas use of the term "toxic" peaked briefly in the 1910s (though it was still less commonly used than "poison") and then surpassed use of the term "poison" around 1960, with a sharp and steady increase in use from 1970 to the early 1990s. It is also worth noting that in the many sources I reviewed, lay use of the terms "toxic" and "toxin" not infrequently conflated the meaning of the two words. Both words were often used to refer to chemical contaminants when in fact "toxin" connotes a poison derived from a living organism.

46. Hays and Hays, *Beauty, Health, and Permanence*.

47. Grace Girdwain, "Immunizations for Public Schools and Passports," *Mothering*, Spring 1979, 10.

48. Roxanne Bank, "A Mother Researches Immunization," in *Vaccinations*, ed. Peggy O'Mara (Santa Fe, NM: Mothering Magazine, 1988), 18–20.

49. Cynthia Cournoyer, *What about Immunizations?*, 4th ed. (Grants Pass, OR: Cynthia Cournoyer, 1987). Cournoyer began publishing this tract in 1983; later editions are available in many public libraries, but the earlier editions are in Cournoyer's own collection, some of which she generously shared with me.

50. McBean, *The Poisoned Needle*, 3rd ed. See also Durbach, *Bodily Matters*.

51. Carol Horowitz, "Immunizations and Informed Consent," *Mothering*, Winter 1983, 37–41.

52. Cournoyer, *What about Immunizations?*, 1st ed.

53. Lora Little, *Crimes of the Cowpox Ring: Some Moving Pictures Thrown on the Dead Wall of Official Silence* (Minneapolis: Liberator Publishing, 1906); Hale, *The Medical Voodoo*.

54. McBean, *The Poisoned Needle*, 3rd ed., 21.

55. Ibid., 7.

56. Mendelsohn, *How to Raise a Healthy Child*, 211; Cournoyer, *What about Immunizations?*, 5th ed., 30.

57. Mendelsohn, *The Risks of Immunization*, 2.

58. Mendelsohn, *How to Raise a Healthy Child*, 211.

59. See chapter 9.

60. Harold E. Buttram and John Chriss Hoffman, "Bringing Vaccines into Perspective," *Mothering*, Winter 1985, 42.

61. Ibid. Buttram was no traditional allopathic practitioner and had personal ties to the anti-immunization groups of the early part of the century, including the Anti-Vaccination Society of America. Biographical details on Buttram appear in Allen, *Vaccine*, 338–42.

62. J.M., "Reader Letter," in *The Risks of Immunization and How to Avoid Them*, ed. Robert S. Mendelsohn (Evanston, IL: The People's Doctor, 1988), 90.

63. Harris L. Coulter and Barbara Loe Fisher, *DPT: A Shot in the Dark* (New York: Harcourt, Brace, Jovanovich, 1985). For more details on *DPT: A Shot in the Dark*, see chapter 5. The book was reissued in 1991. Harris L. Coulter and Barbara Loe Fisher, *A Shot in the Dark: Why the P in the DPT Vaccination May be Hazardous to Your Child's Health* (Garden City Park, NY: Avery Publishing, 1991).

64. Ibid., 110.

65. Ibid., 124.

66. Harris Coulter, *Vaccination, Social Violence, and Criminality: The Medical Assault on the American Brain* (Berkeley, CA: North Atlantic Books, 1990).

67. Walene James, *Immunization: The Reality Behind the Myth* (South Hadley, MA: Bergin & Garvey, 1988); Neil Z. Miller, *Vaccines: Are They Really Safe and Effective?: A Parent's Guide to Childhood Shots* (Santa Fe, NM: New Atlantean Press, 1992); Jamie Murphy, *What Every Parent Should Know about Childhood Immunization* (Boston: Earth Healing Products, 1994); Randall Neustaedter, *The Immunization Decision: A Guide for Parents* (Berkeley, CA: North Atlantic Books, Homeopathic Educational Services, 1990); Viera Scheibner, *Vaccination: 100 Years of Orthodox Research Shows that Vaccines*

Represent a Medical Assault on the Immune System (Santa Fe, NM: New Atlantean Press, 1993). Such titles were on the whole published by small specialized alternative presses.

68. Whorton, *Nature Cures*, ix.

69. Murphy, *What Every Parent Should Know*, 20.

70. Neustaedter, *The Immunization Decision*, 8.

71. Miller, *Vaccines*, 49.

72. Scheibner, *Vaccination*, 92.

73. Murphy, *What Every Parent Should Know*, 53. Largent has argued that popular worries about thimerosal were linked to the public health community's "aggressive campaign against environmental toxins like lead, mercury, and arsenic" from the 1970s through the 1990s. Largent, *Vaccine*, 12.

74. Randall Neustaedter, *The Vaccine Guide: Making an Informed Choice*, 2nd ed. (Berkeley, CA: North Atlantic Books, 1996), 51–52.

75. Associated Press, "High Mercury Levels Found in Everglades Fish," *Washington Post*, March 13, 1989, A16.

76. Jane Brody, "Personal Health: Safety Questions about Eating Fish," *New York Times*, June 12, 1991, C10.

77. K. R. Mahaffey et al., *An Assessment of Exposures to Mercury in the United States: Mercury Study Report to Congress* (Washington, DC: U.S. Environmental Protection Agency, 1997).

78. Associated Press, "Minnesota Bans Flashing Sneakers as Toxic," *New York Times*, May 10, 1994, A18.

79. Nancy Beth Jackson, "Groups Debate Safety of Canned Tuna for the Very Young," *New York Times*, May 11, 1999, F14.

80. "Ending Mercury Madness," *Mothering*, July/August 1999, 11. See note 45 regarding use of the word "toxin."

81. Public Health Service and American Academy of Pediatrics, "Notice to Readers: Thimerosal in Vaccines—a Joint Statement of the American Academy of Pediatrics and the Public Health Service," *Morbidity and Mortality Weekly Report* 48, no. 26 (1999): 563–65.

82. Leslie Ball, Robert Ball, and R. Douglas Pratt, "An Assessment of Thimerosal Use in Childhood Vaccines," *Pediatrics* 107, no. 5 (2001): 1147–54.

83. Environmental Protection Agency, *Mercury Study Report to Congress*, 1997, http://www.epa.gov/hg/report.htm (accessed August 2012).

84. Jane M. Hightower, *Diagnosis Mercury: Money, Politics, and Poison* (Washington, DC: Island Press/Shearwater Books, 2009), 37–38.

85. Paul Offit, "Letter to the Editor: Preventing Harm from Thimerosal in Vaccines," *Journal of the American Medical Association* 283, no. 16 (2000): 2104.

86. Stanley Plotkin, "Letter to the Editor: Preventing Harm from Thimerosal in Vaccines," *Journal of the American Medical Association* 283, no. 16 (2000): 2104–5. The hepatitis B vaccine was one of three vaccines, along with the DPT and Hib vaccines, that contained thimerosal.

87. Neal Halsey, "Limiting Infant Exposure to Thimerosal in Vaccines and Other Sources of Mercury," *Journal of the American Medical Association* 282, no. 18 (1999): 1763–66.

88. Arthur Allen, "The Not-So-Crackpot Autism Theory," *New York Times*, November 10, 2002, F66.

89. Ball, Ball, and Pratt, "An Assessment of Thimerosal Use in Childhood Vaccines."

90. Bruce G. Gellin, Edward W. Maibach, and Edgar K. Marcuse, "Do Parents Understand Immunizations?: A National Telephone Survey," *Pediatrics* 106, no. 5 (2000): 1097–102.

91. National Vaccine Information Center, "Autism and Vaccines," *NVIC Newsletter*, Spring 2000, http://www.nvic.org/nvic-archives/newsletter/autismandvaccines.aspx (accessed August 2010). The now-infamous report in the *Lancet* is further discussed in chapters 9 and 10. Andrew J. Wakefield et al., "Ileal-Lymphoid-Nodular Hyperplasia, Non-Specific Colitis, and Pervasive Developmental Disorder in Children," *Lancet* 351, no. 9103 (1998): 637–41.

92. Thomas R. Dunlap, *Faith in Nature: Environmentalism as Religious Quest*, Weyerhaeuser Environmental Books (Seattle: University of Washington Press, 2004), 4, 149.

93. Coulter and Fisher, *DPT: A Shot in the Dark*, 408. See also Harris Coulter and Barbara Loe Fisher, *A Shot in the Dark: Why the P in the DPT Vaccination May Be Hazardous to Your Child's Health* (Garden City Park, NY: Avery Publishing, 1991), 220.

94. Shannon Henry, "A Pox on My Child: Cool!," *Washington Post*, September 20, 2005, F1.

95. Jon Bowen, "Germs Invited to the Party," *Chicago Sun-Times*, August 6, 2000, 8.

96. Brian Wimer, Jacquelyn Emm, and Deren Bader, "A Few Tips for Chickenpox Parties," *Mothering*, January/February 2004, 35.

97. Fran Pado, "Where the Play Dates Are Fun, but Itchy," *New York Times*, August 21, 2005, B5.

98. Richard Halstead, "Marin Fights Outbreak of Chicken Pox," *Marin Independent Journal*, November 15, 2006, 15.

99. In his introduction to Dunlap's book, environmental historian William Cronon noted that environmentalism and religion share a predilection for predictions of disaster as "a platform for critiquing the moral failings of our lives in the present." Dunlap, *Faith in Nature*, xii.

100. Roderick Nash, *The Rights of Nature: A History of Environmental Ethics*, History of American Thought and Culture (Madison: University of Wisconsin Press, 1989), 12.

101. Whorton, *Nature Cures*; Dunlap, *Faith in Nature*.

102. Moskowitz, "Immunizations."

103. Coulter and Fisher, *DPT: A Shot in the Dark*, 407.

CHAPTER SEVEN

1. Office of the Press Secretary, "Remarks by the President at Reading of Immunization Proclamation, April 12, 1993," William J. Clinton Presidential Library, http://clinton6.nara.gov/1993/04/ (accessed February 2010).

2. Early in his first term, the speech was not technically a State of the Union address. See William J. Clinton, "Address before a Joint Session of Congress on Administration Goals," February 17, 1993, American Presidency Project, UCSB.

3. A. R. Hinman, W. A. Orenstein, and L. Rodewald, "Financing Immunizations in the United States," *Clinical Infectious Diseases* 38, no. 10 (2004): 1440–46.

4. Ibid. Today the Vaccines for Children Program provides 43 percent of all childhood vaccines. Jason T. Shafrin and John M. Fontanesi, "Delivering Vaccines: A Case

Study of the Distribution System of Vaccines for Children," *American Journal of Managed Care* 15, no. 10 (2009): 751–54.

5. CDC, "Goal to Eliminate Measles from the United States," *Morbidity and Mortality Weekly Report* 27 (1978): 391.

6. Inspired by the achievements at home and abroad, they also began to make a case for eliminating measles globally. A. R. Hinman et al., "Progress in Measles Elimination," *Journal of the American Medical Association* 247, no. 11 (1982): 1592–95; CDC, "Elimination of Indigenous Measles—United States," *Morbidity and Mortality Weekly Report* 31, no. 38 (1982): 517–19; D. R. Hopkins et al., "The Case for Global Measles Eradication," *Lancet* 1, no. 8286 (1982): 1396–98.

7. CDC, "Measles—United States, 1983," *Morbidity and Mortality Weekly Report* 33, no. 8 (1984): 105–8.

8. CDC, "Current Trends Measles—United States, 1990," *Morbidity and Mortality Weekly Report* 40, no. 22 (1991): 369–72; "Memorandum for the President from Donna E. Shalala," February 11, 1993, Folder 7, Box 14, Domestic Policy Council, Rasco Subject File, William J. Clinton Presidential Library.

9. CDC, "Current Trends Measles—United States, 1990."

10. CDC, "Measles Prevention: Recommendations of the Immunization Practices Advisory Committee (ACIP)," *Morbidity and Mortality Weekly Report* 38, no. S9 (1989): 1–18; American Academy of Pediatrics Committee on Infectious Diseases, "Measles: Reassessment of the Current Immunization Policy," *Pediatrics* 84, no. 6 (1989): 1110–13.

11. National Vaccine Advisory Committee, "The Measles Epidemic: The Problems, Barriers, and Recommendations," *Journal of the American Medical Association* 266, no. 11 (1991): 1547–52.

12. Philip Hilts, "Panel Ties Measles Epidemic to Breakdown in Health System," *New York Times*, January 9, 1991, A17.

13. Susan Okie, "Vaccination Record in U.S. Falls Sharply," *Washington Post*, March 24, 1991, A1.

14. "The Shame of Measles," *New York Times*, May 22, 1990, A26.

15. Despite this, immunization rates among schoolchildren remained 90 percent or higher through the eighties. A. R. Hinman, "What Will It Take to Fully Protect All American Children with Vaccines?," *American Journal of Diseases of Children* 145, no. 5 (1991): 559–62.

16. Susan Dentzer and Dorian Friedman, "America's Scandalous Health Care," *U.S. News & World Report*, March 12, 1990, 24–28, 30.

17. Janice Castro, "American Health Care Condition: Critical," *Time*, November 25, 1991.

18. Lloyd Grove, "That Other Southern President," *Washington Post*, January 14, 1993, C1.

19. "Health Security for All Americans," address of President William Jefferson Clinton before a Joint Session of Congress, September 22, 1993, William J. Clinton Presidential Library, http://clinton6.nara.gov/1993/09/ (accessed February 2010); Carol H. Rasco, "Special Update on Healthcare Reform from the White House," *Infection Control and Hospital Epidemiology* 16, no. 9 (1995): 526–28.

20. Hart and Teeter Research, "NBC News/Wall Street Journal Poll," January 1993.

21. Dana Priest, "Clinton Names Wife to Head Health Panel," *Washington Post*, January 26, 1993, A1. On the series of events that led to the delay, see David Blumenthal and

James A. Morone, *The Heart of Power: Health and Politics in the Oval Office* (Berkeley: University of California Press, 2009), 355–75; and Theda Skocpol, *Boomerang: Clinton's Health Security Effort and the Turn against Government in U.S. Politics* (New York: Norton, 1996).

22. Clinton, "Address before a Joint Session of Congress on Administration Goals."

23. Ibid.

24. Memo from Ruth Katz and Tim Westmoreland to Josh Steiner, December 9, 1992, Folder 15, Box 204, White House Health Care Interdepartmental Working Group, Participants' Working Papers, William J. Clinton Presidential Library.

25. N. A. Vanderpool and J. B. Richmond, "Child Health in the United States: Prospects for the 1990s," *Annual Review of Public Health* 11 (1990): 185–205.

26. "Mandate for Children," news release, April 19, 1993, Folder 7, Box 9, Domestic Policy Council, Rasco Subject Files, William J. Clinton Presidential Library.

27. List of national health care plan options, not dated, Folder 1: Transitional Issues and Congressional Bills, Box 350, White House Health Care Interdepartmental Working Group, Participants' Working Papers, William J. Clinton Presidential Library; memo from David Nather to Atul Gawande, October 10, 1992, Folder 24, Box 204, White House Health Care Interdepartmental Working Group, Participants' Working Papers, William J. Clinton Presidential Library.

28. "Memorandum for the President from Donna E. Shalala"; "Stories," memo from the Director of the National Vaccine Program Office to Jerry Klepner, February 11, 1993, Folder 8, Box 14, Domestic Policy Council, Rasco Subject Files, William J. Clinton Presidential Library.

29. U.S. Congress, House Committee on Appropriations, *Emergency Supplemental Appropriations for Fiscal Year 1993: Hearings before the Subcommittees of the Committee on Appropriations, House of Representatives, One Hundred Third Congress, First Session* (Washington, DC: U.S. Government Printing Office, 1993).

30. William J. Clinton, "To the Congress of the United States," April 1, 1993, Folder 8, Box 14, Domestic Policy Council, Rasco Subject File, William J. Clinton Presidential Library.

31. "Memorandum to the President," 2/7/93, Folder 7, Box 14, Domestic Policy Council, Rasco Subject File, William J. Clinton Presidential Library; memo from Kevin Thurm to Christine Varney, "Re: Childhood Immunization Initiative," 3/21/93, Folder 8, Box 14, Domestic Policy Council, Rasco Subject File, William J. Clinton Presidential Library.

32. Letter from David Williams to Carol Rasco, March 2, 1993, Folder 8, Box 14, Domestic Policy Council, Rasco Subject File, William J. Clinton Presidential Library.

33. Robert Pear, "Clinton Considers Plan to Vaccinate All U.S. Children," *New York Times*, February 1, 1993, A1.

34. Press release from American Cyanamid Company, March 24, 1993, Folder 8, Box 14, Domestic Policy Council, Rasco Subject File, William J. Clinton Presidential Library.

35. "Memorandum for the President from Donna E. Shalala." Part of the price increase came from health inflation that affected the prices of all drugs; another part of the increase was attributed to the excise tax added to vaccines as a result of the National Childhood Vaccine Injury Compensation Act of 1986.

36. Office of the Press Secretary, "Remarks by the President at Reading of Immunization Proclamation."

37. "Statement of the President, Arlington County Department of Human Services," 2/12/93, Folder 7, Box 14, Domestic Policy Council, Rasco Subject File, William J. Clinton Presidential Library.

38. Ibid.

39. Christopher Connell, "Clinton Knocks Drug Prices, Launches Plan for Kids' Shots," *Chicago Sun Times*, February 12, 1993, 1; Spencer Rich and Ann Devroy, "President Blasts Cost of Vaccines," *Washington Post*, February 13, 1993, A1; P-I News Services, "Profits at the Expense of Children; Clinton Calls Pharmacy Prices Shocking," *Seattle Post-Intelligencer*, February 13, 1993, A1.

40. Robert Samuelson, "Nationalize Health Care," *Newsweek*, October 26, 1992, 50.

41. Federal News Service, "News Conference Transcript: Vaccines for Children Program," 7/19/94, Folder 8, Box 33, Domestic Policy Council, Rasco Subject File, William J. Clinton Presidential Library. At the time, half of all children were immunized in public clinics. C. A. Robinson, S. J. Sepe, and K. F. Lin, "The President's Child Immunization Initiative: A Summary of the Problem and the Response," *Public Health Reports* 108, no. 4 (1993): 419–25.

42. "Memo on Immunization Bill," Folder 6, Box 14, Domestic Policy Council, Rasco Subject File, William J. Clinton Presidential Library.

43. "Statement of the President, Arlington County Department of Human Services."

44. Robert Pear, "Clinton, in Compromise, Will Cut Parts of Childhood Vaccine Plan," *New York Times*, May 5, 1993, A1, A20; "Memorandum for the President, Legislative Initiative to Comprehensively Address the Nation's Childhood Immunization Crisis," February 7, 1993, Folder 7, Box 14, Domestic Policy Council, Rasco Subject File, William J. Clinton Presidential Library.

45. Kay Johnson, *Who Is Watching Our Children's Health?: The Immunization Status of American Children* (Washington, DC: Children's Defense Fund, 1987).

46. "Public Clinics Found to Lack Children's Vaccine," *New York Times*, June 19, 1991, C9.

47. "Memo from Marian Wright Edelman to the President and First Lady," February 18, 1993, Folder 7, Box 14, Domestic Policy Council, Rasco Subject File, William J. Clinton Presidential Library.

48. "Memo from Donna Shalala to the President," February 7, 1993, Folder 7, Box 14, Domestic Policy Council, Rasco Subject File, William J. Clinton Presidential Library.

49. Robert Pear, "Bush Announces New Effort to Immunize Children," *New York Times*, May 12, 1992, A18.

50. Felicia Lee, "Immunization of Children Is Said to Lag; Third World Rate Seen in the New York Area," *New York Times*, October 16, 1991, B1; "The Shame of Measles."

51. Jay P. Sanford et al., *The Children's Vaccine Initiative: Achieving the Vision* (Washington, DC: National Academy Press, 1993).

52. Lisa Belkin, "A Resurgence of Plagues and Pestilences of Yesteryear," *New York Times*, January 19, 1992, E4.

53. "Presidential Briefing Book, Population-Based Prevention and Public Health," March 17, 1993, Folder 5, Box 654, Records of the White House Health Care Interdepartmental Working Group, Participants' Working Papers, William J. Clinton Presidential Library.

54. David Satcher, "Emerging Infections: Getting Ahead of the Curve," *Emerging Infectious Diseases* 1, no. 1 (1995): 1–6.

55. Laurie Garrett, *The Coming Plague: Newly Emerging Diseases in a World out of Balance* (New York: Farrar, Straus and Giroux, 1994); Richard Preston, *The Hot Zone* (New York: Random House, 1994).

56. William J. Clinton, "Inaugural Address, January 20, 1993," American Presidency Project, UCSB.

57. "Letters to the First Lady," Folder 6, Box 358, White House Health Care Interdepartmental Working Group, Participants' Working Papers, Ira Magaziner, William J. Clinton Presidential Library.

58. Amy Goldstein and Spencer Rich, "Health Experts Skeptical about Immunization Plan," *Washington Post*, April 2, 1993, B1.

59. C. Everett Koop, "In the Dark about Shots," *Washington Post*, February 10, 1993, A21.

60. "The Drive to Vaccinate," *Washington Post*, May 3, 1993, A18.

61. Spencer Rich, "Childhood Vaccines Program Cut Back by Administration," *Washington Post*, May 6, 1993, A25.

62. Note from Christopher "Kit" Ashby to Bill and Hillary, February 22, 1993, Folder 8, Box 14, Domestic Policy Council, Rasco Subject File, William J. Clinton Presidential Library.

63. Memo from Ruth Katz and Tim Westmoreland to Josh Steiner, December 9, 1992, Folder 15, Box 204, White House Health Care Interdepartmental Working Group—Participants' Working Papers, William J. Clinton Presidential Library.

64. One outbreak occurred in a school with a 99 percent measles vaccination rate. CDC scientists concluded that the children were either vaccinated too young (when maternal antibodies would have interfered with their ability to mount an immune response), or before a new stabilizer was added to the vaccine, in 1979. See R. T. Chen et al., "An Explosive Point-Source Measles Outbreak in a Highly Vaccinated Population: Modes of Transmission and Risk Factors for Disease," *American Journal of Epidemiology* 129, no. 1 (1989): 173–82; E. E. Mast et al., "Risk Factors for Measles in a Previously Vaccinated Population and Cost-Effectiveness of Revaccination Strategies," *Journal of the American Medical Association* 264, no. 19 (1990): 2529–33; and W. A. Orenstein et al., "Appropriate Age for Measles Vaccination in the United States," *Development of Specifications for Biotechnology Pharmaceutical Products* 65 (1986): 13–21.

65. Blumenthal and Morone, *The Heart of Power*, 355, 359–64.

66. "Atlanta Project Pulled It Off," *Atlanta Journal-Constitution*, May 5, 1993, A12.

67. The program required states to amend their Medicaid programs to include "a pediatric immunization distribution program." The cost of the program was $1.4 billion over four years, but it was projected to result in federal and state Medicaid savings of roughly the same amount. The program also made the vaccine taxes used to fund the National Vaccine Injury Compensation Program permanent. The national registry did not survive the final bill. See Omnibus Budget Reconciliation Act of 1993, PL103-66; and Melvina Ford, "Health Care Fact Sheet: Child Immunization Provisions in the Omnibus Budget Reconciliation Act of 1993, P.L. 103-66, September 2, 1993," CRS Report for Congress 93-781 EPW, Congressional Research Service.

68. "How to Help Immunization," *Atlanta Journal-Constitution*, May 5, 1993, A14.

69. Editorial, "Immunity for the Children," *Washington Post*, February 4, 1993, A20.

70. Editorial, "The Drive to Vaccinate," *Washington Post*, May 3, 1993, A18.

71. Jason DeParle, "With Shots, It's Not Only about Costs, but Stories," *New York Times*, May 16, 1993, E18.

72. Talking points, January 1, 1994, Folder 8, Box 13, Domestic Policy Council, Reed Welfare Reform Subject File, William J. Clinton Presidential Library.

73. Ann Devroy, "Bush Announces New Push to Improve Vaccination Programs," *Washington Post*, June 14, 1991, A17.

74. Okie, "Vaccination Record in U.S. Falls Sharply."

75. Devroy, "Bush Announces New Push to Improve Vaccination Programs."

76. Mary G. Fontrier, "Price of Vaccines for Measles Criticized," *New York Times*, September 17, 1989, LI25.

77. Robert M. Goldberg and American Enterprise Institute for Public Policy Research, *The Vaccines for Children Program: A Critique*, AEI Studies in Policy Reform (Washington, DC: AEI Press, 1995).

78. See, for example, Robert Pear, "Clinton Criticized as Too Ambitious with Vaccine Plan," *New York Times*, May 29, 1994, 1; K. A. Johnson, A. Sardell, and B. Richards, "Federal Immunization Policy and Funding: A History of Responding to Crises," *American Journal of Preventive Medicine* 19, no. S3 (2000): 99–112; memo from Sara Rosenbaum to Carol Rasco, July 19, 1994, Folder 7, Box 33, Domestic Policy Council, Rasco Subject Files, William J. Clinton Presidential Library; and letter from Dale Bumpers and John Danforth to Donna Shalala, April 13, 1994, Folder 9, Box 33, Domestic Policy Council, Rasco Subject File, William J. Clinton Presidential Library.

79. William A. Galston and Geoffrey L. Tibbetts, "Reinventing Federalism: The Clinton/Gore Program for a New Partnership among the Federal, State, Local, and Tribal Governments," *Publius* 24, no. 3 (1994): 23–47.

80. Hillary Rodham Clinton, *It Takes a Village: And Other Lessons Children Teach Us* (New York: Simon & Schuster, 1996), 110–15.

81. Hinman, Orenstein, and Rodewald, "Financing Immunizations in the United States."

CHAPTER EIGHT

1. CDC, "National, State, and Urban Area Vaccination Coverage Levels among Children Aged 19–35 Months, United States, 1999," *Morbidity and Mortality Weekly Report* 49, no. 26 (2000): 585–89.

2. World Health Organization, "WHO/UNICEF Estimates of National Immunization Coverage: Estimated Coverage by Country, Year, and Vaccine," *Immunization Surveillance, Assessment, and Monitoring*, http://www.who.int/immunization_monitoring /routine/immunization_coverage/en/index4.html (accessed February 2011).

3. Leon Jaroff, "Vaccine Jitters," *Time*, September 13, 1999, 64–65.

4. See Baruch S. Blumberg, *Hepatitis B: The Hunt for a Killer Virus* (Princeton, NJ: Princeton University Press, 2002), introduction, chap. 11.

5. William Muraskin, "Hepatitis B as a Model (and Anti-Model) for AIDS," in *AIDS and Contemporary History*, ed. Virginia Berridge and Philip Strong (Cambridge: Cambridge University Press, 1993), 108–32.

6. Blumberg, *Hepatitis B*, chap. 5.

7. Ibid., 134–36. See also B. S. Blumberg, A. I. Sutnick, and W. T. London, "Australia

Antigen and Hepatitis," *Journal of the American Medical Association* 207, no. 10 (1969): 1895–96; A. I. Sutnick, W. T. London, and B. S. Blumberg, "Australia Antigen and the Quest for a Hepatitis Virus," *American Journal of Digestive Diseases* 14, no. 3 (1969): 189–94; W. T. London, A. I. Sutnick, and B. S. Blumberg, "Australia Antigen and Acute Viral Hepatitis," *Annals of Internal Medicine* 70, no. 1 (1969): 55–59; and B. S. Blumberg et al., "Australia Antigen and Hepatitis," *New England Journal of Medicine* 283, no. 7 (1970): 349–54.

8. Norman Pastorek, "Hepatitis," *Today's Health*, September 1974, 46–69.

9. Boyce Rensberger, "Sketches of Two Winners of Nobel Prizes in Medicine," *New York Times*, October 15, 1976, 13.

10. "A Wave of Death from Hepatitis," *Newsweek*, August 27, 1979, 72.

11. *CBS Evening News*, September 29, 1980, Vanderbilt Television News Archive (hereafter VTNA).

12. *CBS Evening News*, November 16, 1981, VTNA.

13. Ibid.

14. See, for example, W. Szmuness et al., "Hepatitis B Vaccine in Medical Staff of Hemodialysis Units: Efficacy and Subtype Cross-Protection," *New England Journal of Medicine* 307, no. 24 (1982): 1481–86.

15. William Muraskin, "The Silent Epidemic: The Social, Ethical, and Medical Problems Surrounding the Fight against Hepatitis B," *Journal of Social History* 22 (1988): 277–98.

16. CDC and Prevention, National Center for Health Statistics, Health Data Interactive, available at www.cdc.gov/nchs/hdi.htm (accessed March 2011). See also "Medicine: Cardiac Shocks," *Time*, August 18, 1980.

17. "Medicine: New Plagues for Old?," *Time*, November 24, 1980.

18. The vaccine's development is described in Louis Galambos and Jane Eliot Sewell, *Networks of Innovation: Vaccine Development at Merck, Sharp & Dohme and Mulford, 1895–1995* (New York: Cambridge University Press, 1995), 181–93. Merck listed the plasma-based vaccine among its all-time top discoveries, and Maurice Hilleman, who oversaw its development, was particularly proud of it. *Fifty Years of Research: Merck Sharp & Dohme Research Laboratory* (Rahway, NJ: Merck Sharp & Dohme, 1983), 3; Louis Galambos and Merck & Co., *Values and Visions: A Merck Century* (Rahway, NJ: Merck & Co., 1991).

19. *CBS Evening News*, November 16, 1981.

20. "Hepatitis Hope," *Time*, October 13, 1980; Jean Seligman, "A Vaccine for Hepatitis," *Newsweek*, October 13, 1980, 132; Carl Sherman, "Hepatitis: Why It's So Common," *Glamour*, March, 1981, 268–70.

21. See, for example, Seligman, "A Vaccine for Hepatitis"; and Lawrence Altman, "Tests of Hepatitis B Vaccine Show Nearly Complete Rate of Protection," *New York Times*, September 29, 1980, A1.

22. W. A. Check, "Looks Like Smooth Sailing for Experimental Hepatitis B Vaccine," *Journal of the American Medical Association* 246, no. 19 (1981): 2111–12.

23. "Recommendation of the Immunization Practices Advisory Committee (ACIP) Inactivated Hepatitis B Virus Vaccine," *Morbidity and Mortality Weekly Report* 31, no. 24 (1982): 317–22.

24. Harvey J. Alter, "The Evolution, Implications, and Applications of the Hepatitis B Vaccine," *Journal of the American Medical Association* 247, no. 16 (1982): 2272–75.

25. Lawrence Altman, "New Homosexual Disorder Worries Health Officials," *New York Times*, May 11, 1982, C1. See also Associated Press, "Rare Cancer Found in Gay Men," *Washington Post*, June 5, 1982, A2.

26. David Dickson, "AIDS Fears Spark Row over Vaccine," *Science*, no. 221 (1983): 437.

27. United Press International, "Two Doctors in U.S. Agency Back Hepatitis B Vaccine," *New York Times*, February 11, 1983, A14; "French Doctors Ban American Blood Imports," *New Scientist*, May 26, 1983, 529.

28. This comment, allegedly made in 1985, was recounted in Pat Griffin Mackie, "Hepatitis B Vaccine and Newborn," *National Immunity Information Network Newsletter*, September/October, 1997, 1.

29. CDC, "The Safety of Hepatitis B Virus Vaccine," *Morbidity and Mortality Weekly Report* 32, no. 10 (1983): 134–36. Follow-up data was published the following year: CDC, "Hepatitis B Vaccine: Evidence Confirming Lack of AIDS Transmission," *Morbidity and Mortality Weekly Report* 33 (1984): 685–87.

30. In the mid-1980s, some commentators in the black and gay media in particular speculated on a link between the emergence of HIV and the hepatitis B vaccine. Dermatologist Alan Cantwell blamed AIDS on hepatitis B trials conducted in homosexual men in the 1970s. See, for example, John Fiske, *Media Matters: Everyday Culture and Political Change* (Minneapolis: University of Minneota Press, 1994), 201, 213–14; Alan Cantwell, *AIDS and the Doctors of Death: An Inquiry into the Origin of the AIDS Epidemic* (Los Angeles: Aries Rising Press, 1988); and Alan Cantwell, *Queer Blood: The Secret AIDS Genocide Plot* (Los Angeles: Aries Rising Press, 1993).

31. CDC, "Current Trends Acquired Immunodeficiency Syndrome (AIDS) Update — United States," *Morbidity and Mortality Weekly Report* 32, no. 24 (1983): 309–11; "Recommendation of the Immunization Practices Advisory Committee (ACIP) Inactivated Hepatitis B Virus Vaccine." This pattern has also been noted by historian Gerald Oppenheimer, "In the Eye of the Storm: The Epidemiological Construction of AIDS," in *AIDS: The Burdens of History*, ed. Elizabeth Fee and Daniel Fox (Berkeley: University of California Press, 1988), 267–300.

32. Claudia Wallis, "AIDS: A Growing Threat," *Time*, August 12, 1985.

33. See, for example, Cheryl Sacra, "A Vaccine for Lovers," *Health* 21 (1989): 47.

34. Sanford Kuvin, "Vaccination Can Halt Epidemic of Hepatitis B, Cousin of AIDS," *New York Times*, April 9, 1989, E24.

35. "Hepatitis, Health, and the Hard Sell," *Gay Community News*, September 10, 1983.

36. Natalie Geary, "Health News: Hepatitis B," *Mademoiselle*, April, 1993, 120.

37. Alter, "The Evolution, Implications, and Applications of the Hepatitis B Vaccine."

38. "Uncorking the Genes: Biotech Stocks Just Coming into Own, Analyst Says," *Barron's National Business and Financial Weekly*, May 5, 1986, 10–16.

39. Claudia Wallis, "Made-to-Order Vaccines," *Time*, October 31, 1983; Janice Castro, "A Breakthrough for Biotech," *Time*, August 4, 1986.

40. Harold Schmeck, "The New Age of Vaccines," *New York Times Magazine*, April 29, 1984, 58. Merck, too, made optimistic predictions about the promise of genetically engineered vaccines. *Fifty Years of Research*, 16.

41. See, for example, Judy Packer, "Chiron Nears Sale of New Hepatitis Vaccine," *San Jose Business Journal* 3, no. 41 (1986); and Tom Post et al., "The Year's Best Entrepreneurial Ideas," *Venture* 8, no. 12 (1986): 6. Though the vaccine was developed by Chiron,

Merck brought it to market; it quickly became the company's most profitable vaccine. Company reports suggest that Merck was focused on the global market over the domestic one at the time, given the high prevalence of hepatitis B in Asia in particular. *Annual Report: Innovations for Global Health* (Rahway, NJ: Merck & Co., 1988), 17, 38.

42. Philip Boffey, "U.S. Approves a Genetically Altered Vaccine," *New York Times*, July 24, 1986, A1; Marlene Cimons, "First Human Vaccine Produced by Genetic Engineering Ok'd by FDA," *Los Angeles Times*, July 24, 1986, 1; Joe Davidson, "Lab-Made Vaccine for Hepatitis B Is Cleared by FDA," *Wall Street Journal*, July 24, 1986, 1. Some vaccine critics also saw genetic engineering as the answer to risks posed by vaccination. See National Vaccine Information Center, "Pertussis Vaccine Research Update," *Vaccine News* 5, no. 1 (1990): 5. The first genetically engineered drug, a recombinant form of human insulin, was approved by the FDA in 1982; the approval of recombinant human growth hormone followed in 1985. Suzanne White Junod, "Celebrating a Milestone: FDA Approval of First Genetically-Engineered Product," *Update*, no. 5 (2007), http://www.fdli.org/pubs/update/toc/2007/issue5.html (accessed December 2008).

43. This was a long-anticipated advantage of genetically engineered vaccines, particularly the genetically engineered hepatitis vaccine. However, when it was first introduced, Heptavax B (the plasma-derived vaccine) was the most expensive vaccine ever marketed. See, for example, Alter, "The Evolution, Implications, and Applications of the Hepatitis B Vaccine."

44. Editorial, "Science and Demagoguery," *Wall Street Journal*, July 31, 1986, 1.

45. Dori Stehlin, "Hepatitis B: Available Vaccine Safe and Underused," *FDA Consumer Magazine*, May 1990.

46. Wallis, "Made-to-Order Vaccines."

47. Associated Press, "FDA Approves Gene-Engineered Hepatitis Vaccine," *Dallas Morning News*, July 24, 1986, 5A.

48. CDC, "Surveillance Summary Viral Hepatitis—1984," *Morbidity and Mortality Weekly Report* 36, no. 3 (1987): 42–43.

49. The targeting of women identified through their sexual (in this case reproductive) behavior is reminiscent of countless historical attempts to curb the spread of sexually transmitted infections by detaining and treating women, especially prostitutes, over men. See, for example, Allan M. Brandt, *No Magic Bullet: A Social History of Venereal Disease in the United States since 1880* (New York: Oxford University Press, 1985); and Dorothy Porter and Roy Porter, "The Enforcement of Health: The British Debate," in *AIDS: The Burdens of History*, ed. Elizabeth Fee and Daniel Fox (Berkeley: University of California Press, 1988), 97–120.

50. CDC, "Recommendation of the Immunization Practices Advisory Committee (ACIP): Postexposure Prophylaxis of Hepatitis B," *Morbidity and Mortality Weekly Report* 33, no. 21 (1984): 285–90.

51. CDC, "Recommendations of the Immunization Practices Advisory Committee Prevention of Perinatal Transmission of Hepatitis B Virus: Prenatal Screening of All Pregnant Women for Hepatitis B Surface Antigen," *Morbidity and Mortality Weekly Report* 37, no. 22 (1988): 341–46, 351.

52. Ibid.

53. M. J. Alter et al., "The Changing Epidemiology of Hepatitis B in the United States: Need for Alternative Vaccination Strategies," *Journal of the American Medical Association* 263, no. 9 (1990): 1218–22.

54. CDC, "Hepatitis B Virus: A Comprehensive Strategy for Eliminating Transmission in the United States through Universal Childhood Vaccination: Recommendations of the Immunization Practices Advisory Committee (ACIP)," *Morbidity and Mortality Weekly Report* 40, no. RR-13 (1991): 1–19.

55. Ibid.

56. Dolores Kong, "U.S. to Urge All Children Be Vaccinated for Hepatitis B," *Boston Globe*, June 11, 1991. The official's comment was not entirely accurate; children were vaccinated to protect pregnant women from rubella infection. See chapter 3 and Leslie J. Reagan, *Dangerous Pregnancies: Mothers, Disabilities, and Abortion in America* (Berkeley: University of California Press, 2010).

57. "Hepatitis B 200x More Contagious than AIDS," *Philadelphia Tribune*, December 31, 1991, 4B; Lisa Holland, "The ABC's of Hepatitis," *Good Housekeeping*, April, 1991, 239.

58. Leslie Laurence, "Beware the Quiet Killer," *Redbook*, October 1991, 24, 28, 32; Sandy Coleman, "Q&A with Leslie Hsu, South Cove Health Center in Chinatown," *Boston Globe*, February 15, 1998, 2.

59. Janice Hopkins Tanne, "The Other Plague: Potentially Deadly Hepatitis Is Fifteen Times More Common than AIDS," *New York*, July 11, 1988, 34–40.

60. Ibid., 35.

61. Ann Devroy, "Bush Announces New Push to Improve Vaccination Programs," *Washington Post*, June 14, 1991, A17.

62. Robert Pear, "Proposal Would Tie Welfare to Vaccinations of Children," *New York Times*, November 29, 1990, A1.

63. Phyllis Freeman and Anthony Robbins, "An Epidemic of Inactivity," *New York Times*, July 10, 1991, A19; Jeremy Waldron, "There We Go Again, Punishing the Poor," *New York Times*, December 12, 1990, A22.

64. CDC, "Measles Prevention: Recommendations of the Immunization Practices Advisory Committee (ACIP)," *Morbidity and Mortality Weekly Report* 38, no. S9 (1989): 1–18. College students were presumed to be at risk because the vaccine they received back in the 1960s, scientists had concluded, was too weak to confer lasting immunity.

65. L. McTaggart and D. Zakruczemski, "The MMR Vaccine," *Mothering*, Spring 1992, 56–62. See also "Are Vaccines Generally Detrimental to the Human Defense System?," *Townsend Letter for Doctors & Patients*, February/March 1994.

66. CDC, "Measles Prevention."

67. Joanne Hatem, "MMR Update," *NVIC News* 1, no. 3 (1991): 9.

68. See chapter 7 for an extensive discussion of Clinton's childhood immunization initiative.

69. Office of the Press Secretary, "Remarks by the President at Reading of Immunization Proclamation, April 12, 1993," William J. Clinton Presidential Library, http://clinton6.nara.gov/1993/04/ (accessed February 2010). See also Omnibus Budget Reconciliation Act of 1993, PL103-66; and Gary Freed and Samuel Katz, "The Comprehensive Childhood Immunization Act of 1993," *New England Journal of Medicine* 329, no. 26 (1993): 1957–60.

70. Harold Margolis et al., "Prevention of Hepatitis B Virus Transmission by Immunization: An Economic Analysis of Current Recommendations," *Journal of the American Medical Association* 274, no. 15 (1995): 1201–8.

71. CDC, "Notice to Readers Update: Recommendations to Prevent Hepatitis B

Virus Transmission—United States," *Morbidity and Mortality Weekly Report* 44, no. 30 (1994): 574-75.

72. U.S. House of Representatives, Subcommittee on Criminal Justice, Drug Policy, and Human Resources of the Committee on Government Reform, *Hepatitis B Vaccine: Helping or Hurting Public Health?*, *106th Congress, 1st session, May 18, 1999* (Washington, DC: Government Printing Office, 2000).

73. Editorial, "AIDS and Immigration," *Washington Post*, February 12, 1993, A26; Clifford Krauss, "Senate Opposes Immigration of People with AIDS Virus," *New York Times*, February 19, 1993, A11.

74. Robert Greenberger, "Clinton Team Seeks Policy to Aid Haiti and Avert Feared Surge of Emigration," *Wall Street Journal*, January 4, 1993, A5; Laura Hawkins, "Facing TB—in the Mirror," *Washington Post*, January 12, 1993, 17.

75. Roberto Suro, "Proposition 187 Could Open Pandora's Box for GOP," *Washington Post*, November 11, 1994, A24; Bruce Nelan, David Aikman, and David Jackson, "Not Quite So Welcome Anymore," *Time*, December 2, 1993. As historian Alan Kraut has pointed out, policies like California's Proposition 187 were simply "old wine in new bottles," as throughout American (and human) history, immigrants have often been held to blame for outbreaks and epidemics—real or imagined—and the resources they consume. Alan M. Kraut, *Silent Travelers: Germs, Genes, and the "Immigrant Menace"* (Baltimore: Johns Hopkins University Press, 1995), 3. Kraut's work describes the treatment of Haitian immigrants to the United States during the height of the AIDS epidemic, as well as many other episodes in American history in which immigration concerns have influenced public health policy and perceptions of disease and vice versa. Similar themes are addressed in Marilyn Chase, *The Barbary Plague: The Black Death in Victorian San Francisco* (New York: Random House, 2003); Judith Walzer Leavitt, *Typhoid Mary: Captive to the Public's Health* (Boston: Beacon Press, 1996); Howard Markel, *Quarantine!: East European Jewish Immigrants and the New York City Epidemics of 1892* (Baltimore: Johns Hopkins University Press, 1997); Howard Markel, *When Germs Travel: Six Major Epidemics that Have Invaded America since 1900 and the Fears They Have Unleashed* (New York: Pantheon, 2004); Naomi Rogers, *Dirt and Disease: Polio before FDR* (New Brunswick, NJ: Rutgers University Press, 1992).

76. Tom Majeski, "State Endorses Hepatitis Shots—Minnesota Is First to Recommend Immunizations for All Adolescents," *St. Paul Pioneer Press*, November 6, 1993, 1A.

77. Peter Shinkle, "Vaccination Line Long as School Opening Nears," *Advocate* (Baton Rouge, LA), August 19, 1993, 1A; Reuters, "Hepatitis B Vaccine for Babies Urged; Most of the Affected Americans Are First Infected as Young Adults," *Philadelphia Inquirer*, October 18, 1995, A12; Michael Romano, "Colorado Will Add Hepatitis B to Required Inoculations for Schoolchildren," *Rocky Mountain News*, April 20, 1996.

78. Andrea Vogt, "CDA Schools First in Idaho to Offer Hepatitis B Shots," *Idaho Spokesman-Review*, October 7, 1997, A1; Diane Eicher, "Hepatitis B Vaccine Carries a Quandary; Debate Rages over Necessity of Wide Usage," *Denver Post*, June 27, 1994, F1; Tina Nguyen, "Parents Need to Get Started on Schools' Hepatitis B Mandate," *Los Angeles Times*, December 26, 1998, 1.

79. Paula S. Fass, *Children of a New World: Society, Culture, and Globalization* (New York: New York University Press, 2007), 215.

80. "A Piercing Look," *Prevention*, November 1996, 46.

81. "If Body Piercing Is So Hazardous, Why Is It So Popular?," *Jet*, April 19, 1999, 56.

82. Eicher, "Hepatitis B Vaccine Carries a Quandary."

83. Romano, "Colorado Will Add Hepatitis B."

84. Eicher, "Hepatitis B Vaccine Carries a Quandary."

85. Rick Ansorge, "State Adds Hepatitis B to Immunization List—Shots Required for Schoolkids Beginning 1997," *Gazette* (Colorado Springs, CO), April 26, 1996, 1.

86. Eicher, "Hepatitis B Vaccine Carries a Quandary"; emphasis in original.

87. Diane Eicher, "HBV Shot Series Must Start Now," *Denver Post*, April 14, 1997, F2.

88. Patti Johnson, "Need for Hepatitis B Vaccine Questioned," *Rocky Mountain News*, July 12, 1999, 30A. That same year, Johnson led a separate campaign targeting the use of Ritalin, the widespread use of which she believed was symptomatic of society's overmedication of children. See Marcela Gaviria, "Medicating Kids," *Frontline* (PBS, 2001). Chiron was not a target of these complaints because its vaccine was developed in cooperation with and subsequently marketed by Merck.

89. Al Knight, "The Limits of Mandatory Medicine," *Denver Post*, March 4, 1999, B11; Al Knight, "Pinning Down the Risks of Vaccinations," *Denver Post*, August 5, 1999, B9.

90. Donna Leusner, "Some Bills Perish in Governor's Pocket," *Star-Ledger* (Newark, NJ), January 21, 1998, 15.

91. Todd Hartman, "Vaccinations' Success Backfiring—Thousands in Colorado Forgo Shots," *Gazette* (Colorado Springs, CO), January 19, 1997, 1; Arthur Allen, "Bucking the Herd," *Atlantic Monthly*, September 2002, 40–42. Overall vaccination coverage in this period was generally high in Colorado and the rest of the United States: 75 to 90 percent of children were fully vaccinated (against diphtheria, tetanus, pertussis, polio, measles, Hib, hepatitis B, and varicella) in each state, although at 75.8 percent Colorado's coverage rate was lower than that of most other states. CDC, "National, State, and Urban Area Vaccination Coverage Levels among Children Aged 19–35 Months—United States, 1998," *Morbidity and Mortality Weekly Report* 49, no. S S09 (2000): 1–26. For more details on vaccination mandate exemption clauses, see note 25 in the introduction.

92. National Vaccine Information Center, "CDC Considers Mass Vaccination with Hepatitis B Vaccine," *Vaccine News* 5, no. 1 (1990): 7; National Vaccine Information Center, "AAP Recommends All Newborn Infants Be Vaccinated with Hepatitis B Vaccine," *NVIC News* 2, no. 1 (1992): 12.

93. "Update: Recommendations to Prevent Hepatitis B Virus Transmission—United States," *Morbidity and Mortality Weekly Report* 48, no. 2 (1999): 33–34; "Global Progress toward Universal Childhood Hepatitis B Vaccination, 2003," *Morbidity and Mortality Weekly Report* 52, no. 36 (2003): 868–70.

94. See, for example, Agence France-Presse, "France Ends Program of Hepatitis B Shots," *New York Times*, October 3, 1998, A4.

95. "Who's Calling the Shots?" *20/20*, ABC News, January 22, 1999. See also Anita Manning, "Now Parents Fear Shots; Kids in USA Get 21 Shots before Start of 1st Grade," *USA Today*, August 3, 1999, 1A.

96. U.S. House of Representatives, *Hepatitis B Vaccine: Helping or Hurting Public Health?*, 7–31, 280–84.

97. Ibid., 67.

98. Introduction to Paula S. Fass and Mary Ann Mason, eds., *Childhood in America* (New York: New York University Press, 2000).

99. U.S. House of Representatives, *Hepatitis B Vaccine: Helping or Hurting Public Health?*, 258.

100. Ibid., 94.

101. Ibid., 112.

102. Charles Marwick and Mike Mitka, "Debate Revived on Hepatitis B Vaccine Value," *Journal of the American Medical Association* 282, no. 1 (1999): 15–17.

103. Lisa Suhay, "A Skirmish over the Hepatitis B Vaccination," *New York Times*, July 18, 1999, NJ1.

104. Laura Maschal, "Debating Hepatitis B Vaccine," *New York Times*, July 25, 1999, NJ13.

105. Mary Ann Mason, "The State as Superparent," in *Childhood in America*, ed. Paula Fass and Mary Ann Mason (New York: New York University Press, 2000), 549–54.

106. World Health Organization, "WHO/UNICEF Estimates of National Immunization Coverage: Estimated Coverage by Country, Year, and Vaccine."

107. Elisabeth Rosenthal, "Parents Face Questions on Vaccinating Infacts for Hepatitis B," *New York Times*, March 3, 1993, C12.

108. This articulation borrows from the concept of the "policy window" developed by political scientist John Kingdon. See John W. Kingdon, *Agendas, Alternatives, and Public Policies* (New York: Longman, 1995).

CHAPTER NINE

1. Andrea Rock, "The Lethal Dangers of the Billion-Dollar Vaccine Business," *Money*, December 1, 1996, 148; emphasis in original.

2. The investigation reported that DPT vaccine caused brain damage in 1 in every 62,000 children immunized and one to two deaths per year. Ibid., 150, 164.

3. See chapter 6.

4. Mark Largent, *Vaccine: The Debate in Modern America* (Baltimore: Johns Hopkins University Press, 2012), 1, 157.

5. An analysis by Christopher E. Clarke found that in a representative sample of U.S. media coverage of the vaccine-autism link, a majority of reports—roughly 80 percent—dismissed the link. Christopher E. Clarke, "A Question of Balance: The Autism-Vaccine Controversy in the British and American Elite Press," *Science Communication* 30, no. 1 (2008): 94.

6. CDC, "Intussusception among Recipients of Rotavirus Vaccine—United States, 1998–1999," *Morbidity and Mortality Weekly Report* 48, no. 27 (1999): 577–81.

7. Lawrence Altman, "U.S. in a Push to Bar Vaccine Given to Infants," *New York Times*, July 16, 1999, A1; "Rotavirus Vaccine Pulled after Illnesses," *Gainesville (FL) Sun*, July 18, 1999, 4G; "Doctors Stop Giving Vaccine after Warning," *Virginian-Pilot*, July 17, 1999, C5; "U.S. Recommends Suspension of a Wyeth Vaccine," *Philadelphia Inquirer*, July 16, 1999, C1.

8. CDC, "Withdrawal of Rotavirus Vaccine Recommendation," *Morbidity and Mortality Weekly Report* 48, no. 43 (1999): 1007; Lawrence Altman, "In Turnabout, Federal Panel Votes against a Vaccine," *New York Times*, October 23, 1999, A11. The rotavirus vaccination recommendation was later reinstated.

9. Edward Hooper, *The River: A Journey to the Source of HIV and AIDS* (Boston: Little, Brown, 1999).

10. Robin Weiss, "Is AIDS Man-Made?," *Science* 286, no. 5443 (1999): 1303; John P. Moore, "Up the River without a Paddle?" *Nature* 401, no. 6751 (1999): 325–26. The culprit vaccine's developer, the Philadelphia-based Wistar Institute, invited independent labs to test remaining stores of the vaccine to see if it did indeed contain chimp virus. It did not—but this finding did not conclusively disprove Hooper's theory. On the results of the independent testing, see Rebecca Voelker, "The World in Medicine: No Chimp DNA in Vaccine," *Journal of the American Medical Association* 284, no. 14 (2000): 1777.

11. Lawrence Altman, "New Book Challenges Theories of AIDS Origins," *New York Times*, November 30, 1999, F1.

12. Michael Woods, "How HIV Started Is Debated," *Blade* (Toledo, OH), December 13, 1999, 32.

13. The CDC's ACIP and the AAP had recently begun coordinating their vaccination recommendations; previously, the AAP had continued to issue its own recommendations distinct from those of the ACIP.

14. Committee on Infectious Diseases, "Prevention of Poliomyelitis: Recommendations for Use of Only Inactivated Poliovirus Vaccine for Routine Immunization," *Pediatrics* 104, no. 6 (1999): 1404–6; "Revised Recommendations for Routine Poliomyelitis Vaccination," *Journal of the American Medical Association* 282, no. 6 (1999): 522; CDC, "Recommendations of the Advisory Committee on Immunization Practices: Revised Recommendations for Routine Poliomyelitis Vaccination," *Morbidity and Mortality Weekly Report* 48, no. 27 (1999): 590.

15. Denise Grady, "Doctors Urge Polio Shots to Replace Oral Vaccine," *New York Times*, December 7, 1999; "Polio Vaccine Switch," *Pittsburgh Post-Gazette*, December 8, 1999, A2.

16. "Vaccine Roulette: Weighing the Odds," *Mothering*, November/December 1999, 30. The last point was a bit of an overstatement; the suspension, which was based on the presence of thimerosal in hepatitis B vaccine, was temporary and the agency actually advised that infants whose mothers did not carry the hepatitis B virus be vaccinated at six months instead of right after birth. Public Health Service and American Academy of Pediatrics, "Notice to Readers: Thimerosal in Vaccines—a Joint Statement of the American Academy of Pediatrics and the Public Health Service," *Morbidity and Mortality Weekly Report* 48, no. 26 (1999): 563–65.

17. "A Closer Look (Medicine—Vaccinations)," *ABC Evening News*, February 16, 1998; "Medicine/Childhood Vaccines," *CBS Evening News*, September 28, 1999, VTNA.

18. Tim Vollmer, "Who Should Call the Shots," *San Francisco Chronicle*, October 10, 1999, 1; Huntly Collins, "Life Giver or Life Taker: A Debate on the Value of Vaccines Special Report, Immunizations: A Public-Health Staple Comes under Siege," *Philadelphia Inquirer*, October 3, 1999, A1.

19. The full list of meetings and links to complete reports are available at http://www.iom.edu/Activities/PublicHealth/ImmunizationSafety.aspx (accessed August 2012). Burton requested that none of the committee members be connected in any way to industry, leading some medical professionals to complain that the resulting committees were, as a result, devoid of vaccine experts. Brian Vastag, "Congressional Autism Hearings Continue: No Evidence MMR Vaccine Causes Disorder," *Journal of the American Medical Association* 285, no. 20 (2001): 2567–69.

20. U.S. Congress, House Committee on Government Reform, *Autism: Present Challenges, Future Needs—Why the Increased Rates? Hearing before the Committee on Government Reform, House of Representatives, 106th Congress, 2nd Session, April 6, 2000* (Washington, DC: U.S. Government Printing Office, 2001), 5–6.

21. Andrew J. Wakefield et al., "Ileal-Lymphoid-Nodular Hyperplasia, Non-Specific Colitis, and Pervasive Developmental Disorder in Children," *Lancet* 351, no. 9103 (1998): 637–41.

22. U.S. Congress, House Committee on Government Reform, *Autism*, 20–21. For a fuller discussion of Wakefield and his work, see Largent, *Vaccine*, chap. 4.

23. See, for example, Brian Lavery, "As Vaccination Rates Decline in Ireland, Cases of Measles Soar," *New York Times*, February 8, 2003.

24. Norman Begg et al., "Medicine and the Media: Media Dents Confidence in MMR Vaccine," *British Medical Journal* 316, no. 7130 (1998): 561. Media studies experts attribute the reaction in Britain in part to the nation's recent experience with mad cow disease, which left the media and the public distrustful of authorities and their reassurances. See, for example, Clarke, "A Question of Balance."

25. Quantitative analyses support this observation; see, for example, Clarke, "A Question of Balance."

26. See, for example, "Measles Study Finds No Evidence that Vaccine Causes Autism," *Chicago Tribune*, May 8, 1998, 7; Michael Day, "MMR/Autism: Have We Fully Investigated the Risks?" *New Scientist*, March 7, 1998; "Vital Signs," *State Journal-Register* (Springfield, IL), May 11, 1998, 9; and Collins, "Life Giver or Life Taker."

27. John Wilkens, "A Father's Day Story: The Rimlands Where There's Hope," *San Diego Union-Tribune*, June 21, 1998, D1.

28. Associated Press, "Parents: Vaccine Likely Caused Son's Autism—Mississippi Couple Seeks Their Day in Court," *Advocate* (Baton Rouge, LA), December 13, 1998, 6B; Judy Foreman, "Answers to Mysteries of Autism May Be Starting to Emerge," *Lewiston (ID) Morning Tribune*, December 30, 1998, 2D. A few media analyses have also found that vaccine blame was a common feature of autism reporting. Juanne Nancarrow Clarke, "Representations of Autism in U.S. Magazines for Women in Comparison to the General Audience," *Journal of Children and Media* 6, no. 2 (2012): 182–97; Seok Kang, "Coverage of Autism Spectrum Disorder in the U.S. Television News: An Analysis of Framing," *Disability and Society* 28, no. 2 (2013): 245–59.

29. See, for example, Leslie Koren, "Doctor's Daughters Pass on Some Vaccinations," *Washington Times*, September 4, 1998, C4; and Karyn Miller-Medzon, "Taking a Shot at Vaccines—Mass. Group Wants to Make Immunizations Optional," *Boston Herald*, August 30, 1998.

30. E. Atlee Bender, "Letter: Sylvia Wood Deserves Praise for Vaccine Story," *Times Union* (Albany, NY), September 5, 1998, A6; Raymond Gallup, "Reader Forum: Vaccine Danger," *Star-Ledger* (Newark, NJ), July 31, 1998, 28; Natalie Veit, "To the Editor," *Erie (PA) Times-News*, April 10, 1988.

31. This claim abounded in media reports on parental vaccination fears and childhood vaccination coverage between 2010 and 2012. See also Seth Mnookin, *The Panic Virus: A True Story of Medicine, Science, and Fear* (New York: Simon & Schuster, 2011); and Paul A. Offit, *Deadly Choices: How the Anti-Vaccine Movement Threatens Us All* (New York: Basic Books, 2011).

32. Largent makes a similar argument about autism advocate Jenny McCarthy,

whom health officials also blamed for igniting parental fears of a vaccine-autism link. Largent, *Vaccine*, 151.

33. Sandra Blakeslee, "Increase in Autism Baffles Scientists," *New York Times*, October 18, 2002, A1; Polly Morrice, "What Caused the Autism Epidemic?," *New York Times*, April 17, 2005, F20.

34. "A Mysterious Upsurge in Autism," *New York Times*, October 20, 2002, C10.

35. But it couldn't be ruled out entirely. In a statement included in bold type in the final report, the committee noted that because of limitations inherent to the epidemiological research, it was possible that MMR vaccine could contribute to autism "in a small number of children." Institute of Medicine Immunization Safety Review Committee, *Immunization Safety Review: Measles-Mumps-Rubella Vaccine and Autism* (Washington, DC: National Academies Press 2001).

36. The report focused on MMR and thimerosal-containing vaccines. Institute of Medicine Immunization Safety Review Committee, *Immunization Safety Review: Vaccines and Autism* (Washington, DC: National Academies Press, 2004).

37. "In Depth: MMR Vaccine and Autism," *NBC Evening News*, April 23, 2001, VTNA. See also, for example, "No Link Found between MMR Vaccine and Autism," *CNN Health*, April 23, 2001; and Sandra Blakeslee, "No Evidence of Autism Link Is Seen in Vaccine, Study Says," *New York Times*, April 24, 2001, A16.

38. "Are Vaccines Safe?: Saying No to Immunization," *60 Minutes*, CBS News, December 27, 2004, http://www.cbsnews.com/video/watch/?id=650462n (accessed June 2012).

39. "Lifeline: Childhood Vaccines," *NBC Evening News*, May 18, 2004, VTNA. See also, for example, "Inside Story: Childhood Vaccines," *CBS Evening News*, May 18, 2004, VTNA; Tina Hesman, "No Link between Vaccines and Autism, Study Says—Report Draws Fire from Parents, Other Researchers," *St. Louis Post-Dispatch*, May 19, 2004, A1; and Robyn Shelton, "Vaccines Don't Cause Autism, Panel Agrees—but Some Parents of Autistic Children Said the Experts' Findings Were Severely Flawed," *Orlando Sentinel*, May 19, 2004, A1.

40. These figures are based on a count of U.S. media reports including the words "vaccine(s)" and "autism" in Newsbank and the VTNA. The same trend is evident in other databases of news reports, including Google News Archive (41 results for 2001; 571 results for 2010) and Lexis-Nexis (260 results for 2001; 856 results for 2010). The number of reports increased and decreased from year to year but generally increased over the course of the decade. The pattern also holds for reports mentioning Andrew Wakefield.

41. See, for example, Sandy Kleffman, "Autism, Heavy Metals Linked; Study Suggests Toxin Can Act as a Trigger in Some Children," *Pittsburgh Post-Gazette*, December 14, 2004, A3.

42. See, for example, "Medicine: Autism and Vaccines," *CNN Evening News*, March 4, 2004, VTNA; Mary Ann Roser, "Censured Doctor Says He'll Resume Autism Research," *Austin American-Statesman*, May 20, 2010, B1; and Susan Dominus, "The Denunciation of Dr. Wakefield," *New York Times Sunday Magazine*, April 24, 2011, MM36.

43. Morrice, "What Caused the Autism Epidemic?"; Miriam Falco, "Study: 1 in 110 U.S. Children Had Autism in 2006," *CNN Health*, December 17, 2009, http://www.cnn.com/2009/HEALTH/12/17/autism.new.numbers/ (accessed January 2010); Jonathan Serrie, "1 in 88 Us Kids Have Autism, CDC Reports," *FoxNews.com*, March 29, 2012,

http://www.foxnews.com/health/2012/03/29/1-in-88-us-kids-have-autism-cdc-reports/ (accessed July 2012). Numerous theories have been put forth to account for the startling rise, from new diagnostic methods to greater awareness to increasing paternal and maternal age.

44. See, for instance, Allan Mazur, *The Dynamics of Technical Controversy* (Washington, DC: Communications Press, 1981); and Nick F. Pidgeon, Roger E. Kasperson, and Paul Slovic, *The Social Amplification of Risk* (New York: Cambridge University Press, 2003).

45. Pew Research Center's Project for Excellence in Journalism, "The State of the News Media 2004," www.stateofthemedia.org/2004 (accessed August 2012).

46. Paul Coelho, "The Internet: Increasing Information, Decreasing Certainty," *Journal of the American Medical Association* 280, no. 16 (1998): 1454; Lee Rainie and Susannah Fox, *The Online Health Care Revolution* (Washington, DC: Pew Internet & American Life Project, 2000).

47. Pam Rajendran, "The Internet: Ushering in a New Era of Medicine," *Journal of the American Medical Association/MSJAMA* 285, no. 6 (2001): 804; A. Risk and C. Petersen, "Health Information on the Internet: Quality Issues and International Initiatives," *Journal of the American Medical Association* 287, no. 20 (2002): 2713–15; Scott Lafee, "Mental Blocked Parents and Researchers at Odds over Treatment "*San Diego Union-Tribune*, January 9, 2002, F1.

48. Foreman, "Answers to Mysteries of Autism."

49. Karen Thacker, "Parents Put Spotlight on Autism," *Los Angeles Times*, August 2, 1999, AV4.

50. Scott Hilyard, "Some Are Seeing Correlation between Autism, Immunizations," *Copley News Service*, December 12, 2001.

51. See, for example, May 8, 1999, snapshot of www.littleangels.org and April 16, 2001, snapshot of www.gti.net/truegrit, accessed July 2012 through the Internet Archive Wayback Machine (archive.org/web).

52. See, for example, October 16, 2003, snapshot of www.tacanow.org (Talk About Curing Autism), accessed July 2012 through the Internet Archive Wayback Machine. The same advice was given by Defeat Autism Now! and the National Vaccine Information Center, among others.

53. Steven Waldman and U.S. Federal Communications Commission, *The Information Needs of Communities: The Changing Media Landscape in a Broadband Age* (Durham, NC: Carolina Academic Press, 2011).

54. Jenny McCarthy, *Louder than Words: A Mother's Journey in Healing Autism* (New York: Dutton, 2007), 11, 166.

55. Data taken from the *New York Times* Bestseller List Archive at http://www.hawes.com/pastlist.htm.

56. Wendy Kline, *Bodies of Knowledge: Sexuality, Reproduction, and Women's Health in the Second Wave* (Chicago: University of Chicago Press, 2010).

57. Boston Women's Health Book Collective, *Ourselves and Our Children: A Book by and for Parents* (New York: Random House, 1978), 215.

58. McCarthy, *Louder than Words*, 76.

59. Ibid., 40, 22.

60. Jenny McCarthy, *Mother Warriors: A Nation of Parents Healing Autism against All Odds* (New York: Plume, 2008).

61. "Jenny McCarthy and Holly Robinson Peete Fight to Save Their Autistic Sons," *Oprah Winfrey Show*, September 18, 2007, http://www.oprah.com/oprahshow/Mothers -Battle-Autism (accessed July 2012).

62. An analysis of autism coverage in U.S. women's magazines in the 2000s found that such coverage underscored mothers' unique and all-consuming role in caring for their children, as well as the need for them to trust in the power of their instincts. Clarke, "Representations of Autism in U.S. Magazines."

63. McCarthy, *Louder than Words*, 200.

64. "Jenny Mccarthy's Autism Fight," *Larry King Live*, CNN, April 2, 2008. Transcript available at http://transcripts.cnn.com/TRANSCRIPTS/0804/02/lkl.01.html (accessed April 2009).

65. The parallels to 1982's *DPT: Vaccine Roulette* (see chapter 5) are striking. Jon Palfreman and Katie McMahon, prod., "The Vaccine War," *Frontline* (PBS, 2010). Full episode available online at http://www.pbs.org/wgbh/pages/frontline/vaccines/ (accessed June 2013).

66. "Jenny McCarthy and Holly Robinson Peete Fight to Save Their Autistic Sons." For an extensive discussion of McCarthy's vaccine activism, see Largent, *Vaccine*, chap. 5. Many thanks to the Spring 2013 participants in the Johns Hopkins University History of Science, Technology, and Medicine Colloquium for their insightful observations on this point.

67. McCarthy details some of these appearances in *Mother Warriors*.

68. "Jenny Mccarthy's Autism Fight"; "Jenny McCarthy and Jim Carrey Discuss Autism; Medical Experts Weigh In," *Larry King Live*, CNN, 2009, transcript available at http://transcripts.cnn.com/TRANSCRIPTS/0904/03/lkl.01.html.

69. Pew Research Center's Project for Excellence in Journalism, "The State of the News Media 2004."

70. See, for example, *CNN News*, March 4, 2004; *NBC Evening News*, May 18, 2004; and *CBS Evening News*, May 18, 2004; all VTNA.

71. Amy Wallace, "An Epidemic of Fear: How Panicked Parents Skipping Shots Endangers Us All," *Wired*, November, 2009, 128; reader comments by "Aloisius" and "richardlefew," http://www.wired.com/magazine/2009/10/ff_waronscience/all (accessed June 2011, July 2012).

72. "Are Vaccines Safe?"

73. "Jenny McCarthy's Autism Fight."

74. See chapter 10 for more on this point.

75. Ibid. Gordon worked closely with McCarthy and authored the introduction to her 2008 book, *Mother Warriors*.

76. "The Vaccine War," *Frontline* (PBS, April 2010), available at http://www.pbs.org /wgbh/pages/frontline/vaccines (accessed January 2012).

77. "Selfish," *Law and Order: Special Victims Unit*, season 10, episode 19, April 28, 2009.

78. "The Vaccine War"; "Selfish."

79. Robert F. Kennedy, "Deadly Immunity," *Rolling Stone*, July 14, 2005, 57–66.

80. Jonann Brady and Stephanie Dahle, "Celeb Couple to Lead 'Green Vaccine' Rally," ABC News.com, June 4, 2008, http://abcnews.go.com/GMA/OnCall/story?id=4987758 (accessed August 2012).

81. Mary Douglas and Aaron B. Wildavsky, *Risk and Culture: An Essay on the Selection*

of Technical and Environmental Dangers (Berkeley: University of California Press, 1982), 10, 11.

82. Ulrich Beck, *Risk Society: Towards a New Modernity* (Los Angeles: Sage, 1992).

83. Sharon R. Kaufman, "Regarding the Rise in Autism: Vaccine Safety Doubt, Conditions of Inquiry, and the Shape of Freedom," *Ethos: Journal of the Society for Psychological Anthropology* 38, no. 1 (2010): 22.

84. Ibid.

85. Some scholars have criticized Beck's limited construal of the mass media and its role in the risk society. See, for instance, Simon Cottle, "Ulrich Beck, 'Risk Society' and the Media: A Catastrophic View?" *European Journal of Communication* 13, no. 1 (1998): 5–32.

86. "Summary of Notifiable Diseases — United States, 2010," *Morbidity and Mortality Weekly Report* 59, no. 53 (2012): 1–111. On the decline of vaccine-preventable infections by this time generally, see S. W. Roush and T. V. Murphy, "Historical Comparisons of Morbidity and Mortality for Vaccine-Preventable Diseases in the United States," *Journal of the American Medical Association* 298, no. 18 (2007): 2155–63; and Kate Yandell, "Vital Signs | Prevention: Chickenpox Down 80 Percent since 2000," *New York Times*, August 21, 2012, D6.

87. The term "elimination" was chosen as an admittedly imperfect term by a panel of infectious disease specialists who debated about how to characterize this achievement. S. L. Katz and A. R. Hinman, "Summary and Conclusions: Measles Elimination Meeting, 16–17 March 2000," *Journal of Infectious Diseases* 189, no. S1 (2004): S43–S47; W. A. Orenstein, M. J. Papania, and M. E. Wharton, "Measles Elimination in the United States," *Journal of Infectious Diseases* 189, no. S1 (2004): S1–S3.

88. "Announcements: National Infant Immunization Week—April 21–28, 2012," *Morbidity and Mortality Weekly Report* 61, no. 15 (2012): 278; CDC, "National, State, and Local Area Vaccination Coverage among Children Aged 19–35 Months — United States, 2009," *Morbidity and Mortality Weekly Report* 59, no. 36 (2010): 1171–77.

89. A. E. Barskey, J. W. Glasser, and C. W. LeBaron, "Mumps Resurgences in the United States: A Historical Perspective on Unexpected Elements," *Vaccine* 27, no. 44 (2009): 6186–95; "Mumps Outbreak—New York, New Jersey, Quebec, 2009," *Morbidity and Mortality Weekly Report* 58, no. 45 (2009): 1270–74.

90. "Measles—United States, 2011," *Morbidity and Mortality Weekly Report* 61, no. 15 (2012): 253–57; "Measles Outbreak Associated with an Arriving Refugee — Los Angeles County, California, August–September 2011," *Morbidity and Mortality Weekly Report* 61, no. 21 (2012): 385–89; "Notes from the Field: Measles Outbreak — Indiana, June–July 2011," *Morbidity and Mortality Weekly Report* 60, no. 34 (2011): 1169; "Notes from the Field: Measles Outbreak—Hennepin County, Minnesota, February–March 2011," *Morbidity and Mortality Weekly Report* 60, no. 13 (2011): 421.

91. "Summary of Notifiable Diseases—United States, 2010."

92. K. Winter et al., "California Pertussis Epidemic, 2010," *Journal of Pediatrics* (2012).

93. "Pertussis Epidemic—Washington, 2012," *Morbidity and Mortality Weekly Report* 61, no. 28 (2012): 517–22.

94. "Summary of Notifiable Diseases — United States, 2010"; P. A. Hall-Baker et al., "Summary of Notifiable Diseases — United States, 2008," *Morbidity and Mortality Weekly Report* 57, no. 54 (2010): 1–92.

95. Barskey, Glasser, and LeBaron, "Mumps Resurgences in the United States"; "Update: Mumps Outbreak—New York and New Jersey, June 2009–January 2010," *Morbidity and Mortality Weekly Report* 59, no. 5 (2010): 125–29.

96. P. Rohani and J. M. Drake, "The Decline and Resurgence of Pertussis in the U.S.," *Epidemics* 3, nos. 3–4 (2011): 183–88; B. M. Kuehn, "Reports Highlight New Cause of Pertussis, Tickborne Illness, and Better Food Safety," *Journal of the American Medical Association* 307, no. 17 (2012): 1785, 1787; C. R. Capili et al., "Increased Risk of Pertussis in Patients with Asthma," *Journal of Allergy and Clinical Immunology* 129, no. 4 (2012): 957–63.

97. January Payne, "What Parents Need to Know about the Latest Vaccine News," *U.S. News & World Report*, January 26, 2009, http://health.usnews.com/health-news/family-health/articles/2009/01/26/what-parents-need-to-know-about-the-latest-vaccine-news (accessed August 2012).

98. "Doctors Report Cases of Whooping Cough on the Rise," *Click on Detroit / MSNBC*, January 19, 2009, http://www.msnbc.msn.com/id/20090128684205/ (accessed July 2012).

99. Campbell Brown, "Commentary: Get Your Children Vaccinated for Measles," *CNN Health*, February 13, 2009, http://www.cnn.com/2009/HEALTH/02/12/campbell.brown.vaccine/ (accessed August 2012).

100. Editorial, "The Solution to the Rising Incidence of Pertussis: Vaccination," *Seattle Times*, July 29, 2012, http://seattletimes.nwsource.com/html/editorials/2018791162_edit30pertussis.html (accessed July 2012); Kent Sepkowitz, "The Hack Is Back," *Newsweek*, August 6, 2012, 11.

101. Reuters, "Most U.S. Kindergartners Getting Vaccines, Risks Remain," August 23, 2012, http://www.reuters.com/article/2012/08/23/us-usa-health-vaccines-idUSBRE87M16820120823 (accessed August 2012).

102. Gary L. Freed et al., "Parental Vaccine Safety Concerns in 2009," *Pediatrics* 125, no. 4 (2010): 654–59.

103. Liz Szabo, "Refusing Kid's Vaccine More Common among Parents," *USA Today*, May 3, 2010, http://usatoday30.usatoday.com/news/health/2010-05-04-vaccines04_ST_N.htm (accessed July 2012).

104. CDC, "Vaccination Coverage among Children in Kindergarten—United States, 2011–12 School Year," *Morbidity and Mortality Weekly Report* 61, no. 33 (2012): 647–52.

105. Karin Klein, "Making It a Little Harder to Say No to Vaccination," *Los Angeles Times*, August 24, 2012, http://www.latimes.com/news/opinion/opinion-la/la-ol-vaccine-bill-california-20120824,0,5816567.story (accessed August 2012).

106. Liz Szabo, "Childhood Diseases Return as Parents Refuse Vaccines," *USA Today*, June 14, 2011, http://usatoday30.usatoday.com/news/health/medical/health/medical/story/2011/06/Childhood-diseases-return-as-parents-refuse-vaccines/48414234/1 (accessed July 2012).

107. Associated Press, "1 in 4 U.S. Parents Buys Unproven Vaccine-Autism Link," *FoxNews.com*, March 1, 2010, http://www.foxnews.com/story/2010/03/01/1-in-4-us-parents-buys-unproven-vaccine-autism-link/ (accessed August 2012).

108. Paul Starr, *Remedy and Reaction: The Peculiar American Struggle over Health Care Reform* (New Haven, CT: Yale University Press, 2011).

109. Jill S. Quadagno, *One Nation, Uninsured: Why the U.S. Has No National Health In-

surance (New York: Oxford University Press, 2005); T. R. Reid, *The Healing of America: A Global Quest for Better, Cheaper, and Fairer Health Care* (New York: Penguin, 2009).

110. Mnookin, *The Panic Virus*.

111. "Why the Myth that Vaccines Cause Autism Survives," *On the Media*, WNYC, May 11, 2012, http://www.onthemedia.org/people/seth-mnookin/ (accessed July 2012).

112. In May 2013, *Mother Jones* offered a rare corrective to the view that autism-fearing parents were solely to blame for rampant under-vaccination. Kiera Butler, "The Real Reason Kids Aren't Getting Vaccines," *Mother Jones*, May 2013, http://m.mother jones.com/environment/2013/05/vaccines-whooping-cough (accessed May 2013).

CHAPTER TEN

1. Lena Dunham, "All Adventurous Women Do," *Girls*, season 1, episode 3 (HBO, 2012).

2. Cori Rosen, "'Girls' Fan to Get a Tattoo in Lena Dunham's Handwriting," *Holly wood.com*, June 10, 2013.

3. Not all HPV infections are pre-cancerous, and "scraping out" the cervix is not a treatment for HPV infection. See, for example, Roni Caryn Rabin, "TV Show 'Girls' Adds to the Muddle on HPV Testing," *New York Times*, May 14, 2012, http://well.blogs .nytimes.com/2012/05/14/tv-show-girls-adds-to-the-muddle-on-hpv-testing/?_r=0 (accessed June 2013).

4. The streamlined process is reserved for drugs that represent a major new advance in treatment or that serve an unmet need. Priority review requests are typically made by drug companies. See Food and Drug Administration, "Fast Track, Accelerated Approval and Priority Review," http://www.fda.gov/forconsumers/byaudience/forpatie ntadvocates/speedingaccesstoimportantnewtherapies/ucm128291.htm (accessed March 2011).

5. Committee on Adolescent Health Care and the ACOG Working Group, "Committee Opinion: Human Papillomavirus Vaccination," *Obstetrics and Gynecology* 108, no. 3 (2006): 699–703.

6. Associated Press, "Panel Urges Approval of Vaccine for Cancer," *New York Times*, May 19, 2006, 21.

7. CDC, "STD Prevention Counseling Practices and Human Papillomavirus Opinions among Clinicians with Adolescent Patients—United States, 2004," *Morbidity and Mortality Weekly Report* 55, no. 41 (2006): 1117–20. In the committee's first formal recommendations, they added that the "vaccine can be administered as young as age 9 years." The committee also ruled that the vaccine should be made available to indigent and uninsured girls through Vaccines for Children. CDC, "Vaccines Included in the VFC Program," June 29, 2006, http://www.cdc.gov/vaccines/programs/vfc/downloads /resolutions/0606vaccines.pdf (accessed March 2011); CDC, "Quadrivalent Human Papillomavirus Vaccine: Recommendations of the Advisory Committee on Immunization Practices (ACIP)," *Morbidity and Mortality Weekly Report* 56, no. RR02 (2007): 1–24.

8. Gardiner Harris, "Panel Unanimously Recommends Cervical Cancer Vaccine for Girls 11 and Up," *New York Times*, June 30, 2006, 12.

9. Cynthia Dailard, "The Public Health Promise and Potential Pitfalls of the World's First Cervical Cancer Vaccine," *Guttmacher Policy Review* 9, no. 1 (2006): 6–9.

10. CDC, *Eliminate Disparities in Cancer Screening & Management*, http://www.cdc.gov/omhd/AMH/factsheets/cancer.htm (accessed December 2007).

11. National Conference of State Legislatures (NCSL), "HPV Vaccine: State Legislation and Statutes," http://www.ncsl.org/default.aspx?tabid=14381 (accessed December 2007, January 2009, April 2011, and August 2012).

12. Virginia lawmakers made several subsequent attempts to repeal the mandate; none were successful. NCSL, "HPV Vaccine."

13. Harris Interactive, "Seventy Percent of U.S. Adults Support Use of the Human Papillomavirus (HPV) Vaccine," *Wall Street Journal Online Health-Care Poll* 5, no. 18 (2006): 1–8.

14. A smaller debate followed the announcement that the Department of Homeland Security would require the shot for all immigrant women between the ages of eleven and twenty-six, because immigration law required immigrants to get all immunizations recommended by the ACIP. The requirement was ultimately abandoned. See Associated Press, "Green Card Applicants Mandated to Get HPV Vaccine," *New York Daily News*, October 3, 2008, http://www.nydailynews.com/latino/green-card-applicants-mandated-hpv-vaccine-article-1.299017 (accessed June 2013); and Associated Press, "Immigrant Seekers Won't Have to Get HPV Vaccine," *USA Today*, November 16, 2009, http://usatoday30.usatoday.com/news/health/2009-11-16-immigration-hpv-vaccine_N.htm (accessed June 2013).

15. Barbara Loe Fisher, "Speech Given at the March 8, 2007, Rally Sponsored by the Parents and Citizens Committee to Stop Medical Experimentation in D.C.," National Vaccine Information Center, 2007, http://www.nvic.org/vaccines-and-diseases/HPV/fisherhpv.aspx (accessed April 2013).

16. Sigrid Fry-Revere, "Mandatory Vaccines Help Drug Firms, Not Necessarily Consumers," *Tampa Tribune*, March 20, 2007; Sigrid Fry-Revere, "The Rush to Vaccinate," *New York Times*, March 25, 2007, 9.

17. See, for instance, "Focus on the Family Position Statement: Human Papillomavirus Vaccines," www.family.org/socialissues/A000000357.cfm (accessed December 14, 2007).

18. Laura Schlessinger, "Mandatory Testing for Cervical Cancer for Pre-Teen Girls? I Don't Think So!," February 9, 2007, www.drlaurablog.com/2007/02/09/mandatory-testing-for-cervical-cancer-for-pre-teen-girls-i-dont-think-so/ (accessed June 2013).

19. Gregory Lopes, "CDC Doctor Opposes Law for Vaccine; Cancer-Causing Virus Not Contagious Disease," *Washington Times*, February 27, 2007, A1.

20. See, for example, Allan M. Brandt, *No Magic Bullet: A Social History of Venereal Disease in the United States since 1880* (New York: Oxford University Press, 1985); and Dorothy Porter and Roy Porter, "The Enforcement of Health: The British Debate," in *AIDS: The Burdens of History*, ed. Elizabeth Fee and Daniel Fox (Berkeley: University of California Press, 1988), 97–120.

21. See chapter 8.

22. On the history of struggles to define blame for cancer prevalence in the U.S. population, see James T. Patterson, *The Dread Disease: Cancer and Modern American Culture* (Cambridge, MA: Harvard University Press, 1987); Robert Proctor, *Cancer Wars: How Politics Shapes What We Know and Don't Know about Cancer* (New York: Basic Books,

1995); Gerald Markowitz and David Rosner, *Deceit and Denial: The Deadly Politics of Industrial Pollution* (Berkeley: University of California Press, 2002); and Devra Lee Davis, *The Secret History of the War on Cancer* (New York: Basic Books, 2007).

23. On the rise of the anti-globalization movement, see Luke Martell, *The Sociology of Globalization* (Cambridge: Polity Press, 2010); and Saskia Sassen, *Cities in a World Economy*, 3rd ed. (Thousand Oaks, CA: Pine Forge Press, 2006).

24. Terrance Neilan, "Merck Pulls Vioxx Painkiller from Market, and Stock Plunges," *New York Times*, September 30, 2004, http://www.nytimes.com/2004/09/30/business /30CND-MERCK.html (accessed June 2013); Marc Kaufman, "Merck Found Liable in Vioxx Case," *Washington Post*, August 20, 2005, A1.

25. Fernando Meirelles, dir., *The Constant Gardener* (Focus Features, 2005).

26. John Le Carré, *The Constant Gardener* (New York: Scribner, 2001), 490.

27. Marcia Angell, *The Truth about the Drug Companies: How They Deceive Us and What to Do about It* (New York: Random House, 2004).

28. Greg Critser, *Generation Rx: How Prescription Drugs Are Altering American Lives, Minds, and Bodies* (Boston: Houghton Mifflin, 2005); John Abramson, *Overdosed America: The Broken Promise of American Medicine* (New York: HarperCollins, 2004); Ray Moynihan and Alan Cassels, *Selling Sickness: How the World's Biggest Pharmaceutical Companies Are Turning Us All into Patients* (New York: Nation Books, 2005); Jerry Avorn, *Powerful Medicines: The Benefits, Risks, and Costs of Prescription Drugs* (New York: Knopf, 2004).

29. Stephen S. Hall, "'The Truth about the Drug Companies' and 'Powerful Medicines': The Drug Lords," *New York Times*, November 14, 2004, Sunday Book Review, 1.

30. Jenny McCarthy, *Louder than Words: A Mother's Journey in Healing Autism* (New York: Dutton, 2007). Data taken from the *New York Times* Bestseller List Archive at http://www.hawes.com/pastlist.htm.

31. Sigrid Fry-Revere, "Mandatory HPV Vaccines: Who Benefits?," Cato Institute, December 14, 2007, http://www.cato.org/blog/mandatory-hpv-vaccines-who-benefits (accessed December 2007).

32. Fisher, "Speech Given at the March 8, 2007, Rally."

33. MaryAnna Clemons, "So Why Does the State Want to Require HPV Vaccinations?," *San Francisco Chronicle*, March 12, 2007, 7.

34. Steve Lawrence, "Vote Delayed on Bill Requiring Girls to Be Vaccinated against HPV," *San Francisco Chronicle*, March 13, 2007, 1.

35. Schlessinger, "Mandatory Testing for Cervical Cancer for Pre-Teen Girls?"; U.S. Cancer Statistics Working Group, *United States Cancer Statistics: 1999–2009* (Atlanta, GA: Department of Health and Human Services, Centers for Disease Control and Prevention, and National Cancer Institute, 2013).

36. Courtland Milloy, "District's HPV Proposal Tinged with Ugly Assumptions," *Washington Post*, January 10, 2007, 1.

37. Elisabeth Rosenthal, "Cervical Cancer Vaccine Is Popular, but Fails to Cure Doubts," *New York Times*, August 19, 2008, http://www.nytimes.com/2008/08/19/world /americas/19iht-vaccine.4.15437772.html?pagewanted=all (accessed December 2008).

38. Deborah Kamali, "Requiring a Vaccine for Young Girls," *New York Times*, February 10, 2007, 14.

39. Bradford King, "The HPV Vaccine Debate — Gardasil: Beyond the Scope of Public Health," *Free Lance-Star* (Fredericksburg, VA), April 22, 2007; Victoria Cobb, "HPV

Legislation: A Train Wreck Waiting to Happen," *Richmond (VA) Times-Dispatch*, March 23, 2007, A15; Rita Rubin, "Vaccines: Mandate or Choice?," *USA Today*, February 8, 2007, 6D.

40. Cobb, "HPV Legislation."

41. See, for example, Laura Smitherman, "Drug Firm Pushes Vaccine Mandate," *Baltimore Sun*, January 29, 2007, http://articles.baltimoresun.com/2007-01-29/news/07 01290104_1_vaccine-cervical-cancer-hpv (accessed December 2008); Elisabeth Rosenthal, "Drug Makers' Push Leads to Cancer Vaccines' Rise," *New York Times*, August 20, 2008, A1; Lianne Hart, "Texas HPV Vaccine Mandate Meets Swift Resistance," *Los Angeles Times*, February 27, 2007, A25; and Amanda Terkel and Ryan Grim, "Michele Bachmann Defends HPV Vaccine Comments, Goes after Rick Perry," *Huffington Post*, September 22, 2011, http://www.huffingtonpost.com/2011/09/22/michele-bachmann -rick-perry-hpv-vaccine_n_977139.html (accessed June 2013).

42. Rosenthal, "Drug Makers' Push Leads to Cancer Vaccines' Rise."

43. Merck, "Driving Growth with our Commitment to Vaccines," *2007 Annual Review* (Whitehouse Station, NJ: Merck & Co., 2007), http://www.merck.com/finance /annualreport/ar2007/home.html (accessed April 2011). In 2008 the company swept the Phame Awards, the Academy Awards of the pharmaceutical industry, based on the "creative excellence" behind its Gardasil campaign. Matthew Arnold, "Gardasil Tops at Annual Phame Awards," *Medical Marketing and Media* 43, no. 6 (2008), http://www .mmm-online.com/issue/june/01/2008/823/ (accessed March 2011).

44. FDA guidelines issued in 1997 permitted direct-to-consumer advertising of prescription drugs through electronic media, including television for the first time. Meredith Rosenthal et al., "Promotion of Prescription Drugs to Consumers," *New England Journal of Medicine* 346, no. 7 (2002): 498–505; Julie Donohue, Marisa Cevasco, and Meredith Rosenthal, "A Decade of Direct-to-Consumer Advertising of Prescription Drugs," *New England Journal of Medicine* 357, no. 7 (2007): 673–81.

45. Michael Applebaum, "Life Gard," *Brandweek* 48, no. 36 (2007): 1–7.

46. Deconstructions of Merck's ads and its presentation of "risky girlhood" appear in Laura Mamo, Amber Nelson, and Aleia Clark, "Producing and Protecting Risky Girlhoods," in *Three Shots at Prevention: The HPV Vaccine and the Politics of Medicine's Simple Solutions*, ed. Keith Wailoo et al. (Baltimore: Johns Hopkins University Press, 2010), 121–45; Giovanna Chesler and Bree Kessler, "Re-Presenting Choice: Tune in HPV," in ibid., 146–64.

47. Mary Engel, "Cervical Cancer Vaccine Gains Acceptance in California," *Los Angeles Times*, February 18, 2009, http://articles.latimes.com/2009/feb/18/local/me -hpv18 (accessed April 2013).

48. J. A. Tiro et al., "What Do Women in the U.S. Know about Human Papillomavirus and Cervical Cancer?," *Cancer Epidemiology Biomarkers and Prevention* 16, no. 2 (2007): 288–94.

49. "Merck & Co: The Marketing Machine Behind Gardasil," *PharmaWatch: Cancer* 5, no. 7 (2006): 4.

50. "Gardasil Ads Remain Platonic . . . for Now," *Medical Market & Media* 41, no. 7 (2006): 11.

51. Sheila M. Rothman and David J. Rothman, "Marketing HPV Vaccine: Implications for Adolescent Health and Medical Professionalism," *Journal of the American Medical Association* 302, no. 7 (2009): 781–86.

52. Rosenthal, "Drug Makers' Push Leads to Cancer Vaccines' Rise."

53. John Simons, "From Scandal to Stardom: How Merck Healed Itself," *Fortune*, February 18, 2008, 94–98.

54. Arlene Weintraub, "Making Her Mark at Merck," *BusinessWeek*, January 8, 2007, 64–65.

55. Beth Herskovits, "Brand of the Year: Gardasil," *Pharmaceutical Executive*, February 2007, 58–70. Merck's marketing efforts won widespread recognition; see also Arnold, "Gardasil Tops at Annual Phame Awards."

56. See, for example, David Healy, *The Antidepressant Era* (Cambridge, MA: Harvard University Press, 1997); Jeremy Greene, *Prescribing by Numbers: Drugs and the Definitions of Diseases* (Baltimore: Johns Hopkins University Press, 2007); and Moynihan and Cassels, *Selling Sickness*.

57. Robert Aronowitz, "Gardasil: A Vaccine against Cancer and a Drug to Reduce Risk," in Wailoo et al., *Three Shots at Prevention*, 21–38.

58. FDA guidelines issued in 1997 permitted direct-to-consumer advertising of prescription drugs through electronic media, including television for the first time. Rosenthal et al., "Promotion of Prescription Drugs to Consumers"; Donohue, Cevasco, and Rosenthal, "A Decade of Direct-to-Consumer Advertising of Prescription Drugs."

59. Applebaum, "Life Gard"; Herskovits, "Brand of the Year: Gardasil."

60. Sue Abercrombie, "Requiring a Vaccine for Young Girls," *New York Times*, February 10, 2007, 14.

61. Ajantha Jayabarathan, "What about the Boys?," *Canadian Family Physician* 54, no. 10 (2008): 1375. For a discussion of consumer-led demands to bring attention to HPV's role in anal cancer, see Steven Epstein, "The Great Undiscussable: Anal Cancer, HPV, and Gay Men's Health," in Wailoo et al., *Three Shots at Prevention*.

62. CDC, "Recommended Immunization Schedules for Persons Aged 0 through 18 Years—United States, 2012," *Morbidity and Mortality Weekly Report* 61, no. 5 (2012): 1–4. Gardasil had been licensed for females only, and not until 2009 was an HPV vaccine licensed for use in boys. Moreover, between the ACIP's 2006 decision and its 2011 decision, evidence had accumulated to implicate HPV not only in cervical cancer, but also in vaginal cancer, penile cancer, and, in both sexes, anal cancer and head and neck cancers. Minutes from October 2011 Meeting, Advisory Committee on Immunization Practices, Department of Health and Human Services and Centers for Disease Control and Prevention, http://www.cdc.gov.proxy.library.emory.edu/vaccines/acip/meetings /meetings-info.html (accessed June 2013).

63. Heather Munro Prescott, "'I Was a Teenage Dwarf': The Social Construction of 'Normal' Adolescent Growth and Development in the United States," in *Formative Years: Children's Health in the United States, 1880–2000*, ed. Alexandra Minna Stern and Howard Markel (Ann Arbor: University of Michigan Press, 2005), 170; Heather Munro Prescott, *A Doctor of Their Own: The History of Adolescent Medicine* (Cambridge, MA: Harvard University Press, 1998).

64. Quotes collected from www.myspace.com; www.facebook.com; www.youtube .com, using search terms "gardasil," "cervical cancer," and "HPV," December 2007.

65. Mollysevilfather, "You Could Be One Less," http://www.youtube.com/watch?v =Yj7aSivwgvM (accessed September 2008).

66. wowTHATSfunny954, "A parody commercial for gardasil called shymali," http:// www.youtube.com/watch?v=fa6IEARWpiM (accessed September 2008).

67. logomojo529, "Guard Yourself Commercial," http://www.youtube.com/watch?v=9YkAhdoxWzU (accessed September 2008).

68. The poster's profile information is no longer available, but her comments can still be viewed at http://www.youtube.com/user/ThinkOneMore/feed (accessed June 2013).

69. Mary Jane Horton, "A Shot against Cervical Cancer," Ms., Summer 2005, 65–66.

70. Cindy Wright, "Lifesaving Politics," Ms., Spring 2007, 12–13.

71. Adina Nack, "Why Men's Health Is a Feminist Issue," Ms., Winter 2010, 32–35.

72. Approval Letter—Gardasil, October 16, 2009, BL 125126/1297, http://www.fda.gov/BiologicsBloodVaccines/Vaccines/ApprovedProducts/ucm186991.htm (accessed June 2013).

73. Meeting Minutes, Advisory Committee on Immunization Practices (ACIP) Summary Report, October 21–22, 2009, http://www.cdc.gov/vaccines/acip/meetings/minutes-archive.html (accessed June 2013).

74. J. J. Kim and S. J. Goldie, "Cost Effectiveness Analysis of Including Boys in a Human Papillomavirus Vaccination Programme in the United States," British Medical Journal 339 (2009): b3884, doi: http://dx.doi.org/10.1136/bmj.b3884 (accessed October 2013). Specifically, male vaccination was not cost-effective when female vaccination levels were 80 percent or higher; below 80 percent, studies produced varying predictions of the cost-effectiveness of vaccinating males. See also "FDA Licensure of Quadrivalent Human Papillomavirus Vaccine (HPV4, Gardasil) for Use in Males and Guidance from the Advisory Committee on Immunization Practices (ACIP)," Morbidity and Mortality Weekly Report 59, no. 20 (2010): 630–32.

75. Jacob Goldstein, "Routine Gardasil Vaccination for Boys: Not Recommended," Wall Street Journal Health Blog, October 21, 2009, http://blogs.wsj.com/health/2009/10/21/routine-gardasil-vaccination-for-boys-not-recommended/ (accessed June 2013); quote under "Comments."

76. Jim Edwards, "Girls as Guinea Pigs: What the CDC's 'Gardasil for Boys' Issue Says about Sexism in Medicine," CBS MoneyWatch, November 1, 2010, http://www.cbsnews.com/8301-505123_162-42846288/girls-as-guinea-pigs-what-the-cdcs-gardasil-for-boys-issue-says-about-sexism-in-medicine/ (accessed June 2013).

77. William Saletan, "Sexually Transmitted Injection," Slate, October 15, 2009, http://www.slate.com/articles/health_and_science/human_nature/2009/10/sexually_transmitted_injection.html (accessed June 2013).

78. Amanda Hess, "The Feminist Implications of Male Reproductive Health," Washington City Paper, February 24, 2010, http://www.washingtoncitypaper.com/blogs/sexist/2010/02/24/the-feminist-implications-of-male-reproductive-health/ (accessed December 2010).

79. Saletan, "Sexually Transmitted Injection." See also Epstein, "The Great Undiscussable."

80. Epstein, "The Great Undiscussable."

81. Meeting Minutes, Advisory Committee on Immunization Practices (ACIP) Summary Report.

82. A. B. Moscicki et al., "Chapter 5: Updating the Natural History of HPV and Anogenital Cancer," Vaccine 24, no. S3 (2006): S42–S51.

83. A. B. Moscicki et al., "Updating the Natural History of Human Papillomavirus and Anogenital Cancers," Vaccine 30, no. S5 (2012): F24–F33.

84. A. Jemal et al., "Annual Report to the Nation on the Status of Cancer, 1975–2009, Featuring the Burden and Trends in Human Papillomavirus (HPV)-Associated Cancers and HPV Vaccination Coverage Levels," *Journal of the National Cancer Institute* 105, no. 3 (2013): 175–201. Liver, kidney, and thyroid cancers also increased.

85. Recurrent respiratory papillomatosis is characterized by benign growths in the respiratory tract; it affects men and women.

86. Meeting Minutes, Advisory Committee on Immunization Practices (ACIP) Summary Report. GlaxoSmithKline's Cervarix was approved for use in October 2009. See Food and Drug Administration, "Approval Letter—Cervarix," October 16, 2009, http://www.fda.gov/BiologicsBloodVaccines/Vaccines/ApprovedProducts/ucm186959.htm (accessed December 2009).

87. The research was conducted in collaboration with researchers from the CDC, the National Cancer Institute, and the North American Association of Central Cancer Registries. Jemal et al., "Annual Report to the Nation on the Status of Cancer."

88. H. Mehanna et al., "Oropharyngeal Carcinoma Related to Human Papillomavirus," *British Medical Journal* 340 (2010): c1439, doi: http://dx.doi.org/10.1136/bmj (accessed June 2013).

89. On the history of such collaborations, see, for example, John Patrick Swann, *Academic Scientists and the Pharmaceutical Industry: Cooperative Research in Twentieth-Century America* (Baltimore: Johns Hopkins University Press, 1988).

90. Dan Childs and Radha Chitale, "Farrah Fawcett's Anal Cancer: Fighting the Stigma," *ABC News*, June 27, 2009, http://abcnews.go.com/Health/story?id=7939402 (accessed April 2013).

91. "Farrah Fawcett's Struggle with Anal Cancer," *FoxNews.com*, June 25, 2009, http://www.foxnews.com/story/2009/06/25/farrah-fawcett-struggle-with-anal-cancer/ (accessed April 2013).

92. Val Willingham, "Oral Cancers in Women Rising, HPV Sometimes a Factor," *CNN Health*, November 30, 2009, http://www.cnn.com/2009/HEALTH/11/30/oral.cancer.women/ (accessed April 2013).

93. Xan Brooks, "Michael Douglas on Liberace, Cannes, Cancer and Cunnilingus," *Guardian*, June 2, 2013, http://www.theguardian.com/film/2013/jun/02/michael-douglas-liberace-cancer-cunnilingus; Deborah Kotz, "Throat Cancer and Oral Sex," *Boston Globe*, June 10, 2013, http://www.bostonglobe.com/lifestyle/health-wellness/2013/06/09/michael-douglas-blames-throat-cancer-oral-sex-what-are-risks/Akb38cr5CCvj2HUKXJ5SCP/comments.html (accessed June 2013).

94. Matthew Herper, "At Our Throats," *Forbes.com*, October 15, 2009, http://www.forbes.com/forbes/2009/1102/health-cancer-tonsils-virus-hpv-at-our-throats.html (accessed June 2013).

95. D. Forman et al., "Global Burden of Human Papillomavirus and Related Diseases," *Vaccine* 30, no. S5 (2012): F12–F23.

96. "Recommendations on the Use of Quadrivalent Human Papillomavirus Vaccine in Males—Advisory Committee on Immunization Practices (ACIP), 2011," *Morbidity and Mortality Weekly Report* 60, no. 50 (2011): 1705–8.

97. Jemal et al., "Annual Report to the Nation on the Status of Cancer."

98. In 2010, 32 percent of girls ages thirteen to seventeen were fully vaccinated against HPV; 48.7 percent had received at least one dose of vaccine. "Recommendations on the Use of Quadrivalent Human Papillomavirus Vaccine in Males."

99. Richard Knox, "Why HPV Vaccination of Boys May Be Easier," *Shots—NPR Health Blog*, November 7, 2011, http://www.npr.org/blogs/health/2011/11/07/142030282/why -hpv-vaccination-of-boys-may-be-easier (accessed June 2013).

100. L. E. Markowitz et al., "Reduction in Human Papillomavirus (HPV) Prevalence among Young Women Following HPV Vaccine Introduction in the United States, National Health and Nutrition Examination Surveys, 2003–2010," *Journal of Infectious Diseases* (2013), doi: 10.1093/infdis/jit192.

101. Press Briefing Transcript, CDC Telebriefing on HPV Prevalence among Young Women Following HPV Vaccination Introduction in the United States, NHANES, 2003–2010, June 19, 2012, http://www.cdc.gov/media/releases/2013/t0619-hpv-vacci nations.html (accessed June 2013).

CONCLUSION

1. A comprehensive and up-to-date overview of laws is available at http://www .immunize.org/laws/.

2. "Remarks by the President on Senate Passage of Health Insurance Reform," White House Office of the Press Secretary, December 24, 2009, http://www.whitehouse.gov /the-press-office/remarks-president-senate-passage-health-insurance-reform (accessed June 2013).

3. "The Affordable Care Act: Secure Health Coverage for the Middle Class," White House Office of the Press Secretary, June 28, 2012, http://www.whitehouse.gov/the -press-office/2012/06/28/fact-sheet-affordable-care-act-secure-health-coverage -middle-class (accessed June 2013). "Grandfathered" health plans are exempt from the requirement to provide all ACIP-endorsed vaccines at no cost. Non-grandfathered plans must provide the vaccines no later than one year after they become recommended by ACIP. Vaccines for Children remains in place under the law, as does Section 317, funding for which increased to $620 million in 2012. 317 Coalition, "FY 2013 Senate Labor HHS Appropriations Bill—Centers for Disease Control and Prevention— Immunization," http://www.317coalition.org/update.html (accessed May 2013). For a complete discussion of the Affordable Care Act's potential impact on immunization, see Alexandra M. Stewart et al., *The Affordable Care Act: U.S. Vaccine Policy and Practice* (Washington, DC: Department of Health Policy, School of Public Health and Health Services, George Washington University Medical Center, 2010).

4. CDC, "National, State, and Local Area Vaccination Coverage among Children Aged 19–35 Months—United States, 2009," *Morbidity and Mortality Weekly Report* 59, no. 36 (2010): 1171–77.

5. Saad Omer et al., "Vaccine Refusal, Mandatory Immunization, and the Risks of Vaccine-Preventable Diseases," *New England Journal of Medicine* 360, no. 19 (2009): 1981–88.

6. Tara Parker-Pope, "Vaccination Is Steady, but Pertussis Is Surging," *New York Times*, August 17, 2010, D1; Anemona Hartocollis, "Jewish Youths Are at Center of Outbreak of Mumps," *New York Times*, February 12, 2010, http://www.nytimes.com/2010/02/12 /nyregion/12mumps.html (accessed October 2011); Jennie Lavine, Aaron King, and Ottar Bjornstad, "Natural Immune Boosting in Pertussis Dynamics and the Potential

for Long-Term Vaccine Failure," *Proceedings of the National Academy of Science* 108, no. 17 (2011): 7259–64. Outbreaks of measles and mumps that occurred in the 1980s, by contrast, were attributed to falling immunization rates due to federal and state budget cuts for vaccine programs, or to vaccines that proved less effective in practice than they had in vaccine trials. See Melinda Wharton et al., "A Large Outbreak of Mumps in the Postvaccine Era," *Journal of Infectious Diseases* 158, no. 6 (1988): 1253–60.

APPENDIX

1. The rotavirus vaccine, for example, is made from weakened cow rotavirus that contains segments of human rotavirus. Stanley Plotkin and Susan Plotkin, "A Short History of Vaccination," in *Vaccines*, ed. Stanley Plotkin, Walter Orenstein, and Paul Offit (Philadelphia: Elsevier, 2008), 1–16.

2. Ibid.

3. Dorothy Porter and Roy Porter, "The Enforcement of Health: The British Debate," in *AIDS: The Burdens of History*, ed. Elizabeth Fee and Daniel Fox (Berkeley: University of California Press, 1988), 97–120; Donald Hopkins, *The Greatest Killer: Smallpox in History* (Chicago: University of Chicago Press, 1983).

4. Norman Baylor and Karen Midthun, "Regulation and Testing of Vaccines," in *Vaccines*, ed. Plotkin, Orenstein, and Offit, 1611–28.

5. Baylor and Midthun, "Regulation and Testing of Vaccines."

6. Ibid.

SELECTED BIBLIOGRAPHY

| | | | | | | | | | |

ARCHIVES AND SPECIAL COLLECTIONS

American Philosophical Society Library, Philadelphia, PA

American Presidency Project, University of California, Santa Barbara (UCSB), http://www.presidency.uscb.edu

Baby Books Collection, Louise M. Darling Biomedical Library, History & Special Collections for the Sciences, University of California, Los Angeles (UCLA), Los Angeles, CA

Centers for Disease Control and Prevention (CDC) Museum Collection, Atlanta, GA

General Records of the Department of Health and Human Services, National Archives and Records Administration (NARA), College Park, MD

Jimmy Carter Library and Museum, Atlanta, GA

John F. Kennedy Presidential Library and Museum, Boston, MA

National Library of Medicine (NLM), Bethesda, MD

Othmer Library of Chemical History, Chemical Heritage Foundation, Philadelphia, PA

Records of the Centers for Disease Control and Prevention (CDC), National Archives and Records Administration (NARA), Southeast Region, Atlanta, GA

Vanderbilt Television News Archive (VTNA), Vanderbilt University, http://tvnews.vanderbilt.edu

William J. Clinton Presidential Library, Little Rock, AR

Special thanks to the editorial offices of *Mothering* magazine for access to the collected letters, newsletters, articles, magazines, books, and notes in their Santa Fe, New Mexico, office.

SELECTED BOOKS, ESSAYS, ARTICLES, AND REPORTS

Allen, Arthur. *Vaccine: The Controversial Story of Medicine's Greatest Lifesaver.* New York: Norton, 2007.

Anderson, Patrick. *Electing Jimmy Carter: The Campaign of 1976.* Baton Rouge: Louisiana State University Press, 1994.

Arms, Suzanne. *Immaculate Deception: A New Look at Women and Childbirth in America.* Boston: Houghton Mifflin, 1975.

Beauregard, Robert A. *Voices of Decline: The Postwar Fate of U.S. Cities.* New York: Routledge, 2003.

————. *When America Became Suburban.* Minneapolis: University of Minnesota Press, 2006.

Beck, Ulrich. *Risk Society: Towards a New Modernity.* Los Angeles: Sage, 1992.

Berkowitz, Edward D. *Something Happened: A Political and Cultural Overview of the Seventies.* New York: Columbia University Press, 2006.

Blumberg, Baruch S. *Hepatitis B: The Hunt for a Killer Virus.* Princeton, NJ: Princeton University Press, 2002.

Blumenthal, David, and James A. Morone. *The Heart of Power: Health and Politics in the Oval Office.* Berkeley: University of California Press, 2009.

Bourne, Peter G. *Jimmy Carter: A Comprehensive Biography from Plains to Post-Presidency.* New York: Scribner, 1997.

Brandt, Allan M. *No Magic Bullet: A Social History of Venereal Disease in the United States since 1880.* New York: Oxford University Press, 1985.

Brewer, Gail Sforza, and Tom Brewer. *What Every Pregnant Woman Should Know: The Truth about Diet and Drugs in Pregnancy.* New York: Penguin, 1977.

Brownmiller, Susan. *In Our Time: Memoir of a Revolution*. New York: Dial, 1999.

Califano, Joseph. *Inside: A Public and Private Life*. New York: Public Affairs, 2004.

Carroll, Peter N. *It Seemed Like Nothing Happened: The Tragedy and Promise of America in the 1970s*. New York: Holt, Rinehart and Winston, 1982.

Clinton, Bill. *My Life*. New York: Knopf, 2004.

Clinton, Hillary Rodham. *It Takes a Village: And Other Lessons Children Teach Us*. New York: Simon & Schuster, 1996.

Coates, Peter A. *Nature: Western Attitudes since Ancient Times*. Berkeley: University of California Press, 1998.

Colgrove, James. *State of Immunity: The Politics of Vaccination in Twentieth-Century America*. Berkeley: University of California Press, 2006.

Colgrove, James, and Ronald Bayer. "Could It Happen Here?: Vaccine Risk Controversies and the Specter of Derailment." *Health Affairs* 24, no. 3 (2005): 729–39.

Commoner, Barry. *Science and Survival*. New York: Viking, 1966.

Corea, Gena. *The Hidden Malpractice: How American Medicine Treats Women as Patients and Professionals*. New York: William Morrow, 1977.

Coulter, Harris. *Vaccination, Social Violence, and Criminality: The Medical Assault on the American Brain*. Berkeley, CA: North Atlantic Books, 1990.

Coulter, Harris, and Barbara Loe Fisher. *DPT: A Shot in the Dark*. New York: Harcourt, Brace, Jovanovich, 1985.

———. *A Shot in the Dark: Why the P in the DPT Vaccination May Be Hazardous to Your Child's Health*. New York: Avery, 1991.

Cournoyer, Cynthia. "What about Immunizations?" Canby, OR: Concerned Parents for Information, 1983.

———. *What about Immunizations?* Grants Pass, OR: Cynthia Cournoyer, 1987.

———. *What about Immunizations?: Exposing the Vaccine Philosophy*. Santa Cruz, CA: Nelson's Books, 1991.

Cunningham, Hugh. *Children and Childhood in Western Society since 1500*. New York: Longman, 1995.

Dallek, Robert. *An Unfinished Life: John F. Kennedy, 1917–1963*. Boston: Little, Brown, 2003.

Duffy, John. *The Sanitarians: A History of American Public Health*. Urbana: University of Illinois Press, 1990.

Dumbrell, John. *The Carter Presidency*. New York: St. Martin's Press, 1995.

Dunlap, Thomas R. *DDT: Scientists, Citizens, and Public Policy*. Princeton, NJ: Princeton University Press, 1981.

———. *Faith in Nature: Environmentalism as Religious Quest*. Weyerhaeuser Environmental Books. Seattle: University of Washington Press, 2004.

Dupree, A. Hunter. *Science in the Federal Government: A History of Policies and Activities*. Baltimore: Johns Hopkins University Press, 1986.

Durbach, Nadja. *Bodily Matters: The Anti-Vaccination Movement in England, 1853–1907*. Durham, NC: Duke University Press, 2005.

Ehrlich, Paul R. *The Population Bomb*. New York: Ballantine, 1968.

Etheridge, Elizabeth W. *Sentinel for Health*. Berkeley: University of California Press, 1992.

Fass, Paula S. *Children of a New World: Society, Culture, and Globalization*. New York: New York University Press, 2007.

Fass, Paula S., and Mary Ann Mason. *Childhood in America*. New York: New York University Press, 2000.

Fee, Elizabeth. "Public Health and the State: The United States." In *The History of Public Health and the Modern State*, edited by Dorothy Porter, 224–75. Atlanta: Rodopi, 1994.

Fee, Elizabeth, and Daniel Fox, eds. *AIDS: The Burdens of History*. Berkeley: University of California Press, 1988.

Fiske, John. *Media Matters: Everyday Culture and Political Change*. Minneapolis: University of Minneota Press, 1994.

Fox, Stephen R. *The Mirror Makers: A History of American Advertising and Its Creators*. Urbana: University of Illinois Press, 1997.

Gaillard, Frye. *Prophet from Plains: Jimmy Carter and His Legacy*. Athens: University of Georgia Press, 2007.

Galambos, Louis, and Jane Eliot Sewell. *Networks of Innovation: Vaccine Development at Merck, Sharp & Dohme and Mulford, 1895–1995*. New York: Cambridge University Press, 1995.

Garrett, Laurie. *The Coming Plague: Newly Emerging Diseases in a World out of Balance*. New York: Farrar, Straus and Giroux, 1994.

Gostin, Lawrence. *Public Health Law: Power, Duty, Restraint*. Berkeley: University of California Press, 2000.

Greene, Jeremy. *Prescribing by Numbers: Drugs and the Definitions of Diseases*. Baltimore: Johns Hopkins University Press, 2007.

Greenough, Paul. "Intimidation, Coercion and Resistance in the Final Stages of the South Asian Smallpox Eradication Campaign, 1973–1975." *Social Science and Medicine* 41, no. 5 (1995): 633–45.

Hale, Annie Riley. *The Medical Voodoo*. New York: Gotham House, 1935.

Halpern, Sydney A. *American Pediatrics: The Social Dynamics of Professionalism, 1880–1980*. Berkeley: University of California Press, 1988.

Hammonds, Evelynn Maxine. *Childhood's Deadly Scourge: The Campaign to Control Diphtheria in New York City, 1880–1930*. Baltimore: Johns Hopkins University Press, 2002.

Harris, John F. *The Survivor: Bill Clinton in the White House*. New York: Random House, 2005.

Hawes, Joseph M., and N. Ray Hiner. *American Childhood: A Research Guide and Historical Handbook*. Westport, CT: Greenwood Press, 1985.

———, eds. *Children in Historical and Comparative Perspective*. Westport, CT: Greenwood Press, 1991.

Hays, Samuel, and Barbara Hays. *Beauty, Health, and Permanence: Environmental Politics in the United States, 1955–1985*. New York: Cambridge University Press, 1987.

Healy, David. *The Antidepressant Era*. Cambridge, MA: Harvard University Press, 1997.

Heller, Jacob. *The Vaccine Narrative*. Nashville: Vanderbilt University Press, 2008.

Henderson, Donald A. *Smallpox: The Death of a Disease*. New York: Prometheus Books, 2009.

Hightower, Jane M. *Diagnosis Mercury: Money, Politics, and Poison*. Washington, DC: Island Press/Shearwater Books, 2009.

Hoffman, Beatrix. *Health Care for Some: Rights and Rationing in the United States since 1930*. Chicago: University of Chicago Press, 2012.

Honorof, Ida, and Eleanor McBean. *Vaccination: The Silent Killer*. Sherman Oaks, CA: Honor Publications, 1977.

Hopkins, Donald. *The Greatest Killer: Smallpox in History*. Chicago: University of Chicago Press, 1983.

———. *Princes and Peasants: Smallpox in History*. Chicago: University of Chicago Press, 1985.

Illich, Ivan. *Medical Nemesis: The Expropriation of Health*. New York: Pantheon, 1976.

Jackson, Kenneth T. *Crabgrass Frontier: The Suburbanization of the United States*. New York: Oxford University Press, 1985.

James, Walene. *Immunization: The Reality Behind the Myth*. South Hadley, MA: Bergin & Garvey, 1988.

Jenkins, Philip. *Decade of Nightmares: The End of the Sixties and the Making of Eighties America*. New York: Oxford University Press, 2006.

Johnston, Robert, ed. *The Politics of Healing: Histories of Alternative Medicine in Twentieth-Century North America*. New York: Routledge, 2004.

———. *The Radical Middle Class: Populist Democracy and the Question of Capitalism in Progressive Era Portland, Oregon*. Princeton, NJ: Princeton University Press, 2003.

Kaufman, Martin. "The American Anti-Vaccinationists and Their Arguments." *Bulletin of the History of Medicine* 41, no. 5 (1967): 463–78.

Kempton, Willett, James S. Boster, and Jennifer A. Hartley. *Environmental Values in American Culture*. Cambridge, MA: MIT Press, 1995.

Kingdon, John W. *Agendas, Alternatives, and Public Policies*. New York: Longman, 1995.

Kirby, David. *Evidence of Harm: Mercury in Vaccines and the Autism Epidemic — a Medical Controversy*. New York: St. Martin's Press, 2005.

Kline, Wendy. *Bodies of Knowledge: Sexuality, Reproduction, and Women's Health in the Second Wave*. Chicago: University of Chicago Press, 2010.

Kraut, Alan M. *Silent Travelers: Germs, Genes, and the "Immigrant Menace."* Baltimore: Johns Hopkins University Press, 1995.

Lander, Daniel A. "Immunization: An Informed Choice." Glen Cove, ME: Dr. Daniel Lander, Family Chiropractor, 1978.

Largent, Mark. *Vaccine: The Debate in Modern America*. Baltimore: Johns Hopkins University Press, 2012.

Leavitt, Judith Walzer. *The Healthiest City: Milwaukee and the Politics of Health Reform*. Madison: University of Wisconsin Press, 1996.

————. *Typhoid Mary: Captive to the Public's Health.* Boston: Beacon Press, 1996.

Little, Lora. *Crimes of the Cowpox Ring: Some Moving Pictures Thrown on the Dead Wall of Official Silence.* Minneapolis: Liberator Publishing, 1906.

Lombardo, Paul A. *Three Generations, No Imbeciles: Eugenics, the Supreme Court, and Buck V. Bell.* Baltimore: Johns Hopkins University Press, 2008.

Lord, Alexandra M. *Condom Nation: The U.S. Government's Sex Education Campaign from World War I to the Internet.* Baltimore: Johns Hopkins University Press, 2010.

Luker, Kristin. *When Sex Goes to School: Warring Views on Sex — and Sex Education — since the Sixties.* New York: Norton, 2006.

Maraniss, David. *First in His Class: A Biography of Bill Clinton.* New York: Simon & Schuster, 1995.

Markel, Howard. *When Germs Travel: Six Major Epidemics That Have Invaded America since 1900 and the Fears They Have Unleashed.* New York: Pantheon, 2004.

Markowitz, Gerald, and David Rosner. *Deceit and Denial: The Deadly Politics of Industrial Pollution.* New York: Milbank Memorial Fund, 2002.

Martell, Luke. *The Sociology of Globalization.* Cambridge: Polity Press, 2010.

Martin, Emily. *Flexible Bodies: Tracking Immunity in American Culture from the Days of Polio to the Age of AIDS.* Boston: Beacon Press, 1994.

McBean, Eleanor. *The Poisoned Needle.* Mokelumne Hill, CA: Health Research, 1974.

————. *Swine Flu Exposé.* Los Angeles: Better Life Research Center, 1977.

————. *Vaccinations Do Not Protect.* Yorktown, TX: Life Science, 1980.

McCarthy, Jenny. *Louder than Words: A Mother's Journey in Healing Autism.* New York: Dutton, 2007.

McCarthy, Jenny. *Mother Warriors: A Nation of Parents Healing Autism Against All Odds.* New York: Plume, 2008.

Melling, Joseph, and Christopher Sellers, eds. *Dangerous Trade: Histories of Industrial Hazard across a Globalizing World.* Philadelphia: Temple University Press, 2011.

Mendelsohn, Robert S. *Confessions of a Medical Heretic*. Chicago: Contemporary Books, 1979.

———. *How to Raise a Healthy Child... In Spite of Your Doctor*. Chicago: Contemporary Books, 1984.

———. *Male Practice: How Doctors Manipulate Women*. Chicago: Contemporary Books, 1981.

———. *The Risks of Immunization and How to Avoid Them*. Evanston, IL: People's Doctor, 1988.

Miller, Neil Z. *Vaccines: Are They Really Safe and Effective? A Parent's Guide to Childhood Shots*. Santa Fe: New Atlantean Press, 1992.

Mitman, Gregg. *Breathing Space: How Allergies Shape Our Lives and Landscapes*. New Haven, CT: Yale University Press, 2007.

Mitman, Gregg, Michelle Murphy, and Christopher Sellers. "Introduction: A Cloud over History." In *Landscapes of Exposure: Knowledge and Illness in Modern Environments*, Osiris 19 (2004): 1–20.

Mnookin, Seth. *The Panic Virus: A True Story of Medicine, Science, and Fear*. New York: Simon & Schuster, 2011.

Morgen, Sandra. *Into Our Own Hands: The Women's Health Movement in the United States, 1969–1970*. New Brunswick, NJ: Rutgers University Press, 2002.

Moynihan, Ray, and Alan Cassels. *Selling Sickness: How the World's Biggest Pharmaceutical Companies Are Turning Us All into Patients*. New York: Nation Books, 2005.

Muraskin, William. "Hepatitis B as a Model (and Anti-Model) for AIDS." In *AIDS and Contemporary History*, edited by Virginia Berridge and Philip Strong, 108–32. Cambridge: Cambridge University Press, 1993.

Murphy, Jamie. *What Every Parent Should Know about Childhood Immunization*. Boston: Earth Healing Products, 1994.

Murphy, Michelle. *Sick Building Syndrome and the Problem of Uncertainty: Environmental Politics, Technoscience, and Women Workers*. Durham, NC: Duke University Press, 2006.

Nash, Roderick. *The Rights of Nature: A History of Environmental Ethics*. Madison: University of Wisconsin Press, 1989.

Neustadt, Richard E., and Harvey V. Fineberg. *The Swine Flu Affair: Decision-Making on a Slippery Disease*. Washington, DC: U.S. Department of Health, Education, and Welfare, 1978.

Neustaedter, Randall. *The Immunization Decision: A Guide for Parents*. Berkeley, CA: North Atlantic Books, Homeopathic Educational Services, 1990.

———. *The Vaccine Guide: Making an Informed Choice*. Berkeley, CA: North Atlantic Books, 1996.

Noll, Steven, and James W. Trent. *Mental Retardation in America*. New York: New York University Press, 2004.

Offit, Paul. *Autism's False Prophets: Bad Science, Risky Medicine, and the Search for a Cure*. New York: Columbia University Press, 2008.

———. *The Cutter Incident: How America's First Polio Vaccine Led to the Growing Vaccine Crisis*. New Haven, CT: Yale University Press, 2005.

———. *Deadly Choices: How the Anti-Vaccine Movement Threatens us All*. New York: Basic Books, 2011.

———. *Vaccinated: One Man's Quest to Defeat the World's Deadliest Diseases*. Washington, DC: Smithsonian Books, 2007.

Olmsted, Dan, and Mark Blaxill. *The Age of Autism: Mercury, Medicine, and a Man-Made Epidemic*. New York: St. Martin's Press, 2010.

O'Mara, Peggy, ed. *Vaccinations*. Santa Fe: Mothering, 1988.

Oshinsky, David M. *Polio: An American Story*. New York: Oxford University Press, 2005.

Parmet, Wendy. *Populations, Public Health, and the Law*. Washington, DC: Georgetown University Press, 2009.

Patel, Kant, and Mark E. Rushefsky. *Health Care Politics and Policy in America*. Armonk, NY: M.E. Sharpe, 2006.

Patterson, James T. *The Dread Disease: Cancer and Modern American Culture*. Cambridge, MA: Harvard University Press, 1987.

Plant, Rebecca Jo. *Mom: The Transformation of American Motherhood*. Chicago: University of Chicago Press, 2010.

Plotkin, Stanley, Walter Orenstein, and Paul Offit, eds. *Vaccines*. Philadelphia: Elsevier, 2008.

Porter, Dorothy. *Health, Civilization, and the State*. London: Routledge, 1999.

Proctor, Robert. *Cancer Wars: How Politics Shapes What We Know and Don't Know about Cancer*. New York: Basic Books, 1995.

Quadagno, Jill. *One Nation, Uninsured: Why the U.S. Has No National Health Insurance*. New York: Oxford University Press, 2005.

Reagan, Leslie J. *Dangerous Pregnancies: Mothers, Disabilities, and Abortion in America*. Berkeley: University of California Press, 2010.

Rogers, Naomi. *Dirt and Disease: Polio before FDR*. Health and Medicine in American Society. New Brunswick, NJ: Rutgers University Press, 1992.

Rosenberg, Charles. "Disease in History: Frames and Framers." *Milbank Quarterly* 67, Supplement no. 1 (1989): 1–15.

———. "Framing Disease: Illness, Society, and History." In *Framing Disease: Studies in Cultural History*, edited by Charles Rosenberg and Janet Golden, xxi–xxvi. New Brunswick, NJ: Rutgers University Press, 1997.

———. "What Is Disease?" *Bulletin of the History of Medicine* 77 (2003): 491–505.

Rozario, Diane. *The Immunization Resource Guide*. Burlington, IA: Patter, 1994.

Scheibner, Viera. *Vaccination: 100 Years of Orthodox Research Shows That Vaccines Represent a Medical Assault on the Immune System*. Santa Fe: New Atlantean Press, 1993.

Sealander, Judith. *The Failed Century of the Child: Governing America's Young in the Twentieth Century*. New York: Cambridge University Press, 2003.

Sears, Robert. *The Vaccine Book: Making the Right Decision for Your Child*. New York: Little, Brown, 2007.

Shorter, Edward. *The Health Century*. New York: Doubleday, 1987.

———. *The Kennedy Family and the Story of Mental Retardation*. Philadelphia: Temple University Press, 2000.

Skocpol, Theda. *Boomerang: Clinton's Health Security Effort and the Turn against Government in U.S. Politics*. New York: Norton, 1996.

Smith, Jane S. *Patenting the Sun: Polio and the Salk Vaccine*. New York: Morrow, 1990.

Starr, Paul. *Remedy and Reaction: The Peculiar American Struggle over Health Care Reform*. New Haven, CT: Yale University Press, 2011.

———. *The Social Transformation of American Medicine*. New York: Basic Books, 1982.

Tomes, Nancy. *The Gospel of Germs: Men, Women, and the Microbe in American Life*. Cambridge, MA: Harvard University Press, 1998.

Tone, Andrea. *The Age of Anxiety: A History of America's Turbulent Affair with Tranquilizers*. New York: Basic Books, 2009.

Valencius, Conevery Bolton. *The Health of the Country: How American Settlers Understood Themselves and Their Land*. New York: Basic Books, 2002.

Wailoo, Keith, et al., eds. *Three Shots at Prevention: The HPV Vaccine and the Politics of Medicine's Simple Solutions*. Baltimore: Johns Hopkins University Press, 2010.

Wallace, Alfred Russel. *The Wonderful Century: Its Successes and Its Failures*. New York: Dodd, Mead, 1898.

Watkins, Elizabeth. *The Estrogen Elixir: A History of Hormone Replacement Therapy in America*. Baltimore: Johns Hopkins University Press, 2007.

———. *On the Pill: A Social History of Oral Contraceptives, 1950–1970*. Baltimore: Johns Hopkins University Press, 1998.

Whorton, James C. *Nature Cures: The History of Alternative Medicine in America*. New York: Oxford University Press, 2002.

Willrich, Michael. *Pox: An American History*. New York: Penguin, 2011.

INDEX

AAP. *See* American Academy of Pediatrics

ABC Evening News, 207

abortion politics, 65, 76

Abramson, Jon S., 231

ACA. *See* Affordable Care Act, 2010

Advisory Committee on Immunization Practices (ACIP): accusations of sexism in HPV vaccine recommendations, 243–44; formation of, 32; hepatitis B vaccine recommendations, 184, 188, 190–91; HPV vaccine recommendations, 229, 239, 246–47, 317n7, 321n62; influence on state laws, 8, 32; measles vaccine position, 52–53, 190–91, 306n34; mumps vaccine debate, 70–71, 80; polio vaccine recommendation change, 207; rotavirus vaccine recommendation reversal, 205–6. *See also* Centers for Disease Control and Prevention

Affordable Care Act (ACA, 2010), 252, 324n3

African Americans: cervical cancer disparities and, 230; positioning in the connection between poverty and public health, 46–48; urban polio outbreaks and, 43

AIDS: cultural connection to the hepatitis B vaccine, 198, 232; domestic concern over the globalization of disease and, 172; fears of a link between hepatitis B and AIDS, 146, 185–86, 304nn30–31; reactions to Hooper's theory about the origins of, 206;

theory of a link to the polio vaccine, 206; vaccine rejection and, 157

aluminum, 150

Alvarez, Walter, 108, 109, 110

American Academy of Pediatrics (AAP), 8, 24, 32, 151, 207

American Cancer Society, 245

American Drug Manufacturers' Association, 22

American Medical Association (AMA), 22, 36–37, 67, 209

American Public Health Association (APHA), 32, 74, 125

anal cancer, 245–46

Angell, Marcia, 233

anti-vaccinationists: books attacking vaccinations, 134–38, 148–49; contrast with vaccine skeptics, 132; McBean's book and (see *Poisoned Needle, The*); nature-based ideology, 133–34, 141, 142, 143, 144; officials' blaming of disease outbreaks on anti-vaccine mothers, 219; position on vaccines, 4, 11; public forums for, 115. *See also* vaccine safety movement; vaccine skepticism: belief that vaccines cause cancer

APHA. *See* American Public Health Association

Arkansas, 93–94, 108, 109

Arms, Suzanne, 119